Coastal Processes II

WIT*PRESS*

WIT Press publishes leading books in Science and Technology.
Visit our website for the current list of titles.
www.witpress.com

WIT*eLibrary*

Home of the Transactions of the Wessex Institute.
Papers presented at Coastal Processes 2011 are archived in the WIT elibrary in volume 149 of WIT Transactions on Ecology and the Environment (ISSN 1743-3541).
The WIT electronic-library provides the international scientific community with immediate and permanent access to individual papers presented at WIT conferences.
http://library.witpress.com.

SECOND INTERNATIONAL CONFERENCE ON
PHYSICAL COASTAL PROCESSES, MANAGEMENT
AND ENGINEERING

COASTAL PROCESSES II

CONFERENCE CHAIRMEN

G. Benassai
University of Pharthenope, Italy

C.A. Brebbia
Wessex Institute of Technology, UK

G. Rodriguez
University of Las Palmas, Spain

INTERNATIONAL SCIENTIFIC ADVISORY COMMITTEE AND LOCAL COMMITTEE

Organised by

University of Parthenope, Italy
Wessex Institute of Technology, UK
University of Las Palmas (Canary Islands), Spain

Sponsored by

WIT Transactions on Ecology and the Environment
ASCE UK Group

WIT Transactions

Transactions Editor

Carlos Brebbia
Wessex Institute of Technology
Ashurst Lodge, Ashurst
Southampton SO40 7AA, UK
Email: carlos@wessex.ac.uk

Editorial Board

Coastal Processes

Editors

G. Benassai
University of Pharthenope, Italy

C.A. Brebbia
Wessex Institute of Technology, UK

G. Rodriguez
University of Las Palmas, Spain

WIT*PRESS* Southampton, Boston

Editors

G. Benassai
University of Pharthenope, Italy

C.A. Brebbia
Wessex Institute of Technology, UK

G. Rodriguez
University of Las Palmas, Spain

Published by

WIT Press
Ashurst Lodge, Ashurst, Southampton, SO40 7AA, UK
Tel: 44 (0) 238 029 3223; Fax: 44 (0) 238 029 2853
E-Mail: witpress@witpress.com
http://www.witpress.com

For USA, Canada and Mexico

Computational Mechanics Inc
25 Bridge Street, Billerica, MA 01821, USA
Tel: 978 667 5841; Fax: 978 667 7582
E-Mail: infousa@witpress.com
http://www.witpress.com

British Library Cataloguing-in-Publication Data

A Catalogue record for this book is available
from the British Library

ISBN: 978-1-84564-534-2
ISSN: 1746-448X (print)
ISSN: 1743-3541(online)

The texts of the papers in this volume were set individually by the authors or under their supervision. Only minor corrections to the text may have been carried out by the publisher.

Printed in Great Britain by Quay Digital, Bristol.

Preface

Coastal regions present a complex dynamic web of natural and human related processes. Although coastal zones are narrow areas extending a few kilometres on either side of the shoreline, and occupying small strip of ocean and land, they play a very important role as they account for nearly a quarter of all oceanic biological production, which in turn supplies approximately 80% of the world's fish. About 60% of the human population lives in the coastal zone, and around 70% of big cities are placed in this narrow area. Concomitantly, more than 90% of the pollutants generated by human economic activities end up in the coastal zone.

The unstoppable demand on the coast for recreational and tourism activities has increased the need for shore and beach protection, as well as the construction of artificial beaches, ports and harbours. Most of the coastlines are subjected to the direct impact of wind waves, swell and storm wave activity. As a result, wind waves and wave driven currents are the dominant mechanisms controlling littoral sand transport and determining the nearshore morphology. In addition, many other physical phenomena, such as tides and associated currents, long waves and storm surges, among others, can play a significant role in the dynamic behaviour of the coastal zone.

Due to its great socio-economic importance, the physical aspects of coastal processes have been of concern for decades, but recent advances in a number of areas, including satellite remote sensing, are giving rise to significant progress in this field. In particular, the use of satellite and imaging systems has significantly enhanced the monitoring and understanding of coastal processes. Accordingly, it has become clear that the ocean side of the coastal zone represents a very sensitive and particularly vulnerable sector of the ocean to any kind of man-made action or natural extreme events. Consequently, the problem of environmental protection and conservation takes special relevance in this zone, and any decision concerning its viability must be preceded by a forecast of its consequences. Their adequate prediction is only possible on the basis of a clear understanding and careful analysis of the fundamental dynamic processes occurring in such areas.

A greater knowledge of sediment transport mechanisms at beaches may permit to avoid some common mistakes of the past, consisting in uncontrolled development of groin fields and seawalls, dam constructions on rivers that reduce sand supply to the coast, hydrocarbon and groundwater extraction inducing local ground subsidence and associated inundation and erosion of coastal areas.

In order to reach satisfactory solutions for the demands imposed on the coastal areas and the protection of its environment, one needs to understand very different aspects and their interaction. The problems are essentially interdisciplinary and scientists need to be able to exchange ideas with colleagues from other disciplines with a variety of different experiences. Thus, an acceleration of research is needed to improve the quality of the coastal processes prediction, together with an adequate level of scepticism about the model results and a continuous comparison with well-documented case studies and field experiments.

This is the reason why, after the successful experience of the first Conference in Malta, the Wessex Institute of Technology, the University Parthenope of Naples and the University of Las Palmas, Gran Canaria decided to convene a second International Conference in Coastal Processes. This book comprises the edited papers of that meeting grouped into the following topics:

- Coastal Management
- Coastal Processes and GIS
- Coastal Geomorphology
- Extreme Events and Sea Level Rise
- Coastal Processes and Navigation
- Sediment Transport and Erosion
- Interaction between Coastal Defences and Processes
- Pollution and dispersion
- Hydrodynamics

The Editors are grateful to all the authors for their excellent contributions as well as to the members of the International Scientific Advisory Committee for the review of both the abstracts and the papers included in this book. The quality of the material makes this volume a most valuable and up-to-date tool for professionals, scientists and managers to appreciate the state-of-the-art in this important field of knowledge.

The Editors
Naples 2011

Contents

Section 7: Interaction between coastal defences and processes

Section 8: Pollution and dispersion

Section 9: Hydrodynamic forces

Section 1
Coastal management

A sustainability index for offshore wind farms and open water aquaculture

G. Benassai[1], C. Stenberg[2], M. Christoffersen[2] & P. Mariani[2]
[1]*Department of Applied Sciences, University Parthenope, Naples, Italy*
[2]*National Institute of Aquatic Resources,*
Technical University of Denmark, Lyngby, Denmark

Abstract

One of the possible multi uses of marine space is offshore wind farms and open water aquaculture plans. Unlike coastal installations, offshore wind farms allow for increased availability of wind power and wind persistence, as well as lower visual impact of the turbines. Existing offshore wind farms offer important advantages for aquaculture plans especially in terms of lack of major physical constrains, e.g., navigation routes, submarine cables, marine protected areas. Moreover, enhanced current velocity due to the presence of the piles and to the air fluxes of the turbines may increase the environmental suitability of aquaculture plans in these areas. In addition, the transmission of localized depleted water masses or waste material towards near-shore zones can be avoided, excluding potential impact close to the coast. On the other hand, other environmental constrains (e.g. temperature and salinity variability, dissolved oxygen concentrations, phytoplankton dynamics) need also to be considered when planning aquaculture activities.

In this context, the present paper gives a contribution towards the definition of a sustainability index for the large scale localization of marine areas as offshore wind farms and aquaculture plans. This index is developed on the basis of a Multi-Criteria Evaluation (MCE) technique, already used in open water aquaculture localization: raw data were converted to suitability scores, which were combined using additive models, in order to define the overall suitability. A detailed analysis of the environmental suitability is performed for few specific test cases at offshore wind farms located in the Denmark coastal zone, for which some projects of offshore aquaculture plans have been proposed.

WIT Transactions on Ecology and the Environment, Vol 149, © 2011 WIT Press
www.witpress.com, ISSN 1743-3541 (on-line)
doi:10.2495/CP110011

The use of this sustainability index will perform quantitative evaluations of the degree of compatibility between the offshore wind farms and the open water aquaculture, highlighting the main problems which can rise by the integration of both technologies.
Keywords: sustainability index, offshore wind farms, open water aquaculture.

1 Introduction

Compared to onshore wind, offshore wind farms are more costly to install and maintain but also have a number of key advantages: stronger and more stable wind; larger wind turbines; less conflict with neighbouring citizens and other stakeholders unless they interfere with competing maritime activities or impact negatively on important marine environmental interests.

In Europe, over 100 GW of offshore wind projects are already in various stages of planning [1]. If realised, these projects would produce 10% of the EU's electricity whilst avoiding 200 million tonnes of CO^2 emissions each year. There are currently 830 wind turbines now installed and grid connected, totalling 2,063 MW in 39 wind farms in nine Northern European countries. During 2009, 201 wind turbines installed and grid connected totalled 584 MW, up 56% from 2008 installations [1].

Several large projects are planned in other countries, including Southern Europe where wind energy productions are not as mature as in the Northern European Countries [2]. In Italy, Enel (national body for the electric energy) deposited the project design for one of the first offshore wind farms in the Mediterranean Sea, proposing the installation of 115 large generators with a capacity between 3 and 5 MW in the waters of the Gulf of Gela. Moreover, another offshore wind project is based on floating wind turbines in higher depths (more than 110m), located 20 km offshore the Apulia coastline (Tricase offshore wind farm) [2]. This project follows the technology of Hywind project in Norway (see http://www.statoil.com/en/technologyinnovation/newenergy/renewablepowerproduction/onshore/pages/karmoy.aspx).

The overall high-competitive use of the coastal zones highlights the need to optimize co-management of the resources, which raises some advantages in the potential combination of marine space utilization. Many of the smaller wind farms are located relatively close to the shore (<5 km) while the larger parks are positioned more offshore (>20km). The trend seems to be bigger turbines and parks and more offshore position. An example of this is the Dogger Bank wind farm which is on the drawing board these years. This park is planned to be on 9000 MW, consist of more than 1500 turbines, cover an area of 2000 km^2 and is to be 200 km from the shore.

The most used foundations for the turbines are monopile and gravity ones. In order to prevent sediment erosion large boulders are placed around the monopile turbine for scour protection. While windmills use the wind above the surface to produce energy, their fixed pylons may offer a basis to connect systems used in aquaculture or create artificial nursery areas for wild fish species. Additionally, the multi-use concepts of these two industries can assist to avoid user conflicts with fishermen, who may lose part of their traditional fishing grounds [3].

The advantage of moving aquaculture further away from coastal waters consists in an enhanced water quality. Open ocean waters are in general less exposed to anthropogenic impacts and might provide a continuous supply of clean water having satisfactory levels of dissolved oxygen and minor concentrations of pollutants, pesticides and near-surface agents. However, open ocean hydrodynamic conditions can represent further challenges for aquaculture infrastructures and require resistant species that can withstand strong currents and large wave heights. There are however candidates [4] that can be cultivated in these conditions, such as blue mussels. The growth of these mussels is largely controlled by food availability and phytoplankton dynamics. Several studies have identified strong direct linkage (R^2=0.77) between upwelling indices and cultured shellfish production and quality [5]. In fact upwelling typically provides a rich source of nutrients to enhance phytoplankton growth and large volumes of shellfish are cultured in areas of high phytoplankton concentrations. Sea surface Chl-a concentrations is often considered a good proxy to identify areas of high biological production [6].

Biological productivity in marine areas is generally the result of oceanographic processes such as upwelling, gyres or eddies, which can transport cold, nutrient-rich water from below the pycnocline to the euphotic zone where photosynthesis of autotrophs organisms takes place. The cold water signature of the nutrient-rich water in these productivity hot spots is typically associated with low SSTs. The spatial integration of normalized SST and Chl-a anomalies indicates areas of productive processes such as upwelling, gyres, etc. [5, 6]. The use of climatological (long-term) datasets allow the identification of persistently productive regions, independent of short-term variability.

2 Methods and data

2.1 Structure of the SI

We defined the Sustainability Index (SI) as the geometric mean of all environmental indexes relevant to our study. Those environmental indexes were based on climatological values of: sea surface temperature (SST), chlorophyll concentration, topography, wind power and SST anomalies [8] which are then modified by their Parameter Specific Suitability Functions and subsequent restrictions by the Boolean constraints layer:

$$SI = \prod_{i=1}^{n} PSSF_i \qquad (1)$$

The Parameter Specific Suitability Function (PSSF) for each environmental data map was defined on an arbitrary scale between 0 and 1, where 0 defined a non-suitable area and 1 the most suitable. The Boolean constraints layer allows an element to belong to a 'crisp' set (0 or 1) taking into account the existing technology and environmental restrictions.

2.2 Environmental data

Depth data in the North Sea area were derived from ETOPO2 seabed topography (Figure 1) which showed a quite shallow and uniform depth in the southern part of the basin and along the entire coastline. In particular close (< 30 Km) to the Danish coast, depth was always lower than 20-25m.

The Sea Surface Temperature (SST) and SST anomalies in the North Sea (Figures 2 and 3) were obtained from the Advanced Very High Resolution Radiometer (AVHRR) Pathfinder data produced by NOAA and the University of Miami's Rosenstiel School of Marine and Atmospheric Science (http:// pathfinder.nodc.noaa.gov). SST could have ecological consequences on the growth of organisms with higher temperatures allowing faster growth of the aquaculture species. Moreover, additional potential environmental changes in response to changes in SST could be measured as the frequency of temperature anomalies [7]. Data for SST anomalies included anomaly frequency between 2000-2005 and 1985-1990.

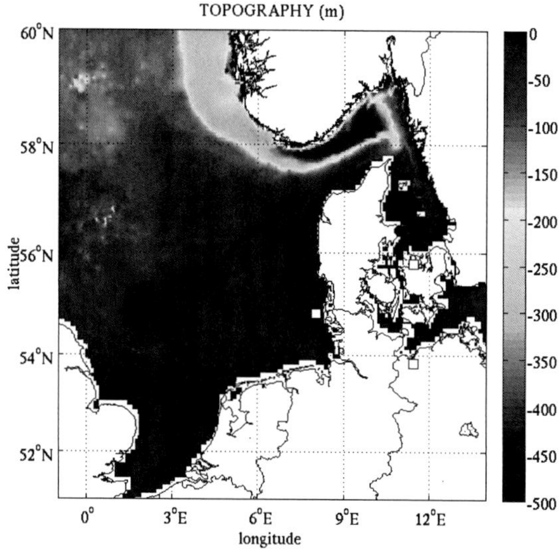

Figure 1: Map of topography in the study area with indication (dots) of the three locations analyzed for multi-use purpose platforms in Denmark.

Those data were derived from Halpern et al. [7] and provided a baseline for determining when temperatures were unusually warm having negative consequences on aquaculture species. Lower anomalies occurred in the South-Eastern part of the North Sea and further East, into the Skagerrak and Kattegat. The Southern and Eastern North Sea, from the German Bight into the Skagerrak, were under the influence of the Jutland current which in the Northern part of Denmark branches into two currents, one staying in the North Sea and the other

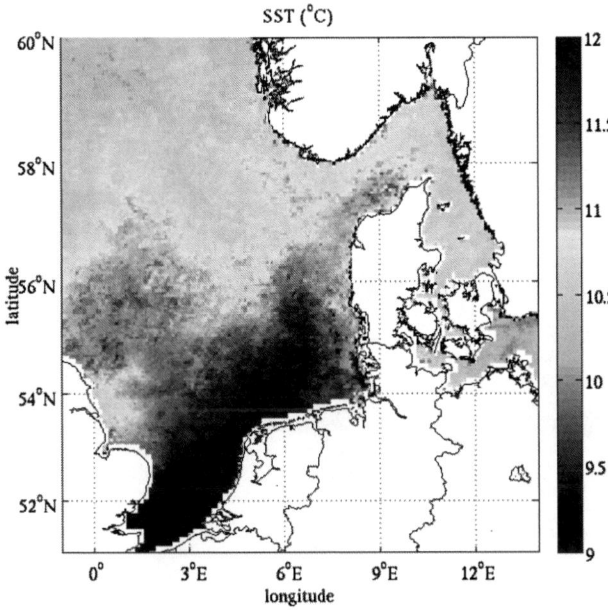

Figure 2: Map of the distribution of Sea Surface Temperature.

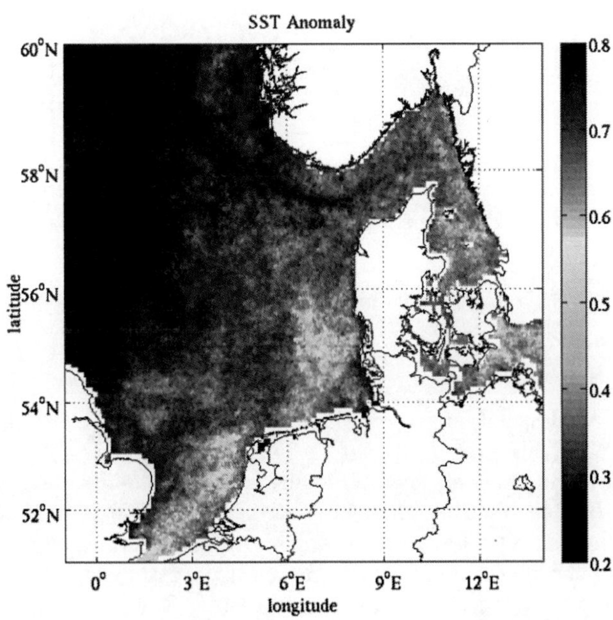

Figure 3: Map of the distribution of SST anomaly.

entering the Skagerrak. In the Kattegat the influence of oceanic circulation was weaker than in the North Sea since dominating factors were river run-off and larger atmospheric variability above land [8, 9].

The Chl-a climatology in the North Sea (Figure 4), exhibited a similar variation: in the South-Eastern part of the North Sea Chl-a was influenced by the shallow water and the strong tidal currents which transported high sediment loads. In general the North Sea circulation was anti-clockwise, so that German Bight waters, which were nutrient-rich due to river discharge in relatively low depth, were transported up along the Jutland west coast, into the Skagerrak and even spilled into the Kattegat [9].

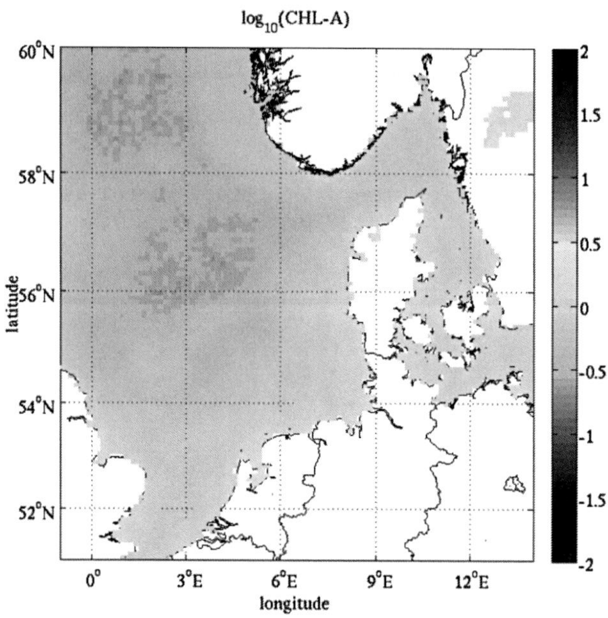

Figure 4: Map of the distribution of chlorophyll-a concentration (in log scale).

Finally, wind speed data were obtained from Scatterometer Climatology of Ocean Winds (SCOW) [10] and were based on a yearly average of climatological monthly values (122 months included) at ~26Km resolution in the study area (Figure 5). This relatively low resolution did not allow a detailed description of the Kattegat area. In order to have a complete coverage in the definition of the PSSF of the study area we then extrapolated the wind speed values in the Skagerrak into the Kattegat area. The resulting map showed average wind speed ranging between 7–10 m/s with higher values in the Northern zones.

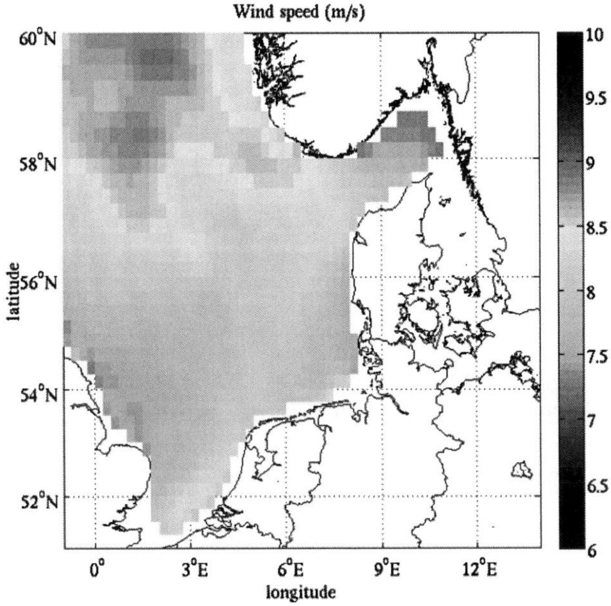

Figure 5: Maps of the distribution of wind speed.

2.3 Computation of PSSF for offshore wind farms and open aquaculture plans

Raw data of the environmental conditions described above were re-binned in the study area using relatively high horizontal resolution (1Km) and nearest values interpolation method. An ad-hoc normalization procedure was then applied to compute the PSSF ranging between 0 and 1.

Topography and wind speed were considered to derive PSSF of the wind farm activities in the North Sea. Due to the different cost percentage between the turbine and the monopile structure in relatively low depths (< 30 m) a step cost function for the depth was assigned, with five steps ranging from below 20 m (most favourable), to higher than 80 m (least favourable). A Boolean constraint was assigned to the benthic sediment type, which can be only constituted by sand or gravel, due to the limits of the monopile technology. A conversion between wind speed and energy was obtained using the standard power curve for the Vestas V112, 3MW turbine which was particularly suited for offshore sites [2]. Then a normalization procedure (values/max value) ensured PSSF varying between 0 and 1. The maximum value of 1 (most favourable) was given to the mean wind speed corresponding to 100% of the power transfer function (12 m/s), while the other PSSF values were assigned consistently. Thus the PSSF was not linear with wind speed but shaped accordingly to the physical energy transfer from the wind.

SST, SST anomaly and Chl-a were the three environmental factors assumed to affect the PSSF of aquaculture activities. The calculation of PSSF for the SST and CHL-a was made on the basis of a normalization technique over the environmental data (values/max value). Chlorophyll values changed over many orders of magnitude, so in order to weight similarly all the environmental parameters an additional rescaling was applied to the logarithmic (*log10*) of Chl-a in order to obtain the PSSF ranging from 0 and 1. The SST anomalies were transformed in anomaly indexes suitable for aquaculture, attributing 1 to low values of temperature anomaly and 0 to maximum positive anomaly, because large SST anomalies are detrimental for aquaculture activities.

3 Results

The Suitability Index (SI) for co-location of wind farm and aquaculture plans (Figure 6) showed values ranging between ~0, in the Northern deepest zones of the study area, and higher values (0.6 – 0.8) in the Southern part of the North Sea and all around Denmark, with the highest values on the Western coast.

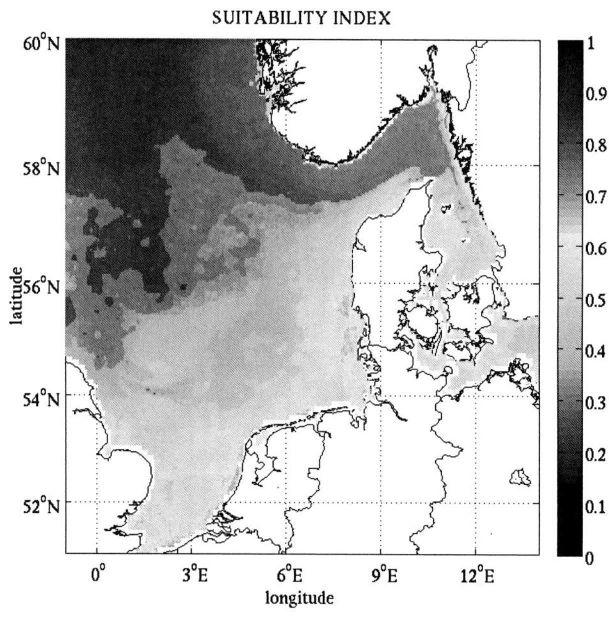

Figure 6: Map of the suitability index for the North Sea and adjacent seas.

As shown in figures 6 and 7, relatively high SI values were also evident in the Central North Sea close to the Dogger bank. The SI appeared to be consistent with shallow areas having high SST and low SST anomalies while Chl-a and wind speed appeared to have a secondary effect due to their relatively uniformity in the study area.

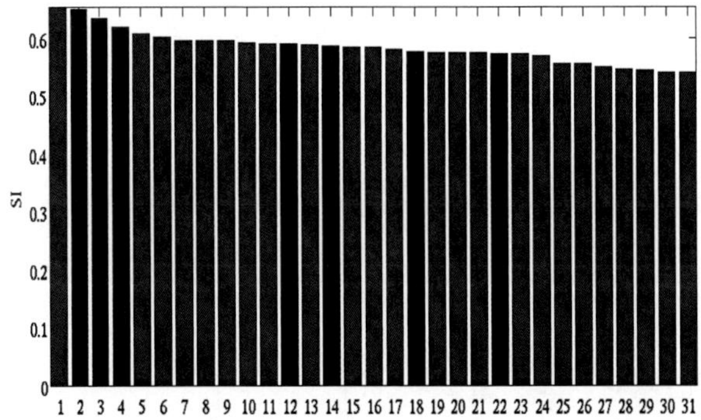

Figure 7: Suitable Index values at the locations (proposed=light grey, operating=dark grey) of Danish offshore wind farm. The highest value is for the proposed location Horn Rev (7.9E 55.7N), the 2nd and 3rd highest are the Horn Rev1 (7.8E 55.5N) while the 4th is the Nysted (11.2E 55.0N). The site called Anholt (11.3E 56.5N) is at position 23.

4 Discussions

In Denmark several sites were used or planned to be used for offshore wind farms installations (Figure 8). According to our SI most of those sites appeared to be suitable locations for a multi-use purpose combining wind energy and aquaculture activities. The SI defined above provided a first and large scale evaluation of the feasibility of setting aquaculture activities in offshore wind farm sites. However, the index as defined above did not include other relevant aspects which needed to be accounted for in making the aquaculture activity feasible.

Offshore aquaculture has drawn a lot of attention because of the large areas available offshore and thus its huge potential. However, there are today worldwide few, if any, aquaculture facilities that are positioned offshore in exposed conditions. The rough conditions demands special solutions and equipment that at present are not fully developed. Fitting offshore aquaculture into wind farms is another and addition challenge. Operation and maintenance of the turbines requires frequent visits of smaller ships and yearly call of large ship crane [4]. Aquaculture facilities in a WF must therefore adapt to these constrains and challenges. Solutions could be only to occupy a part of the WF area for aquaculture and/or to have mobile facilities that can be towed away when needed.

Stenberg et al. [4] estimated that up to 25% of the area in a WF could be used for pelagic mussel productions. Another solution for mussels could be to collect mussel seeds elsewhere and use the seafloor in WF for mussel beds thereby avoiding lines, anchors, floats etc in the WF area that are in potential conflict

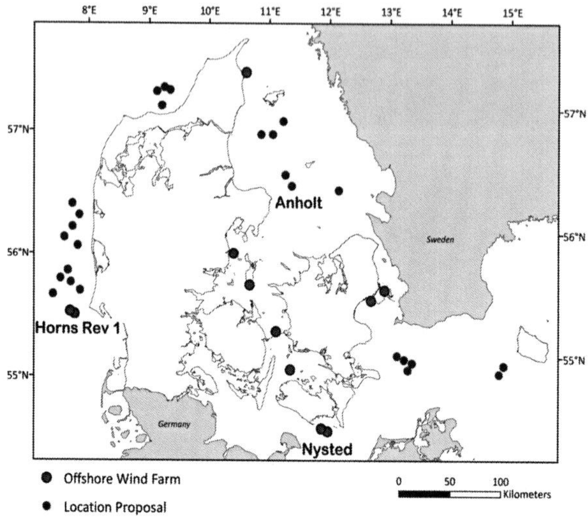

Location	*Horns Rev 1*	*Anholt*	*Nysted*
Number of turbines	80	112	72
Area covered in WF (km²)	27,5	88	24
Distance between turbines (m)		450-550	
Depth in WF (m)	2-9	14-20	6-9,5
Sediment type in WF	Sand/gravel	Sand	Moraine/sand
Distance to nearest port (km)		10-15	

Figure 8: Distribution of offshore WF (operational/proposed) in Denmark.

with other WF activities. An analysis of the economical feasibility is associated with great uncertainty but estimate a break-even price for blue mussels between 1.5–2.0€ kg-1 [4]. The market price in 2010 was less than 1€ kg^{-1}. However, the mussel production from a single WF would be the same as the total production of mussels in Denmark at present. The large scale would probably bring down operating costs thereby reducing the break-even price.

For fish aquaculture, a solution could be smaller cages that are easily moveable and only to farm fish in the growth season (spring to autumn), e.g. juvenile to consumer size fish allowing a time window for turbine maintenance and reducing risk of sea ice damages [11].

Another approach could be to use the WF areas for extensive aquaculture as sea ranching and stock enhancement solutions. A suitable species for this could be the European lobster (H. gammarus). Lobsters are economical valuable and can be artificially raised for release of the juveniles in WF areas. The scour protection around each turbine is believed to be a very suitable lobster habitat for juvenile and adult lobster. Due to lack of suitable habitat in the adjacent areas is hypothesized that the lobsters will remain in the WF area throughout their entire life. This would allow an effective sea ranching concept with harvest 3–5 years after the release of the juveniles. An increase in suitable habitat could be combined with innovative cable protection or artificial lobster reefs within the WF.

5 Conclusions

The application of MCE technique for co-location of offshore wind farms and open water mussel cultivation was collected in a definition of the Suitability Index (SI), which is a possible response to the actual and future need of a smart tool for multiple marine space utilization. This SI is easily implemented in GIS in order to do a first order selection of the most promising areas to be more specifically studied in a second order approach, to be based on local field data.
The SI defined above provided a first and large scale evaluation of the feasibility of aquaculture activities in offshore wind farm sites. However, the index as defined above did not include other relevant aspects, such as the wave and current loads on the long-lines of mussels. Moreover, the present results rely heavily on the data resolution, which could be probably increased for the wind velocity and the CHL-a which should be extend to values other than the surface. However, the relative comparisons upon which the method is based are susceptible of further on-site detailed studies.

References

[1] EWEA 2009 Annual Report. http://www ewea org.
[2] Floating Wind Turbines, Deep Water Installation Offshore Industry 2 (4): 48–51. 2009. www.wikipedia org.
[3] Buck B (2004) Extensive open ocean aquaculture development within wind farms in Germany: the prospect of offshore co-management and legal constraints. Ocean and Coastal Management 47: 95-122.
[4] Stenberg C, Christoffersen M, Mariani P, Krog C, Dolmer P, Maar M, Tørring D. ICES Conference 2010 ICES CM 2010: O:12.
[5] Espinosa-Carreon TL, Strub PT, Beier E, Ocampo-Torres F, Gaxiola-Castro G (2004) Seasonal and interannual variability of satellite-derived chlorophyll pigment, surface height, and temperature off Baja California. Journal of Geophysical Research-Oceans 109.
[6] Valavanis VD, Kapantagakis A, Katara I, Palialexis A (2004) Critical regions: A GIS-based model of marine productivity hotspots. Aquatic Sciences 66: 139-148.
[7] Halpern BS, Walbridge S, Selkoe KA, Kappel CV, Micheli F et al. (2008) A global map of human impact on marine ecosystems. Science 319: 948-952.
[8] Becker GA, Pauly M (1996) Sea surface temperature changes in the North Sea and their causes. Ices Journal of Marine Science 53: 887-898.
[9] EC-FP5 REVAMP (2005) Atlas of Chlorophyll-a concentration for the North Sea based on MERIS imagery of 2003. Project Report 1-121.

[10] Risien CM, Chelton DB (2008) A Global Climatology of Surface Wind and Wind Stress Fields from Eight Years of QuikSCAT Scatterometer Data. Journal of Physical Oceanography 38: 2379-2413.

[11] Møller and Bregnballe (2009) Assessing the possibilities of establishing mariculture at Nysted (in Danish). 3-26, in: Vurdering af mulighederne for opdræt af fisk og skaldyr i områder med havvindmøller - pilotprojekt Nysted Havmøllepark, editor Carsten Krog

Recent evolution and the present-day conditions of the Campanian Coastal plains (South Italy): the case history of the Sele River Coastal plain

G. Pappone[1], I. Alberico[2], V. Amato[3], P. P. C. Aucelli[1] & G. Di Paola[3]
[1]*DiSAm, Università Parthenope, Napoli, Italy*
[2]*CIRAM, Università Federico II, Napoli, Italy*
[3]*DiSTAT, Università degli Studi del Molise, Pesche (IS), Italy*

Abstract

The low coasts of the Campania are generally located in the main alluvial coastal plains of the region. These coasts have been affected, during late Quaternary, by strong progradation and more recently by erosion and episodic flooding particularly during storm events. The causes are essentially to be sought in the decrease in sedimentary discharge due to forest hydraulic engineering works but especially to the construction of many artificial dams along the main rivers, coupled whit subsidence and or increases in sea levels. Such events generally occur in coastal plains where the sectors close to the present-day dune ridge are morphologically lower (in general no more than 1–3 m a.s.l.). The main goal of this study was to provide a synopsis of the coastal vulnerability and present a new semi-quantitative method to assess coastal erosion. Eight factors describing the current system state of the beaches and the effects of the wave climate and human activity were combined to assess the potential erosion of the Sele River coastal plain. The method shows the high erosion potential at the mouths of the Sele, Picentino and Tusciano Rivers, while the areas south of the Sele river mouth and the zone stretching from Lido Lago to the Asa channel showed low hazard levels.
Keywords: coastal vulnerability, coastal erosion potential, Campanian coastal plains, South Italy.

WIT Transactions on Ecology and the Environment, Vol 149, © 2011 WIT Press
www.witpress.com, ISSN 1743-3541 (on-line)
doi:10.2495/CP110021

1 Introduction

In the last six millennia the plains in the southern Italian region of Campania have experienced coastal progradation amply documented by several dune systems. Since the 20th century this trend has been interrupted and many stretches of the coastline are now affected by erosion, at times severe. This has serious implications both for public safety and of a socio-economic nature. The causes are essentially to be sought in the decrease in sedimentary discharge due to forest hydraulic engineering works but especially to the construction of many artificial lakes along the main water courses. Clear evidence of this is the transformation of the mouths of the main water courses from fluvial-dominated to wave-dominated. A further factor is intense urbanisation, which took place especially after World War II in the wake of tourist development. These trends may be further amplified by the rise in sea level due to climate change [1]. This could increase the possibility of storms producing coastal erosion, temporary or episodic inundation. For much of the coastline in northern Campania, the state of the coastline and the relative coastal hazard have been extensively described by De Pippo *et al.* [2]. To assess the erosion vulnerability of the coastline in southern Campania, especially the plain of the River Sele (fig. 1), we used a new semi-quantitative method.

2 Geomorphological setting

The Tyrrhenian flanks of the Campanian Apennines are characterized by a very articulated topography due to alternating mountainous transverse ridges, valleys and alluvial plains. From NW to SE the Campanian coastal area presents the structural high of Monte Massico, the graben of the Campana Plain, the structural high of the Sorrento and Amalfi peninsula, the graben of the Sele plain and the structural high of the Cilento (fig. 1). The presence of two topographic highs, the Phlegraean Fields and Vesuvius volcanoes, causes the zoning of the Campana plain, from NW to SE, into the Volturno coastal plain, the alternating high and low coasts of the Phlegrean fields, the Sebeto coastal plain, the alternating high and low coasts of the Vesuvian area, and the Sarno River coastal plain. The plains along the coastal strip of Campania result from the aggradation of structural depressions of differing width and geometry whose bed is as much as several thousand metres lower than the Apennine massifs and other hills in between [3–6]. These coastal plains are the terminal point of the major water courses in the region, with their catchment areas comprising almost the whole of Campania, especially that of the River Volturno and the River Sele. At the mouth, both rivers, especially the Volturno, have a marked delta cusp that interrupts the straight coastline. It is precisely such sectors that have been – and still are – subject to sudden phases of coastline advance and retreat.

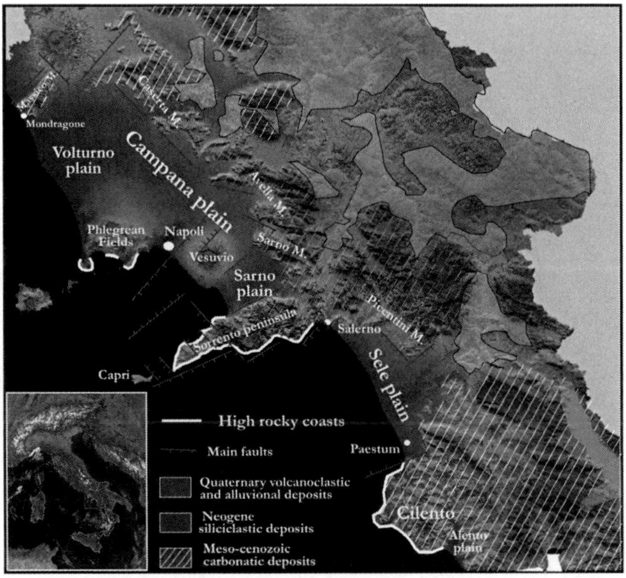

Figure 1: Geological and structural setting of the Campanian coasts.

3 Long- and middle-term evolution of the Campania coastal plain: current state of knowledge

3.1 The Campana plain

The Campana Plain comes from volcano-sedimentary aggradation of the peri-Tyrrhenian graben of the same name, whose fault margins are roughly at the foot slopes of the carbonatic reliefs (fig. 1). This graben is identifiable from the Lower Pleistocene [7, 8]. Thus extensive parts of the area currently occupied by the Campana Plain were invaded by transitional and shallow marine environments: the sedimentation rates, with a significant contribution from the Phlegraean and Vesuvian eruptions, on average managed to offset subsidence. Indeed, the sea-level rise during the Tyrrhenian interglacial (MIS 5) brought the coastline almost right up to the limestone massifs at the margin of the plain (fig. 2A). To confirm this, the coastal deposits of MIS 5 have been found by probes tens of metres below sea level at several points in the plain subsoil [7, 9] and raised along the margins of the Avella Mts [7].

Since the Upper Pleistocene the Campana plain has been divided into two large physiographic units due to the emergence of volcanoes around Naples: the Volturno plain to the NW and the Sarno plain to the SE. Contributing significantly to the forming of the two plains is the thick ignimbrite layer of *Tufo Grigio Campano* (39 ka BP, De Vivo *et al.* [10]) which produces rapid volcanoclastic aggradation and shoreline progradation for several kilometres.

For the Volturno plain, during the final phases of Versilian sea-level rise the shoreline formed a type of gulf reaching some km inland from the present-day terminal sector of the river (fig. 2A).

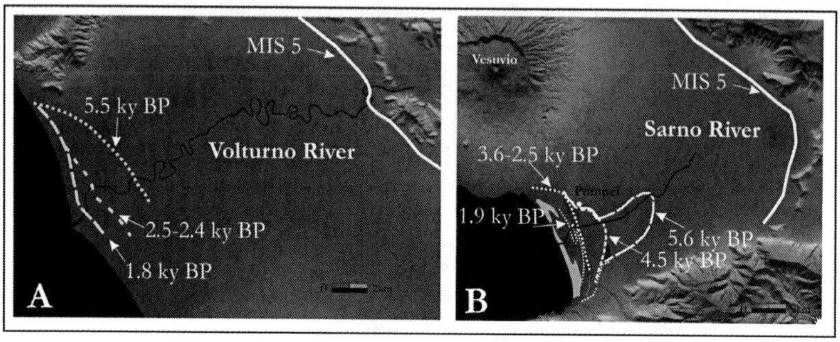

Figure 2: Late Pleistocene and Holocene shorelines of the Campana plain. A. the Volturno river alluvial-coastal plain; B: Sarno river alluvial-coastal plain. The dune ridge positions are shown in dark and light grey.

The last progradation phase of about 2 km occurred during the last 6,000 years when the rates of sea-level rise slowed considerably, leading to the formation of coastal lagoon-dune barrier systems. For the historical and protohistoric period the most morpho-dynamically active sector was that of the river mouth, which presented alternately landward (palaeogulfs) and seaward shorelines (delta cusps). Based on Cocco *et al.* [11], drawing on archaeological data and an analysis of historical maps, it is possible to reconstruct the trend in the shoreline near the delta cusp of the Volturno mouth (fig. 3).

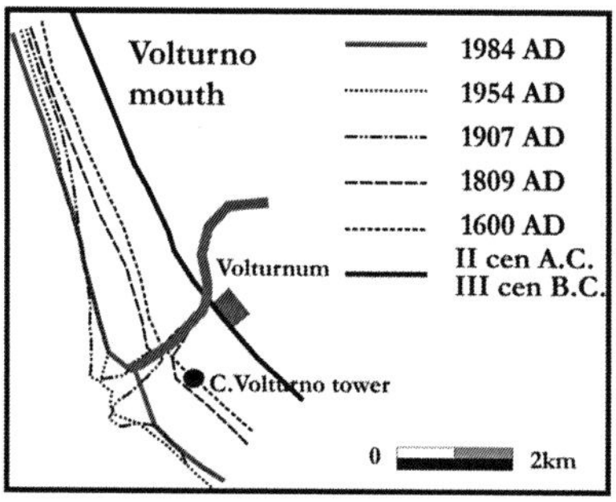

Figure 3: Shoreline variation at Volturno river mouth, from Cocco *et al.* [11].

According to some authors, the coastline has been affected by constant progradation starting from the Roman period (3rd century BC): the delta cusp today extends into the sea over 2.5 km further than the ancient Roman city of Volturnum, founded close to the ancient mouth. The period of peak progradation of the delta cusp occurred during the 19th century. The severe progradation could be related to the well-known cold and wet climatic phase of the Little Ice

Age in the 18th-19th centuries [6, 13]. For the period from 1954 to 1984 the coastline continued to advance between 1.5 m and 2.5 m per year, except for the delta cusp, which was subject to a rapid straightening process, retreating by hundreds of metres, especially in the area on the left. This change could be related to the reduction in river sediment discharge resulting from the end of the Little Ice Age, especially to human intervention in the hydrographic basin (construction of weirs, gravel and sand extraction, "cementification" of extensive sections of the water course). By contrast, in the Sarno plain, south of the Somma-Vesuvius edifice (fig. 2B), subsidence appears to have continued until the historical period [4, 5]: Tyrrhenian coastal deposits are tens of metres lower [13], while the beach linked to the Holocene transgression maximum (more than 6 km beyond the present-day coastline) is about ten metres lower (fig. 2B). This is proof that the rate of Late Quaternary subsidence increased during the Holocene, reaching mean rates of about 2 mm/a. Similar values may also be found for historical times, as shown by the lowering to -4/-5 m of the Roman shoreline near Pompeii [8]. The tendency of the coastline to prograde is only reached when the uplift rate over the base level (given by summing the sea-level variation and subsidence of the plain) is below that of the (sedimentary+pyroclastic) aggradation rate of the plain [14]. Also in this plain progradation has taken place only in the past 5000 years through the juxtaposition of increasingly advanced coastal lagoon-dune barrier systems up to the present-day position (fig. 2B).

3.2 Sele plain

The Sele Plain derives from the aggradation of a Plio-Quaternary depression along the western margin of the southern Apennine chain and known as the Salerno Gulf-Sele Plain graben. Extending about 400 km^2, its triangular shape is delimited seaward by a straight sandy coast, stretching between the towns of Salerno and Agropoli, encircled landwards by a range of calcareous mountains (Lattari, Picentini, Alburni, Soprano-Sottano) as well as by the arenaceous mountains in the Cilento (fig. 1). The southernmost portion of the plain, formed during the Last Interglacial (Tyrrhenian stage, MIS 5 [15–17]), is characterized by beach-dune ridges (Gromola-S. Cecilia-Arenosola-Aversana ridges) (fig. 4).

The present elevation of about 11–15 m a.s.l. of the Tyrrhenian coastal deposits proves that the plain has been moderately uplifted since the last interglacial [3]. A younger coastal sector occurs between the Tyrrhenian sandy-coastal ridge and the present shoreline. This belt represents the evolution of a barrier-lagoon system, shifted alternatively landwards and seawards during the Holocene. It includes a composite sand ridge system, elevated 1-5 metres a.s.l., which is partly exposed along the present coast and disappears inland under a muddy depression, rising about 1 m a.s.l. At the beginning of the Holocene, the progradational trend was interrupted by at least three phases of formation of sandy coastal ridges, knows as the Laura ridge (dating to the interval that spans from 5.3 to 3.6 ky BP) and Sterpina ridges (I and II, dating, respectively, to before 2.6 ky BP and about 2.0 ky BP) [15, 16, 18–21].

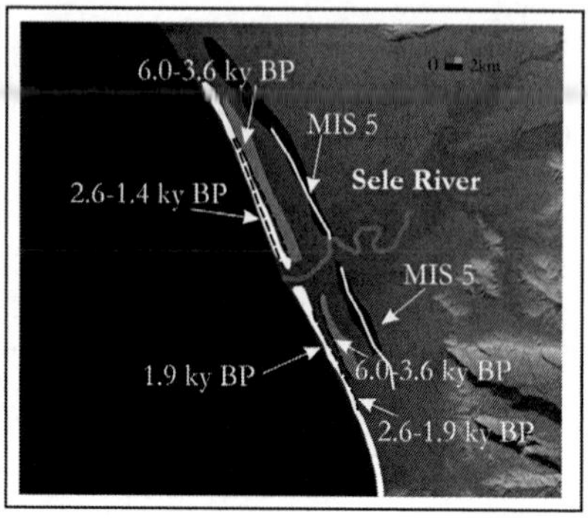

Figure 4: Late quaternary shorelines of the Sele river alluvial-coastal plain. In dark and light grey are the Tyrrhenian and Holocene dune ridges.

The coastal area of the Sele plain could be vulnerable to a relative sea-level rise, storms and flooding due to the presence of large areas with a topographic height lower than 2m. The shoreline from 1200 to 1809 was in a progradation phase, with a mean value of about 0.3-0.4 m/year. Indeed, the coastguard towers built in this period at a short distance from the shoreline are at present located about 200 m from the sea [22, 23]. During this progradation phase, a coastal dune system (Sterpina dunes) was formed. From 1809 to 1908 a strong progradation of the littoral zone occurred over the whole coastal area, reaching its maximum extension at the right of the Sele mouth, while only the Tusciano mouth was affected by erosion [23]. From 1908-1954 the progradational phase is predominant in the areas closest to the Sele mouth, with the exception of its right lobe that has retreated by about 25 m. Since 1954, the coastal area of the Sele plain has experienced an erosional phase [23–25]. Very recent studies using the method of shoreline comparison available in the Digital Shoreline Analysis System (DSAS [26]) have highlighted the shoreline variation from 1870-2009 [27]. Comparison of shoreline variations occurring in all eight investigated time intervals (fig. 5) shows the decrease in the shoreline variation rate from 1870 to 2009 and that the highest rates of coastal retreat are around the river mouth (Sele mouth: maximum erosion −4.8 m/y, Picentino-Tusciano mouths: maximum erosion −1.8 m/y) [28]. This analysis suggests that the shoreline changes along the beaches of the Sele river basin might be mainly associated to a reduction in coarse-grained sediment outflow from the Sele River. Direct measurements of river sediment transport are absent. However, a recent study by Ferrante *et al.* [29] and Vallifuoco *et al.* [30] combining numerical models and core stratigraphy seems to support this hypothesis. The retreat of the shoreline, mainly occurring at the Sele mouth, was also due to the interaction between two longshore currents with opposite directions that generate a cross shore current directed seawards with a direction 220 N. This current coupled with oscillatory motion generates a nested current in the bottom boundary layer that removes sand from the littoral cell and releases it on the inner shelf during major sea storms [29].

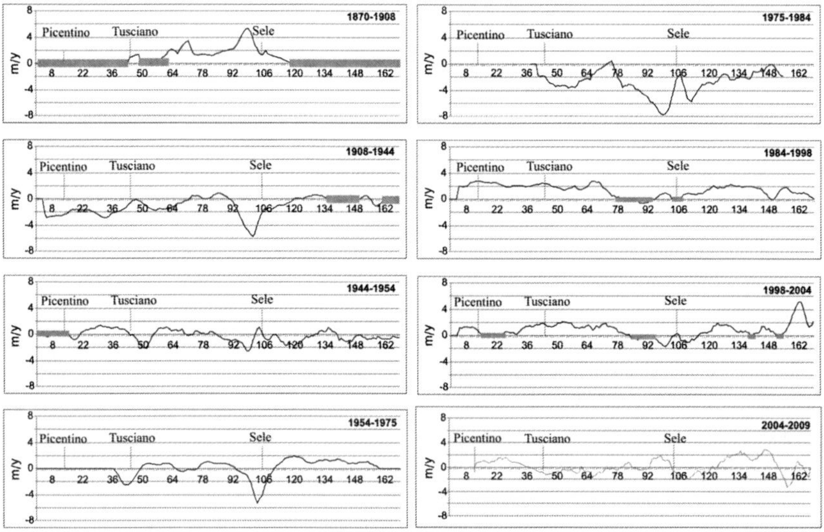

Figure 5: Shoreline variation in different time windows. In the grey areas such variations cannot be defined [28].

4 Erosion potential assessment in the Sele coastal plain

Starting from the method proposed by Gornitz *et al.* [31] to define the coastal vulnerability index as the sum of permanent inundation, episodic inundation and erosion potential, we proposed the following eight factors to define the erosion potential of coastal areas at regional and local scale.

The *condition of the dunal system (DS)* was defined qualitatively, as a relation of three sub-factors:

$$DS = \frac{(DE + DC + DA)_x}{(DE + DC + DA)_{MAX}} \cdot 100 \tag{1}$$

where *DE* is the erosional state of the dune, mainly based on the evaluation of topographic profile and vegetation type; *DC* indicates the spatial continuity of dune system and *DA* the presence or absence of artefacts. The *DS* assumes a normalized percentage with the maximum value in the study area.

The *backshore width (BW)* corresponds to the beach width measured between the dune crest and shoreline position at high tide:

$$BW = \frac{SW_x}{SW_{MAX}} \cdot 100 \tag{2}$$

where SW_x is the maximum backshore width in the study area, while SW_{MAX} is the backshore width along the coastal sub-zones.

The *grain size of the intertidal beach* (*GS*) assesses the influence of the wave climate [32–35] on coastal erosion. It is calculated as follows:

$$GS = \frac{ID50_x}{ID50_{MAX}} \cdot 100 \tag{3}$$

where *ID50* is the maximum grain size at the 50th percentile in the whole study area, $ID50_x$ is the grain size diameter at the 50th percentile calculated for the coastal sub-zones (zoning of the Sele Plain in a 2 km-long sector).

The *rate of shoreline change* (*SC*) was defined, considering both long (100-year) and short term (30-year) variations, as follows:

$$SC = 0{,}75 \cdot SC_1 + 0{,}25 \cdot SC_2 \tag{4}$$

where SC_1 is the short term and SC_2 the long term.

The *effects of the wave climate on the beach* were defined using two variables, the *maximum run-up of storm event* (*XR*) and *tide range* (*XM*), calculated as follows:

$$XR = \frac{R_{2\%}}{SW \cdot sen\beta} \cdot 100 \tag{5}$$

$$XM = \frac{M}{SW \cdot sen\beta} \cdot 100 \tag{6}$$

where$R_{2\%}$ is the run-up due to extreme events as defined by Stockdon *et al.* [36], *M* is the maximum tide range for the study area, *SW* is beach width at each transect and β is the slope of the intertidal beach.

Fluvial system (*FS*) features were defined for each sector as follows:

$$FS = \frac{(MT + LT + AB + RB - DI)_x}{(MT + LT + AB + RB - DI)_{MAX}} \cdot 100 \tag{7}$$

where *MT* is the mouth type, LT indicates the lithology, *AB* the basin area, *RB* is the ratio between the basin area calculated from the dam to the mouth and whole basin area.

The last parameter considered is the *Coastal Erosion Defences (ED)*. This parameter qualitatively describes the conservation state of man-made coastal defences.

The eight factors described above were ranked linearly into five classes (very low, low, moderate, high, and very high) as proposed by [37] Gornitz *et al.* and subsequently adopted by several authors [38–41].

These ranked variables (tab. 1) are combined as the square root of their product and divided by the total number of variables to calculate *coastal erosion potential* (*EP*).

$$EP = \sqrt{\frac{DS \cdot WB \cdot SE \cdot SC \cdot XR \cdot XT \cdot FS \cdot ED}{8}} \tag{8}$$

EP was calculated for 16 zones (labelled from A to P) 2 km long and shown in the erosion potential map (fig. 6). The study area was ranked into five classes, from very high to very low, by using the natural break method which defines the

Table 1: Ranking of factors used to define coastal erosion potential.

Potential coastal erosion	*1*	*2*	*3*	*4*	*5*
Dunal system condition **DS (%)**	< 20	20 - 40	40 - 60	60 - 80	> 80
Backshore width **BW (%)**	> 80	60 - 80	40 - 60	20 - 40	< 20
Grain size of intertidal beach **GS (%)**	< 20	20 - 40	40-60	60 - 80	> 80
Rate of shoreline change **SC (m)**	> 2 Accretion	2 - 1	1- -1	-1- -2	> -2 Erosion
Effects of wave climate on the beach **XR (%)**	< 20	20 - 40	40 - 60	60 - 80	> 80
Effects of tide range on the beach **XT (%)**	< 20	20 - 40	40 - 60	60 - 80	> 80
Fluvial system **FS (%)**	< 20	20 - 40	40 - 60	60 - 80	> 80
Coastal Erosion Defences **ED**	Excellent	Fair	Good	Very poor	Insufficient or absent

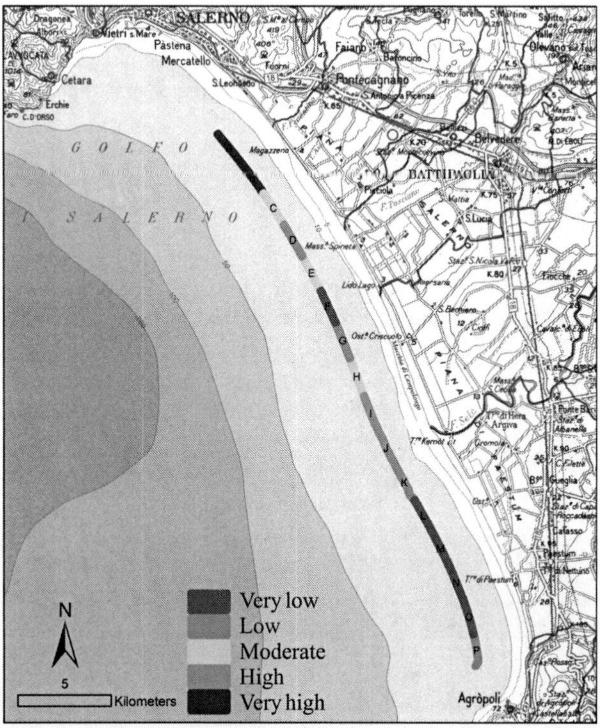

Figure 6: The Sele plain map of coastal erosion potential.

separation points among the frequency values and minimizes the variance in each class. Application of the new method shows that 37% of the coastline of the Sele river plain has a high erosion potential, corresponding to sectors closer to the town of Salerno and the areas on or close to river mouths. By contrast, about 44% of the whole coastline is not subject to erosion, almost all concentrated S of the Sele river mouth. The remaining areas (19%) have an intermediate or moderate erosion potential. High potential erosion is found at the mouths of the main rivers of the Sele plain (Sele, Picentino, Tusciano) and of the Asa stream (transects A, B, D, I, J, K in fig. 6). This is mainly due to the discontinuity or absence of the dune system, the high run-up value and beach width (fig. 6). The opposite features typify the area south of the Sele river and the zone stretching from Lido Lago to the Asa stream (transects F, G, L, M, N, O, P in fig. 6), giving these areas a low and very low erosion potential. Medium hazard potential applies to the zone from the Asa stream and Tusciano river and that close to Lido Lago and Macchia di Capolongo (transects C and H, in fig. 6). This hazard class is mainly due to the condition of the dunal system, the distance from the river mouth and beach width, respectively. The reasons for such differences are to be sought in the set of all elements considered in the calculation model. Human activity, the decrease in sediment transport, and destruction of the dunes over extensive areas are to be considered the prime causes behind such variations.

In conclusion the analysis of the coastal hazard in the Campanian Plains shows that many sectors are exposed to high hazard degrees. De Pippo *et al.* [3] identify for the Volturno Coastal Plain a 55% of coastal area in extreme or high hazard degree, while for the Sarno and Sebeto Plain these hazard classes reach the 80%. On the contrary the Sele Plain is less exposed, in fact only the 37% of coastline is characterized from high to very high hazard degree.

References

[1] IPCC (Intergovernmental Panel on Climate Change) *Fourth Assessment Report – Climate Change 2007*, Cambridge University Press, UK, 2007

[2] De Pippo, T., Donadio, C., Pennetta, M., Petrosino, C., Terlizzi, F., Valente A. *Coastal hazard assessment and mapping in Northern Campania, Italy.* Geomorphology, **97**, 451–466, 2008.

[3] Brancaccio, L., Cinque, A., Romano, P., Rosskopf, C., Russo, F., Santangelo, *N. & Santo, A. Geomorphology and neotectonic evolution of a sector of the Tyrrhenian flank of the southern Apennines (Region of Naples, Italy).* Zeit. Geomorph. N.F., **Suppl. Bd. 82**, 47-58, 1991

[4] Brancaccio, L., Cinque, A., Romano, P., Rosskopf, C., Russo, F. & Santangelo, N. *L'evoluzione delle pianure costiere della Campania: Geomorfologia e Neotettonica.* Mem. Soc. Gegr. It., **53**, 313-336, 1995.

[5] Cinque, A., Aucelli, P. P. C., Brancaccio, L., Mele, R., Milia, A., Robustelli, G., Romano, P., Russo, F., Russo, M., Santangelo, N. & Sgambati, D. *Volcanism, tectonics and recent geomorphological change in the bay of Napoli.* Guide for the excursion 4th Int. Conf. On Geomorphology, 1997.

[6] Amato, V. La risposta di alcuni tipici sistemi morfodinamici della Campania (Italia meridionale) alle variazioni climatiche oloceniche. PhD Thesis. Università degli Studi di Napoli "Federico II". Website www.fedoa.unina/636/., 2006.

[7] Romano, P., Santo, A. & Voltaggio, M., *L'evoluzione morfologica della pianura del fiume Volturno (Campania) durante il tardo Quaternario.* Il Quaternario, **7(1)**, 41–56, 1994.

[8] Cinque, A. & Irollo, L., *Il vulcano di Pompei: Nuovi dati geomorfologici e stratigrafici.* Il Quaternario, **17(1)**, 101-116, 2004.

[9] Santangelo, N., Ciampo, G., Di Donato, V., Esposito, P., Petrosino, P., Romano, P., Russo Ermolli, E., Santo, A., Toscano, F. & Villa, I., *Late Quaternary buried lagoons in the northern Campania plain (southern Italy): evolution of a coastal system under the influence of volcano-tectonics and eustatism.* Boll. Soc. Geol. It., **129(1)**, 156-175, 2010.

[10] De Vivo, B., Rolandi, G., Gan, P.B., Calvert, A., Bohrson, W.A., Spera, F.J., Belkin, H.E., *New constraints on the pyroclastic eruptive history of the Campanian volcanic Plain (Italy).* Mineralogy and Petrology, **73**, 47-65, 2001.

[11] Cocco, E, Crimaco, L; de Magistris, M. A., *Dinamica ed evoluzione del litorale campano-laziale: variazioni della linea di di riva dall'epoca romana ad oggi nel tratto compreso tra la foce del Volturno e Torre S. Limato (Mondragone).* Atti 10° congresso AIOL 1992, 543-555.

[12] Ortolani, F. & Pagliuca, S., *Cyclical climatic-environmental changes in the Mediterranean Area (2500 BP- Present day).* Pages, **11(1)**, 15-17, 2003.

[13] Barra, D., Bonaduce, G., Brancaccio, L., Cinque, A. Ortolani, F., Pagliuca, S. & Russo, F. *Evoluzione geologica olocenica della piana costiera del Fiume Sarno.* Mem. Soc. Geol. It., **42**, 255-267, 1989.

[14] Cinque, A., *La trasgressione versiliana nella piana del Sarno (Campania).* Geogr. Fis. Din. Quat., **14**, 63-71, 1991.

[15] Brancaccio, L., Cinque, A., Belluomini, G., Branca, M. & Delitala, L. *Isoleucine epimerization dating and tectonic significance of upper Pleistocene sea-level feature of the Sele Plain (Southern Italy).* Z. Geomorphol. N.F., **Suppl. 62**, 159-166, 1986.

[16] Brancaccio, L., Cinque, A., Russo, F., Santangelo, N., Alessio, M., Allegri, M. Improta, S., Belluomini, G., Branca, M. & Delitala, L., *Nuovi dati cronologici sui depositi marini e continentali della piana del F.. Sele e della costa settentrionale del cilento (Campania, Appennino Meridionale).* Atti del 74.mo Congr. Naz. della Soc. Geol. It., **A**, 55 – 62, 1988.

[17] Barra, D., Calderoni, G., Cipriani, M., De La Genier`E, J., Fiorillo, L., Greco, G., Mariotti Lippi, M., Mori Secci, M, Pescatore, T., Russo, B., Senatore, M.R., Tocco Sciarelli, G. & Thorez, J.;. *Depositional history and palaeogeographic reconstruction of sele coastal plain during magna grecia settlement of Hera argiva (southern Italy).* Geologica Romana, **35**, 151–166, 1999.

[18] Barra, D., Romano, P., Santo, A., Campajola, L., Roca, V. & Tuniz, C., *The versilian transgression in the Volturno River Plain: palaeoenvironmental history e chronological data.* Il Quaternario, **9**, 445-458, 1996.

[19] Barra, D., Calderoni, G., Cinque, A., De Vita, P., Rosskopf, C. & Russo Ermolli, E., *New data on the evolution of the sele river coastal plain (Southern Italy) during the holocene.* Il Quaternario, **11**, 287–299, 1998.

[20] Amato, V., Avagliano, G., Cinque, A., Cipriani, M., Di Paola, G., Pontrandolfo, A., Rosskopf, M. C. & Santoriello, A., *Geomorphology and geoarchaeology of the Paestum area: modifications of the physical environment in historical times*, Mediterranee, **112**, 2009, 129-135, 2009.

[21] D'Argenio, B., Amato, V., Anzalone, E., Aucelli, P.P.C., Cesarano, M., Cinque, A., Da Prato, S., Di Paola, G., Ferraro, L., Pappone, G., Petrosino, P., Rosskopf, C.M. & Russo Ermolli, E. (2011) *Holocene palaeo-geographical evolution of the Sele river alluvial-coastal plain: new morphosedimentary data from Poseidonia-Paestum area.* Rassegna delle attività di ricerca scientifica e tecnologica del CNR nell'ambito del mare e delle sue risorse, Volume Mare, 2011.

[22] De Pippo, T., Cocco, E., De Magistris, M.A., Efaicchio, M.T., Giulivo, I., Tarallo, F. & Valente, A., *Evoluzione e dinamica litoranea delle spiagge della Piana del Sele (Campania,Italia).* Pact., **25**, 271 – 277, 1989.

[23] Cocco, E. & De Magistris, M.A., *Evoluzione storica e recente del litorale di Paestum (Golfo di Salerno).* Memorie della Società Geologica Italiana, **41**, pp. 697-702, 1988.

[24] Cocco, E. & Iuliano, S., *L'erosione del litorale in sinistra foce Sele (Golfo di Salerno): dinamica evolutiva e proposta di intervento a difesa e tutela della spiaggia e della pineta litoranea di Paestum.* Il Quaternario, **12(2)**, pp. 125-140, 1999.

[25] D'Acunzi, G., De Pippo, T., Donadio A., Peduto, F., Santoro U., Sessa F., Terlizzi, F. & Turturiello, M.D., *Studio dell'evoluzione della linea di costa della piana del Sele (Campania) mediante l'uso della cartografia numerica.* Studi Costieri, **14**, pp. 55-67, 2008.

[26] Thieler, E.R., Himmelstoss, E.A., Zichichi, J.L. & Miller, T.L., *Digital Shoreline Analysis System (DSAS) version 3.0: An ArcGIS extension for calculating shoreline change.* U.S. Geological Survey Open-File Report 2005-1304, 2005.

[27] D'Argenio, B., Alberico, I., Amato, V., Aucelli, P.P.C., Di Paola, G., Pappone, G. & Giordano, L. *Historical change and trend evolution assessment of the coastline between Salerno and Agropoli, Sele Plain (Southern Italy).* Rassegna delle attività di ricerca scientifica e tecnologica del CNR nell'ambito del mare e delle sue risorse, Volume Mare, 2011.

[28] Alberico, I., Amato, V., Aucelli P.P.C., D'Argenio, B., Di Paola, G., Pappone G. *Historical change and trend evolution assessment of coastline between Salerno and Agropoli, Sele Plain (Southern Italy).* Journal of Coastal Research, *submitted.*

[29] Ferrante, V., Budillon, F., Iorio, M., Lubritto, C., Monaco, M., Pugliese Carratelli, E., Russo, S.D. & Vicinanza, D., *Tracing possible offshore sand movement during extreme sea storm events: an example from the Salerno Bay (Southern Tyrrhenian Sea) by comparing field data with numerical model results.* Rassegna delle attività di ricerca scientifica e tecnologica del CNR nell'ambito del mare e delle sue risorse, Volume Mare, 2011.

[30] Vallefuoco, M., Lirer, F., Ferraro, L., Sprovieri, M., Capotondi, L., Bellucci, L., Albertazzi, S., Giuliani, S., Angelino, A., Iorio, M. &

Iavarone, M. *Evidences of natural environmental changes and human impact reflected by Foraminifera in the southern Tyrrhenian Sea for the last 520 years.*, 2010
[31] Gornitz, V.M., Daniels R.C., White T.W. & Birdwell K.R., *The development of a coastal risk assessment database: vulnerability to sea-level rise in the U.S.A. southeast.* Journal of Coastal Research, **Special Issue 12**, pp. 327 – 338, 1994.
[32] Bascom, W.N., *The relationship between sand size and beach-face slope.* Am. Geophys. Union Trans. 32(6), pp. 866-874, 1959
[33] Komar, P.D., *Beach Process and Sedimentation.* Prentice-Hall, Englewood Cliff: New jersey, 1998.
[34] Wiegel, R.L., *Oceanographical Engineering* (II ed.). Dover, Mineola: New York, 2005.
[35] Williams, A. & Micallef, A., *Beach management: Principles & Practice.* Earthscan: London, 2009.
[36] Stockdon, H.F., Holman, R.A., Howd, P.A. & Sallenger, A.H., *Empirical parametrization of setup, swash, and runup.* Coastal Engineering **53(7)**, pp. 573 – 588, 2006
[37] Gornitz, V.M., Beaty, T.W., Daniels, R.C., *A coastal hazards data base for the U.S. West Coast.* ORNL/CDIAC-81, NDP-043C, Oak ridge national laboratory, Oak ridge, Tennessee, 162 pp., 1997.
[38] Diez, P.G., Perillo, G.M.E. & C., Piccolo, *Vulnerability to Sea-Level Rise on the Coast of the Buenos Aires Province.* Journal of Coastal Research, **23(1)**, pp. 119-126, 2007.
[39] Doukakis, E., *Coastal vulnerability and risk parameters.* Eur Water **11(12)**, pp.3–7, 2005
[40] Pendleton, E.A., Thieler, E.R., Williams, S.J. & Beavers R.L., *Coastal Vulnerability Assessment of Padre Island National Seashore (PAIS) to Sea-Level Rise* U.S. Geological Survey Open-File Report 2004-1090, Electronic Book, 2004.
[41] Thieler, E. R. & Hammar-Klose, E.S., *National Assessment of Coastal Vulnerability to Sea-Level Rise: Preliminary Results for the U.S. Pacific Coast.* U.S. Geological Survey Open-File Report 00-178, Woods Hole, Massachusetts, 2000

Integrated coastal management for local municipalities in Latvia: sustainability governance and indicator system

R. Ernsteins, J. Kaulins, E. Lice & A. Štals
Department of Environmental Management, University of Latvia, Latvia

Abstract

Municipal audit for integrated coastal management (ICM) development for local municipality includes analysis of stakeholders, document frame and planning process analysis, vertical governance assessment combined with sectorial analysis for all four dimensions of sustainability; nature environment, economics environment, social environment and also governance and communication dimension in Saulkrasti coastal municipality. ICM guidelines model frame were elaborated during this collaboration research project by University of Latvia performed in R&D cooperation with local authority and with involvement of all local stakeholder groups. Further on related indicator system based on these sustainability dimensions and designed ten main integrated work directions were elaborated, including 24 thematic groups with 55 indicators.
Keywords: sustainable development dimensions, governance and communication dimension, collaboration principle, indicators system.

1 Introduction of ICM from international to local level

Sustainable coastal development, as widely acknowledged [1–4], has to be implemented employing integrated coastal (zone) management frame as for many years introduced and here are generic definition and additional explanation available from European Commission [1, 2]. Integrated Coastal Zone Management (ICZM) is a dynamic, multi-disciplinary and iterative process to promote sustainable management of coastal zones. "Integrated" in ICZM refers to the integration of objectives and also to the integration of the many instruments needed to meet these objectives. It means integration of all relevant

WIT Transactions on Ecology and the Environment, Vol 149, © 2011 WIT Press
www.witpress.com, ISSN 1743-3541 (on-line)
doi:10.2495/CP110031

policy areas, sectors, and levels of administration. It means integration of the terrestrial and marine components of the target territory, in both time and space.

Also eight key principles for successful ICZM are officially delivered in 2000 [1]: broad "holistic" perspective; long term perspective; adaptive management during a gradual process; reflect local specificity; work with natural processes; participatory planning; support and involvement of all relevant administrative bodies; use of a combination of instruments.

All this has to be taken into account when planning for local level ICM as well as appropriate system of indicators [3–5, 7] shall be introduced, e.g. like United Nations (UNCSD, 2001) has developed indicators for sustainable development in order to: translate physical and social science knowledge into manageable units of information that can facilitate the decision-making process; help to calibrate and measure progress towards sustainable development goals; provide early warning to prevent damage; and communicate ideas, thoughts and values.

2 ICM model frame for Saulkrasti municipality

Following are ICM development studies [6] in Latvia, particularly using Saulkrasti municipality case study research example, based on previous experiences [7–9] and approaches designed and implemented [10]. Description of Saulkrasti municipality in brief would include following. Geography – area 48 km^2, town area is 6,8 km^2, shoreline 17 km, 45 km from capital city of Riga, four small rivers. Demography and habitat structure – 6105 habitants (2009) in four historical parts (Pabaži, Pēterupe, Neibāde, Zvejniekciems) and almost 13 000 summer houses inhabitants in season. Entrepreneurship and factors, having impact on area development – mainly Skulte port with cargo turnover 451 thsd. tons at 2008. Tourism facilities – main resources of nature and environment are 17 km sand beaches, statutory designed nature park area "Piejura". The most important risks – coastal erosion, transport risks. Saulkrasti coastal municipality sustainability audit and further on ICZM guidelines were elaborated during collaboration research project by University of Latvia (with involvement of environmental management master program students – ViPa16 group) during 2009-2010 in R&D cooperation with local authority and with involvement of fall local stakeholder groups [6]. These guidelines are based on studies of legal framework, national, regional and local planning and management documents and case study research field work: seminars, interviews, questionnaires, etc. Municipal sustainability audit was performed taking into account three main sustainable development capitals - nature environment, economic environment (particularly, emphasizing tourism environment (reviewed separately, meaning the great importance for Saulkrasti municipality development), social environment (incl. culture, health, education etc. subsectors) – as well as adding fourth important (even horizontal and cross-sectoral) capital as governance and communication. In the table 1 we can see all mentioned sustainability dimensions and, subsequently, recognized and structured frame of 10 main integrated problem areas to be seen as main work directions, as well as ICZM work sub-directions.

Table 1: Saulkrasti ICM system frame.

Dimension of sustainability	Integrative problem area – directions	Work sub-directions
Nature environment **D**	The protective zone for coastal dunes: erosion, managing, biodiversity	Permanent managing of coastal dunes protection area; Realisation of conservation for biodiversity; Supporting of collaboration and dialogue among different stakeholder groups; Corresponding construction at coastal dunes protection area; Corresponding activities for tourism and recreation
	Strategic management in public services sector	Decrease waste in nature environment; Decrease emissions of sewage; Decrease emissions from fossil fuel from industry and public services;
Economical environment **E**	Port complex	Further development of port aquatoria and landings; Rational, poly-functional use of port territory; Development of access roads corresponding to perspective needs
	The development planning	Social partnership, involving all stakeholder groups; Strengthening of municipal planning capacity; Elaboration of planning documents and actual amendments
Economical environment: tourism **ET**	Resources of nature, cultural history and recreation as preconditions for tourism development	Resources of nature and cultural environment for tourism development; Development of infrastructures for nature and cultural environment resources; Development of human resources; Information system and forming of environmental awareness; Strategic planning for using the nature and cultural environment resources
	Entrepreneurship in promotion of tourism development	Strategic planning for tourism at municipal level; Project management and development in the tourism branch; Education for tourism entrepreneurship; Marketing of tourism
Social environment **S**	Quality of life for inhabitants	Further development of infrastructures for the public services; Improving the households' comfort and energetic efficiency; Development of public transport and transport infrastructures; Improving the town environment quality;
	Polycentrics or existing of several centers with equivalent dominance in the municipality area	Improve and intensify the communication between spatial parts of the community and inhabitants from different parts; Even up accessibility of public services in different parts of municipality; Promote tourism and entrepreneurship activities in all areas of the municipality;

Table 1: Continued.

Dimension of sustainability	Integrative problem area – directions	Work sub-directions
Governance and communication **P**	Collaboration governance for coastal municipality	Development of collaboration among governance stakeholder groups; Development of vertical and horizontal integrative thematic collaboration; Development of tools for collaboration; Development of assessment collaboration;
	ICZM coastal communication	Develop system of co-ordination for coastal communication, Promote internal and external communication of stakeholders; Design and develop unitary space for coastal communication;

ICZM at Saulkrasti municipality can be given a look as case of good praxis and a model case for other coastal municipalities in Latvia with following recognized adavantages : auditing all sustainability sectors and their interlinkage, particularly, within complex coastal territory; definition then of integrated problem areas (see table 1) as principial stage at integrated coastal planning and management process; both auditing and preparing guidelines for whole sustainable governance/management cycle; measuring coastal sustainable development with indicator method – full scale sustainability indicator system as for the first time in Latvia. Further on indicator system based on sustainability dimensions and designed 10 integrated work directions were elaborated, including 24 thematic groups with 55 indicators.

3 Indicator system for coastal management in Saulkrasti

3.1 General characteristics and structure

An indicator system for measuring coastal sustainability differs from the general case by its spatial specifics: the coastal zone is formed by a coastal line with the related set of other geospatial elements [1]. The indicator system though which coastal sustainability is assessed should therefore be able to at least differ the coastal zone from the inland and provide a comparison, to establish the origin of impact factors on the coastal status and development trends, and to create understanding of the distribution of coastal impacts within the governance territory. Ideally, the term `coastal zone` should apply to a territory where the specific coastal impacts can be detected, and vice versa – a territory which impacts the developments on the coast and its proximity, as these impacts [10]: may in advance be unknown precisely enough; may change over time; differ for different factors; the specifics of spatial distribution of the data used may prevent their correct differentiation. In practice, the term 'coastal zone' is therefore applied to a relative territory which – within a single system - may in addition be applied in a number of ways depending on the data character.

Saulkrasti county, which is comparatively small, relatively urbanised (for Latvian conditions) Baltic Sea coastal territory (Fig. 1), had the coastal integrated development guidelines elaborated for it in 2009–2010 within the University of Latvia COBWEB project [6], as the central component of municipal integrated governance. Special attention was devoted to measuring sustainability, and for this purpose, a system of indicators was developed (see Table 2). The system was elaborated based on the analysis of four dimensions of sustainability (natural, social and economic environment, governance and communication) and integrative problem areas as defined on their points of intersection. In Saulkrasti municipality, the coast may be defined in the following ways based on the character of indicator data spatial distribution: coastal geometrical zone, which is formed as a coastal geometrical buffer; point-shaped measurement locations on the coast; coastal dune protection zone with adjacent protected biotopes; coastal 5 km-wide zone of limited economic activity, which can be adjusted depending on local geographical conditions [9, 10]. For the last two, it is defined with the help of municipal spatial planning, in Saulkrasti its spatial plan is still under preparation.

The nature of data determines not only coastal definition types but also the spatial relations to the coast by indicators themselves. Here, the following cases can be distinguished [10]: special coastal indicators directly characterise some values characteristic of the coast only; coastal discernible indicators which characterise elements not directly coast and coastal relatively discernible indicators; indicators non-applicable to the coast, which characterise a factor in the overall territory. In the indicator system for Saulkrasti 18% of indicators are

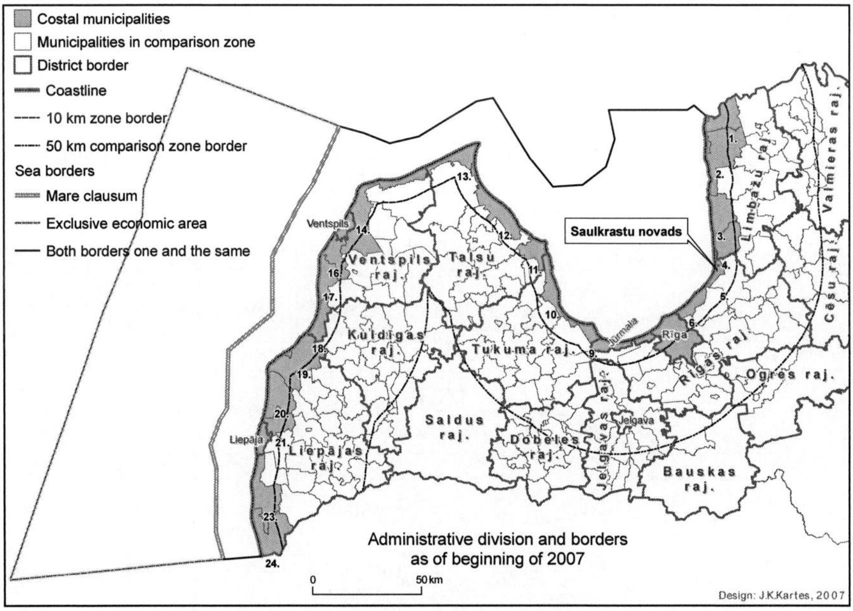

Figure 1: Saulkrasti and coasts in Latvia.

Table 2: The system of sustainability indicators in Saulkrasti.

Dimension of sustainability	Thematic subdivision	Indicator
I. Nature environment **D**	**D1.** Green frame status	D1.1. Cut and restored wooden areas
		D1.2. Cutting permissions in non-wooden lands
		D1.3. Land transformation from non-developed to developed types
	D2. Potential loads on the environment from public utilities	D2.1. Total waste amount and coastal litter
		D2.2. Satisfaction of inhabitants with waste management
		D2.3. Providing of households by centralised drinking water support and sewerage
		D2.4. Emissions from sewage treatment plants
		D2.5. Financial resources for public utilities
	D3. Air quality and climat change factors	D3.1. Using of environmental-friendly fuel at public and industrial sector
		D3.2. Emissions of greenhouse-effect gases at public and industrial sector from fossil fuel
		D3.3. Snow cover condition
	D4. Surface water quality	D4.1. Bathing water quality
	D5. Land use development	D5.1. Permissions for building
	D6. Nature risks	D6.1. Number of stormy days
		D6.2. Coastal erosion
II. Economic environment **E**	**E1.** Economically active people	E1.1. Working-age inhabitant proportion from declared inhabitants
	E2. Municipal budget	E2.1. Structure of municipality budget incomes and expensions
	E3. Traffic routes	E3.1. Proportion of hard-covered roads in all state and municipal roads and density of network of roads
		E3.2. Public transport traffic
	E4. Skulte port development	E4.1. Cargo turnover in Skulte and Salacgrīva ports
		E4.2. The ship visiting in Skulte port
	E5. Tourism characteristics **ET**	E5.1. Number of tourism services and distribution by types of them
		E5.2. Bed number in tourism accommodation
		E5.3. Bed space occupancy in tourism accommodation

Table 2: Continued.

Dimension of sustainability	Thematic subdivision	Indicator
II. Economic environment **E**	**E5.** Tourism characteristics **ET**	E5.4. Personnel at tourism industry
		E5.5. Financial resources for Tourism Information Center and numbers of attendance
		E5.6. Environment friendly tourism accommodation
III. Social environment **S**	**S1.** Health care characteristics	S1.1. Provision by health care personnel
		S1.2. Loading for family doctors
	S2. Supporting for cultural environment	S2.1. Municipal funding for supporting of cultural heritage
		S2.2. Municipal funding for supporting of cultural events
		S2.3. Number of cultural events
	S3. Employment and entrepeneurship	S3.1. Employment and their characteristics
		S3.2. Unemployment
	S4. Social care and social security	S4.1. Degree of social exclusion
		S4.2. Crime statistics
	S5. Education system characteristics	S5.1. Number of pupils in "key" classes
		S5.2. Mutual payments among the municipalities
	S6. Social life quality	S6.1. Habitat comfort level
		S6.2. Public services accessibility
		S6.3. Average incomes *per capita*
IV. Governance and Communi-cation **P**	**P1.** Activities for environment manintenance	P1.1. Number of people involved in environment maintainance and municipal financial resources for it
	P2. Information of society about environmental events	P2.1. Publications in local mass media about environmental questions
		P2.2. Satisfaction of people by municipal communication with society
	P3. Activities in nongovernmental sector	P3.1. Local environment-oriented NGOs
		P3.2. Number of environment friendly activities, organized by all NGOs
V. Integral indicators **I**	**I1.** Number of inhabitants	I1.1. Number of declared and seasonal inhabitants
	I2. Area development index	I2.1. Area development index

Table 2: Continued.

Dimension of sustainability	Thematic subdivision	Indicator
V. Integral indicators **I**	**I3.** Area attractivity index	I3.1. Area attractivity index
	I4. Opinions of population	I4.1. Opinion for municipal management
		I4.2. Opinion for environment quality
		I4.3. Opinion for coastal zone status and facilities
		I4.4. Opinion for municipal planning
		I4.5. Personal attitude and connection with sea

special coastal, 16% – coastal discernible, 32% – relatively coastal discernible 0and 34% has no coastal description function. It is estimated that, by improving data collection and storage quality, the proportion of coastal discernible indicators may increase. Within the small municipality extended along the coast the majority of key development factors, problems and opportunities have a directly relation to the coast.

The integrated environmental management cycle approach applied in Saulkrasti municipality was based on municipal situation analysis in sustainability dimensions and on segmentation of priority integrative problem areas at dimension intersection points. The indicators were selected separately for characterising both the sustainability dimensions and integrative problem areas. By way of combining both resultant systems and assessing how the indicators refer to sustainability components, we divided all indicators into 4 groups [10]: sub-sectoral indicators and sectoral indicators – describe governance level-specific aspect or the whole sector of the respective sustainability dimension; integrative indicators – describe integrative problem areas and other processes which concerns at least two sustainability dimensions; integral indicators – describe the key, more general pointers of the governed system that characterise a given governance system in its entirety and/or compared to other similar systems.

Sectorial indicators are of 20%, integral ones – 16% of the system. Others are integrative ones and almost 2/3 of them are related directly to the integrative problem areas of ICZM in Saulkrasti. Classical dimensions of sustainability are almost equally represented in the system, with the governance and communication dimensions less represented, as pointers characterising these and meeting indicator requirements are more difficult to find. The prevalence of integrative indicators in the system stems from the approach to planning itself, which is based exactly on an integrative perspective of seeing sustainability dimensions in their interactions. Separate sectors (e.g., tourism) are singled out when the related issues is significant enough for the development and welfare of the entire territory. The integral indicators also include separate indicators selected to characterise particular dimensions, as these bear a considerably larger informative load, but are overall designed to characterise resident attitudes and opinions. Their number is comparatively small; in case of a bigger proportion, there is a risk of obtaining too general information, which provides an insufficiently detailed picture for the purposes of practical action and decision-making.

3.2 System building, implementation and documentation

The building and implementation of an indicator system is a process consisting of several stages, which have now been largely completed in Saulkrasti.

First, the development of an indicator system according to the results of sectoral and integrative problem analysis took place. Indicator selection for the system was carried out in a multiple-level scheme. Initially, all proposals by experts and working groups were collected without a critical evaluation; the number of proposed potential indicators reached over 100. Then, the values were dropped which were impossible to measure. A number of parameters were rejected where it was clear that no possibility would exist to obtain the required data, or where compliance with the indicator technical requirements was insufficient. The most significant exceptions were a number of pointers that can be obtained through opinion polls or voluntary monitoring; these were included in the indicator list won the condition that the measurements required would in future be done periodically.

Evaluation expert workshop took place in Saulkrasti (September, 2010), in which the indicator system was presented to municipality experts and the general public. This was followed by work in focus groups to evaluate the proposed system. The participants split into groups according to the interest and competence principle, with one group analysing indicators in the governance and social environments, and the other – in the economic and natural environments. Both groups concluded that the proposal is sufficient and adequately substantiated; the proposed additions were more concerned with the methodological approach in the calculation and interpretation of results. In addition, a proposal was made to apply indicators in the evaluation of sports and life-long learning events as well, which can, in fact, be included as additional parameters among existing indicators.

Summarising the conclusions of both groups, an assertion can be made that by introducing an indicator system for measuring sustainability, a municipality gets: comprehensive and well-arranged information on development and sustainability processes taking place in its territory, and an obvious comparison to its neighbours and competitors; review on the coastal processes and impacts, also in comparison to the inland part of the territory; effective instrument which allows for assessment of success in planning document implementation; forecasting instrument for planning further action, information on resident opinions and opinion changes. Indicators also serve as a powerful communication instrument in demonstrating governance effectiveness to the public and convincing the public of investments or other measures required.

Currently, the development of several indicator methodologies is in different stages of elaboration. During indicator calculations, reports are prepared for each indicator individually. This is done by using a template, the key requirements for which are indicated in the methodologies of the respective indicators. A review of the system values in general is, of course, also prepared. The review is to contain a public part as well, which might be part of the municipality's annual report. It includes fact sheets for individual indicators and the overall sustainability assessment. Indicator system's documentation, which is prepared in the implementation stage, is a relatively independent component of the municipal development strategy. It should be arranged as a text-book consisting of brief overall description of the indicator system in the form of a table, as well as the set of indicator calculation methodologies.

4 Saulkrasti municipality: short indicator assessment

In the planning practice, a number of coastal issues are regulated through national legislation; this, besides quite well keeping general public interests, however, limits planning flexibility for local government in particular local sustainability circumstances, but wide integration work possibilities still do exist. The coastal dune protection zone is in Latvia defined (generally) as a 150m-wide belt in residential areas and a 300m-wide belt beyond residential areas, in which mainly construction and other anthropogenic loads have been limited. A zone of limited economic activity is 5 km wide (in Saulkrasti municipality case covering almost all territory) and has limitations for some types of industrial production, extraction of mineral deposits and placement of waste management objects.

The developed indicator system in Saulkrasti cannot be said to function with maximum efficiency yet. This is due to difficulties in obtaining information for a number of agreed indicators; parts of indicator measurements are carried out for the first time, which does not yet allow evaluating existing trends. However, when analysing the values already obtained, we have come to a number of sustainability governance characterizing aspects concerning both sectoral and integrative approach for coastal management in Saulkrasti county, which have to be further studied, discussed with all stakeholders and implemented, but complementary with established ICM framework and main work directions (see, table 1) planned:

1. Governance environment and communication. The coastal area is the key geographical spatial factor influencing the development of Saulkrasti, which means highest potential as well as creating significant problems at the same time. In existing municipal planning documents of Saulkrasti municipality, coastal issues have been integrated relatively poorly – in fact, only as much as is required under the national regulatory framework. This means that the coastal dune protection zone and coastal zone with limited economic activity have been established, providing for relevant activity limitations.

2. Natural environment. Seasonal pollution, damage inflicted by vehicles and tourism pressure in the dune zone and in other forests are of real threat. Renovation and extension of water and sewerage systems required and started as the systems currently cover the built-up areas insufficiently and are in a poor technical condition, but the level of public knowledge on the status and problems can be considered as very good.

3. Social environment. Symbolism of town and the whole municipality takes out significance of seashore and internal waters. Opportunities afforded by the coast are insufficiently reflected in the territory's cultural environment, except part of mass scale events during summer.

4. Economic environment. Business (except tourism) is relatively little affected by proximity to the sea; however, it can use it to its advantage. The key facilitating factor for business is Skulte port. It is relatively little affected by seasonality. Tourism infrastructure is not functionally closely integrated with the coast, although its activities are largely determined by proximity to the sea and seasonality. Only some tourist objects are direct coastal elements.

Aware of the drawbacks and possibilities, Saulkrasti county municipality is currently working on a new set of development documents, setting coastal impacts and opportunities as one of the key tasks. There are no coastal experts in the municipal administration, however, the importance of these issues has been duly acknowledged and their integration is consistently requested from the spatial and development planning experts involved.

5 Conclusions

Sustainability indicators play a vital role in the sustainability integrated planning and management model as a multiple instrument of situation analysis, prognosis and development strategy. In case of the coastal area, indicators acquire an additional importance as they allow for distinguishing and understanding coastal influences, the extent of their expansion within the territory, and the intensity of the influences in its different parts. The assessment of the spatial factor is even more effective due to the use of geographical information systems both during the indicator calculation process and as a communication instrument – to reflect the results in a form understandable not only to experts but also to the public at large and to decision-makers.

In Latvian circumstances, this sustainability governance and indicators system model approach is new, both in theoretical developments and definitions as well as being local practice oriented, in fact applied in the local municipal planning practice for the first time as one of the key analysis and governance assessment and communication instruments. Saulkrasti municipality have evaluated this jointly elaborated approach and acknowledged the developed indicator system as very significant. The municipality is planning allocation of financing to the further development of the indicator system and implementation in the municipality's everyday work during spring-summer 2011, and to the training of municipal specialists in the practical use of the indicator system.

References

[1] European Commission. Communication From The Commission To The Council And The European On Integrated Coastal Zone Management: A Strategy For Europe Parliament. Brussels, 2000
[2] European Commission, Communication from the Commission - Report to the European Parliament and the Council: an evaluation of Integrated Coastal Zone Management (ICZM) in Europe. Brussels, 2007
[3] Shipman, B., Stojanovic, T. Facts, Fictions, and Failures of Integrated Coastal Zone Management in Europe, Coastal Management, 375-398, 2007
[4] Pickaver A., Salman A. EU Policies and ICZM implementation in Europe: furthering and measuring sustainability. Proceedings of the 9th International Conference Littoral 2008, November, Venice, Italy, 8p, 2008
[5] Pickaver, A., Further testing of the approved EU indicator to measure the progress in the implementation of integrated coastal zone management in Europe. Proceedings. International Symposium on Integrated Coastal Management, Arendal, Norway, 2008
[6] EC Interreg project COBWEB: http://www.hyria.fi/cobweb/about_cobweb
[7] Indicator Guidelines - To adopt an indicators-based approach to evaluate Coastal sustainable development. EC Interreg project DEDUCE Consortium, Government of Catalonia, Barcelona, 2007 – *www.deduce.eu*
[8] Ernsteins R. Integrated environmental communication management for coastal regions. E-Proceedings. 9th International Conference "Littoral 2008" – A Changing Coast: Challenge for the Environmental Policies, European Union for Coastal Conservation, Venice, Italy, 2008
[9] R.Ernšteins, J.Kauliņš, I.Kudreņickis. Sustainable development indicator system studies in Latvia, Sustainable Development and Planning IV, WIT Press, UK, 2009
[10] J.Kauliņš, R.Ernšteins, I.Kudreņickis. Sustainable development indicators for integrated coastal management: definition area, spatial properties and system, Sustainable Development and Planning, WIT Press, in print, 2011

Coastal adaptation to sea level rise along the Nile delta, Egypt

Y. Eldeberky
Helwan University, Department of Civil Engineering, Faculty of Engineering, Cairo, Egypt

Abstract

Global warming and climate change processes are expected to raise sea levels. This paper represents a contribution to the efforts of investigating the impact of sea level rise (SLR) on the Nile delta coastal zone and identifies anticipatory and adaptation measures that may be appropriate today in spite of current SLR uncertainties. SLR was calculated by applying the quadratic equation for 10-year intervals using 1980 as the base year. Despite large variation between SLR predictions, the results indicated that SLR is accelerating, with estimated value for the present (21st) century in the order of 0.6 m, although with a margin of as much as ±0.4 m. Local land subsidence in the Nile delta would exacerbate the impacts of rising seas. It was estimated that with a 1.0m SLR, about 4500 square kilometers representing 19% of the delta area will be submerged and that 6.1 million people will be affected. These potential impacts of SLR would be serious but manageable if appropriate actions are taken. The following adaptation measures to the impact of SLR in the Nile delta coastal zone were identified: maintaining and building coastal protection structures, restoration of sand dunes along the shore, preserving existing wetlands, setting regulations to control development in vulnerable areas, change of land use, relocation of infrastructures in the landward side, and development of comprehensive monitoring and early warning systems.
Keywords: sea level rise, coastal zone, impact assessment, adaptation, the Nile delta.

1 Introduction

Global warming causes significant threats to the environment, human welfare and socio-economic activities, IPCC [1]. It could cause sea levels to rise by

WIT Transactions on Ecology and the Environment, Vol 149, © 2011 WIT Press
www.witpress.com, ISSN 1743-3541 (on-line)
doi:10.2495/CP110041

expanding ocean water, melting mountain glaciers, and perhaps eventually causing polar glaciers to melt and slide into the oceans. As seas rise, many areas of the coasts will be submerged, with increasing wave attacks. Shoreline retreat will be accelerated. In addition to expected inundation events, severe and frequent storms will cause storm surges on coastal areas, Dasgupta *et al.* [2]. Fortunately, many of the adverse consequences can be avoided by taking timely measures in anticipation of SLR.

The impacts of SLR will not be globally uniform, because of local variations in topography, wave climate, longshore currents, and storm frequencies. Low gradient coastal landforms most vulnerable to inundation include deltas and estuaries. SLR is expected to threat the Nile delta coastal zone, water resources, agricultural productivity and population (El-Raey [3]). Therefore, it is important to adapt to SLR in the Nile delta coastal zone not because there is expected threats, but because there are opportunities to avoid adverse impacts by acting now. Such consequences can be avoided by implementing measures in anticipation of SLR. This paper examines the implications of future SLR on Nile delta coastal zone and identifies anticipatory and adaptation options that may be appropriate today in spite of current uncertainties.

The paper is arranged as follows. The Nile delta coastal zone is described in section 2. Potential implications of SLR on the Nile delta coast are summarized in section 3. Coastal adaptation options to SLR and their implications, environmentally and socio-economically, are discussed in section 4. Adaptation measures in the Nile delta coastal zone are identified in section 5. Finally, section 6 presents the summary and conclusions.

2 The Nile delta coastal zone

The Northern coastal zone of Egypt extends along the Mediterranean coast (995km of coastline). The dominant feature of Egypt's northern coast is the low lying delta of the River Nile, with its large cities, industry, agriculture and tourism. About 25% of the populations (total 80 million) live in the low elevation coastal zone. Most of these people live in and around a number of very important and highly populated industrial and commercial cities: Alexandria, Port Said, Damietta, and Rosetta, fig. 1. The Nile delta and the Mediterranean coast include 30-40% of Egypt's agricultural production, and half of Egypt's industrial production. The three main delta lagoons are: Lake Idku, Lake Burullus and Lake Manzala produce over 60% Egypt's fish catch.

The Nile delta coastal zone is most vulnerable to SLR due to its relatively low elevation and local land subsidence. Most of the 50 km wide land strip along the coast is less than 2 m above sea-level and is protected from flooding by a coastal sand belt formed by the sediment discharge of the Rosetta and Damietta branches of the Nile. This protective sand belt is facing rapid erosion, which has been a serious problem since the construction of the Aswan dam. Many studies, [4–6], have shown that the entrapment of sediment in Lake Nasser behind the High Dam, and effects of barrages and river control structures have contributed to erosion along the Nile delta coast.

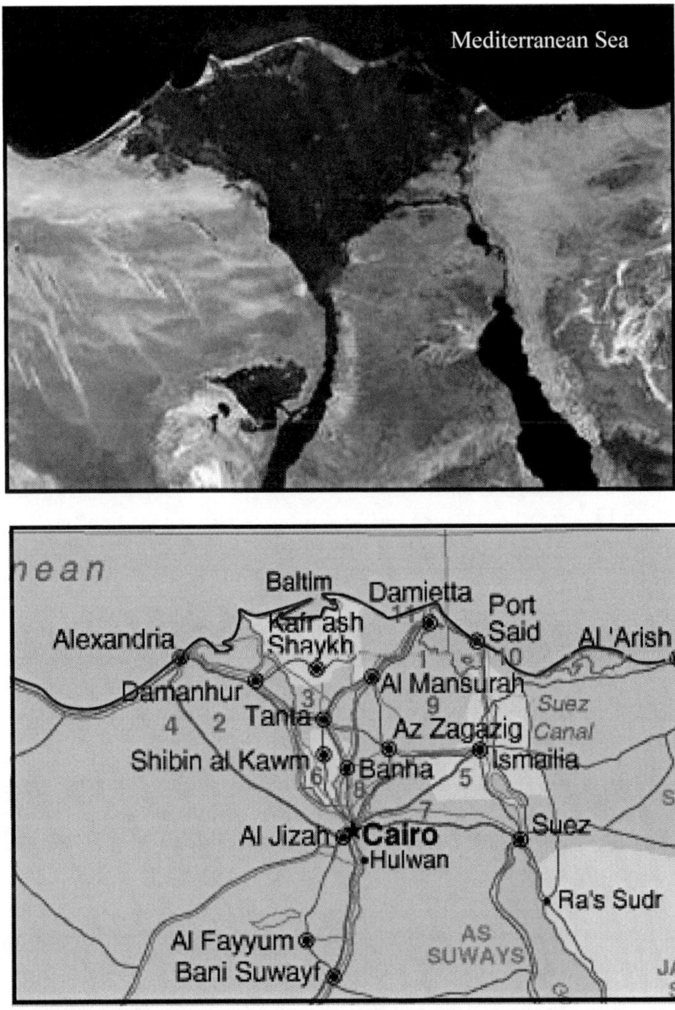

Figure 1: Top: Satellite image of the Nile delta (Source: Google) Bottom: map of the Nile delta and main cities.

3 Potential impacts of sea level rise on the Nile delta

The Nile delta is subject to shoreline changes, subsidence, and sea level rise resulting from climate change. The present global trend of SLR receives much attention worldwide. New insights in the earth climate change appear and new models are being developed. There is general agreement that global SLR is accelerating, with estimated values for the present century between 9 and 88 cm between 1990 and 2100, IPCC [7].

3.1 Prediction of SLR

Egypt is considered one of the top five countries expected to be mostly impacted with about 1.0 m SLR in the world (Dasgupta *et al.* [8]). The accelerated SLR is due to few factors: thermal expansion of the oceans, glacier melting and geological subsidence due to natural or human induced causes (especially in the delta coastal zone).

Local (or relative) changes in sea level depart from the global mean trend due to regional variations in oceanic level change and geological subsidence; it is relative sea-level change that drives impacts and is of primary concern to coastal managers. Relative SLR is expressed as follows:

$$R(t) = Rg(t) + Ro(t) + mt \tag{1}$$

where $R(t)$ = local (or relative) SLR (in meter) in year t, $Rg(t)$ = global SLR in year t, $Ro(t)$= regional sea level change in meter induced by oceanic changes in year t, m = vertical land movement, i.e., geological uplift/subsidence, in meter per year, and t = number of years in the future with respect to a base year.

Global SLR scenarios in year t are expressed using a quadratic equation, NRC [9],

$$Rg(t) = at + bt^2 \tag{2}$$

where a = incremental SLR in meter per year, and b = acceleration factor $(\text{m/year})^2$. The coefficients a and b can be modified to reflect new scientific information on global SLR (Nicholls and Leatherman [10]).

Global SLR $Rg(t)$ could be obtained directly from scenario generators that connect simple climate models with sea level models. Given the uncertainties about global SLR, it is important to select a realistic value for the analysis. Therefore, a maximum scenario in which $Rg(t)$ equals 1.0 m in year 2100 is quite appropriate for such analysis. Currently, there is insufficient understanding to develop detailed scenarios about regional sea level change. Therefore, the value for $Ro(t)$ could be assumed to be zero until more firm information appears.

Values for the vertical land movement m, due to geological uplift/subsidence, can be assessed from geological analysis, geodetic surveys, and the analysis of long-term tide gauge measurements (El-Nahry and Doluschitz [11]). It should be noted that the vertical land movement m, in eqn. (1), determines the departure of relative SLR from global SLR and that m is assumed to be linear and stays unchanged with respect to time in the future.

Some estimates of vertical land movement indicate that the northern delta region is subsiding at a rate that varies from about 2 mm/year at Alexandria to about 2.5 mm/year at Port Said, in addition to tide range about 20 cm.

The computed scenarios of SLR along the Nile delta coast over the present century are shown in fig. 2. In the calculations, it was assumed that 1980 is the base year. The results show three scenarios representing high, average and low trends of SLR scenarios. The expected average SLR scenario by the end of this century is 0.6 m with a margin of as much as ±0.4 m.

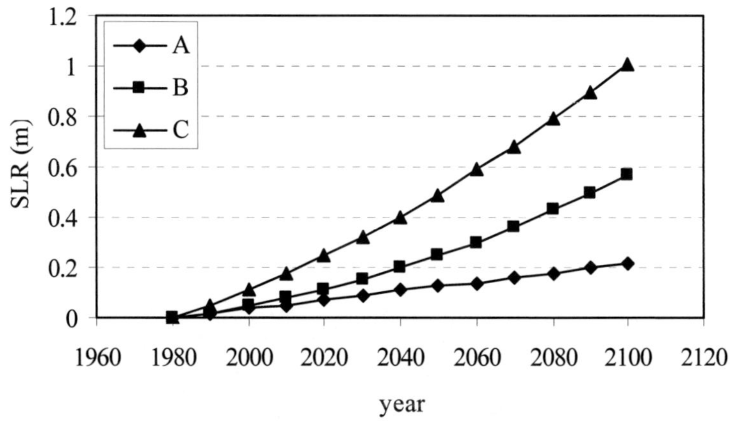

Figure 2: Computed SLR scenarios over the present century (A: high trend; B: average trend; C: low trend).

3.2 Potential land loss due to SLR scenarios

Various scenarios of SLR will have adverse consequences on the Nile delta. Satellite maps of the Nile delta, fig. 3, show potential land loss due to SLR scenarios. In the 0.5 m SLR scenario, recreational areas and beach facilities will be inundated. About 1800 square kilometers of cropland, wetland and fish ponds, representing 7.5% of the total delta area (23,850.76 square kilometres), will be submerged. The 1.5m SLR scenario will have a catastrophic impact on the Nile delta as coastal zone citizens of the major cities such as Port Said and Alexandria would be displaced. The total loss in area if this scenario occurred would reach 5700 square kilometers representing 24% of the delta area.

Potential land loss and people affected due to SLR scenarios (0.5, 1.0, and 1.5m SLR) are summarized in table 1.

Several analyses of the potential impact of SLR on the Nile delta coast have been carried out. These analyses, [12, 13], focused on areas of high vulnerability and possible socio-economic impacts. Most vulnerable areas include parts of Alexandria, Port Said, Damietta, and Suez governorates. Potential impacts of SLR on the Nile delta coastal zone are summarized in the following section.

Table 1: Potential land loss and people affected due to SLR scenarios.

SLR Scenario	Land submerged (square kilometers)	Percentage of land submerged	No. of People affected
0.5m SLR	1800	7.5%	4 million
1.0m SLR	4500	18.9%	6.1 million
1.5m SLR	5700	23.9%	8 million

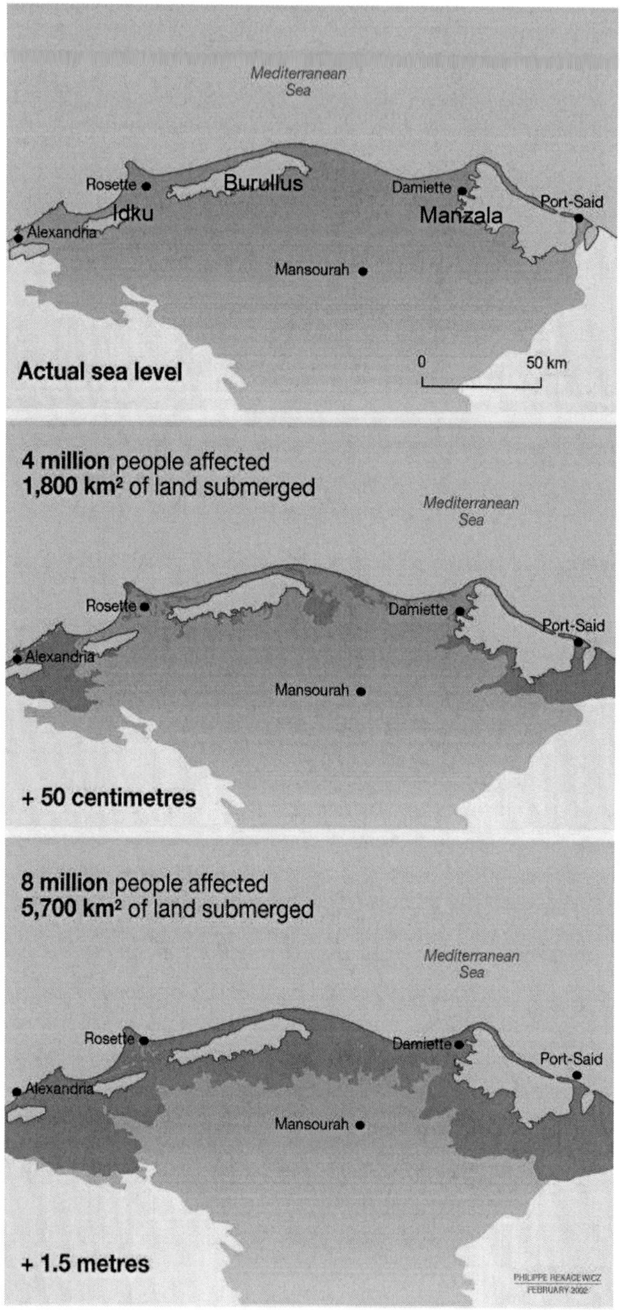

Figure 3: Expected impacts of SLR for two scenarios on the Nile delta. Source: Rekacewicz and Simonett [14].

3.3 The impact of SLR on coastal zone

The Nile delta shoreline is most vulnerable to SLR due to its relatively low elevation and local land subsidence. In the past, the Nile delta shore was in a dynamic equilibrium between sediment supplied by the river and the transport along the coast. After the construction of the High Aswan dam, sediment input in the delta has been strongly reduced. This resulted in serious shoreline erosion and salt-water intrusion. Currently, the Nile delta is exposed to erosion due to the wave action and the currents of the east Mediterranean gyre. Moreover, the construction of waterways for irrigation and transportation has trapped sediment supply to the Nile delta. This entrapment of sediment is a key contributor to coastal erosion and land loss occurring in the Nile delta and the Nile's two branches, Rosetta and Damietta (e.g. [4–6]). This vulnerability of the coastal zone is now further raised by the threat of accelerated SLR. Rising sea levels are expected to destroy weak parts of the coastal sand belt formed by the sediment discharge of the Rosetta and Damietta branches of the Nile. This protective sand belt is essential for the protection of lakes and the low-lying lands.

In summary, SLR could cause major impacts on the Nile delta coast in terms of: increasing erosion rates due to increasing wave action in the nearshore area; inundation of wetlands and other low-lying lands; increasing risk of flooding; accelerating coastal retreat, including erosion of sand dunes and coastal sand belt, breaching of coastal barriers, and damage of coastal inlets.

3.4 The impact of SLR on water resources

The most outstanding impact of SLR on the water resources is that it will increase the occurrence of saline intrusion into coastal aquifers with contamination of groundwater resources in the upper Nile delta. SLR is expected to cause a landward shift of the salt wedge and to increase the rate of saline seepage to the topsoil of the delta. In addition, the salinity in Lake Manzala may increase because of the stronger influence of tidal flows penetrating the lake. El-Raey *et al.* [15] mentioned that changes in the salinity conditions of the lake may affect its ecology and fisheries and the accelerated SLR will increase the salinity.

3.5 The impact of SLR on inhabitants

SLR is expected to affect Egypt coastal population in many ways (Dasgupta *et al.* [8]). Loss of beaches will reduce the number of tourists in coastal areas, forcing tourism dependent individuals and communities to abandon their settlements and look for jobs elsewhere. This will lead to increased unemployment in these areas.

El-Raey [3] used available land-use data, topographic and socio-economic data to calculate the approximate numbers of people expected to be affected by SLR in the future. Estimation of the socio-economic impact due to loss of land and jobs was performed and results of the impact on population and loss of employment for several affected cities were published. He has also estimated that a SLR of 0.5m in Alexandria will cause a displacement of almost 1.5 million people and the loss of about 200,000 jobs by the middle of this century.

3.6 The impact of SLR on agriculture and food resources

Potential implication of SLR on agricultural areas is significant, mainly because Egyptian cropland in surge zones would increase with SLR and intensified storm surges. The increase of frequency of extreme events will reduce crop yield as well as causing changes in the agricultural distribution of crops. Furthermore, it will negatively affect marginal land and force farmers to abandon them and increase desertification. Expected also socio-economic impacts associated with loss of jobs and increase of unemployment.

3.7 The impact of SLR on urban areas and infrastructures

Egypt is ranked the fifth in the world concerning the biggest impact of SLR on the total urban areas (Dasgupta *et al.* [8]). According to The World Bank report, the impact of SLR and intensified storm surges on specific urban centers of the developing world was examined and the top 10 major cities worldwide that are located in storm-surge zones were listed. Port Said city, Egypt, was found to be the first in the Middle East and North Africa region to be mostly affected.

A one-meter SLR in the Mediterranean will possibly submerge Alexandria. Weakness of the erected sea wall on Alexandria shore is shown in fig. 4. This highlights the deadly exposure of its inhabitants, since storm water drainage infrastructure is often outdated and inadequate in low-income urban centers. Generally, fundamental and low-lying installations in Alexandria and Port Said are endangered by SLR and the recreational tourism beach facilities are at risk.

Figure 4: Weakness of the erected sea wall on Alexandria shore.

4 Coastal adaptation to sea level rise

Several measures and adaptation options could be utilized to deal with the SLR impact on the vulnerable coastal segments. This section summarizes responses to SLR and highlights issues that arise with interventions designed to reduce risks to vulnerable coastal areas as a consequence of SLR.

4.1 Responses to SLR

In general, the responses required to protect inhabitants and socio-economic developments against SLR in vulnerable coastal zones can be classified into three options: coastline retreat, accommodation and coastal protection, IPCC [16]. The coastline retreat option requires no effort to protect the coastal area against the sea. In the extreme case, an entire coastal area may be abandoned. This choice can be motivated by excessive economic or environmental impacts of the coastal protection option. However, the retreat option is not feasible in highly populated areas with extensive socio-economic activities.

The accommodation option means that people continue to use the coastal area under the threat of accelerated SLR. This option includes emergency flood measures, converting agriculture to fish farming, or growing salt tolerant crops. The coastal protection option includes hard stabilization solutions and soft stabilization solution. The former involves constructing hard structures such as sea walls, breakwaters, and rock revetments to protect the shoreline against the sea. Soft solutions such as sand dunes are used to stabilize the coast so that existing land uses can continue.

Adaptation options are normally site-dependent. Selecting the most effective measures requires analysis to assess available technology options, environmental and socio-economical implications, and long term as well as short term costs and benefits.

4.2 Environmental implications

The wetlands of the Nile delta represent about 25% of the total area of wetlands in the Mediterranean region. Over 60% of the fish catch of Egypt and many marine species, depend on coastal wetlands (Lake Burullus and Lake Manzala) for their survival. In the retreat option, ecosystems could migrate landward as sea level rises and thus could remain largely intact. Under the protection option, a larger proportion of these ecosystems would be lost. This will be the case especially if coastal protection structures are constructed which will block their landward migration. Coastal structures (such as sea walls) can have a greater impact along coastal areas than soft stabilization solutions (such as dunes and beach nourishment). Hard structures influence beach profiles, sediment deposits and morphology of the coastal zone. Protective coastal structures should be used and designed with great care to avoid negative environmental impacts. Soft stabilization solutions such as beach nourishment retain natural shorelines.

4.3 Socio-economic implications

In case of retreat, relocation of impacted populations may cause political and economical stress of both the displaced people and the host regions. Although the impacts of accommodation and protection would be less stressful, they may still be important. The loss of traditional environments which normally sustain economies could create social instability. Regardless of the response eventually chosen, community participation in the decision making process is important.

The economic impacts of climate change and associated SLR should be included in any response strategy. In the retreat option, coastal landowners would suffer from loss of property, relocation costs and the costs of rebuilding infrastructures. Under the accommodation option, there would be changing property values, increasing damage from storms and costs for modifying infrastructures. Under the protection option, authorities would face the costs for the necessary protective structures. To ensure that coastal development is sustainable, decisions on response strategies should be based on long term as well as short term costs and benefits.

5 Adaptation measures in the Nile delta coastal zone

The appropriate adaptation option in certain coastal areas depends on the associated response. If the land for relocation is available; coastline retreat can be implemented through anticipatory land use regulations and building codes. Accommodation may evolve without governmental action, but could be assisted by strengthening flood preparation programs. Coastal protection can be implemented by the Shore Protection Authority. The current measures for coastal zone protection along the Nile delta are unsatisfactory. The coastline response to protective structures should be considered first before implementation.

Figure 5: Shoreline at the middle delta coast: protection works and sand dunes. Source: El-Shinnawy [17].

Proposed adaptation measures in the Nile delta coastal zone include (but are not limited to):

- Maintaining and building coastal protection structures by the Shore Protection Authority with an emphasis on vulnerable lowlands.
- Restoration and maintenance of sand dunes along the shore, fig. 5, and protective coastal sand belt.
- Preserving existing wetlands (or creating new areas) such as Lake Manzala and Lake Burullus.
- Setting regulations to restrict development in vulnerable areas such as restrictions on set-back line for new developments at low-lying areas.
- Change of land use such as converting agriculture to fish farming or salt tolerant crops.
- Relocation of buildings, roads and other infrastructures in a landward direction.
- Controlling water extraction to minimize saltwater intrusion into ground water in the upper Nile delta.
- Development of comprehensive monitoring program, early warning systems and decision support systems.

6 Summary and conclusions

The Nile delta coastal zone is vulnerable to the impacts of SLR and changes in weather patterns from both the environmental and the socio-economic points of view. SLR was calculated by applying the quadratic equation for 10-year intervals using 1980 as the base year. Despite large variation between SLR model predictions, depending on the modeling parameters and scenarios, the results indicated that SLR is accelerating, with estimated value for the present (21st) century in the order of 0.6 m, although with a margin of as much as ±0.4 m. Local land subsidence in the Nile delta coastal zone would exacerbate the impacts of rising seas. In a 1.0 m SLR scenario, about 4500 square kilometers of cropland, wetland and fish ponds in the Nile delta coastal zone will be submerged. This area represents about 19% of the total delta area. Moreover, it is expected that 6.1 million people will be affected. These potential impacts of SLR would be serious but manageable if appropriate actions are taken. Fortunately, many of the adverse consequences can be avoided by taking timely measures in anticipation of SLR.

The current measures for coastal zone protection along the Nile delta coast are unsatisfactory. Proposed adaptation measures in the Nile delta coastal zone include: maintaining coastal protection structures, restoration of sand dunes along the shore, preserving existing wetlands, setting regulations to control development in vulnerable areas, change of land use, relocation of infrastructures in the landward side, development of comprehensive monitoring program and early warning systems, and technical capacity building. Finally, reactive efforts to reduce SLR impacts on vulnerable coastal areas are less effective than responses which are part of integrated coastal zone management (ICZM).

References

[1] Intergovernmental Panel on Climate Change (IPCC), Fourth Assessment Report. Climate Change 2007: Synthesis Report. www.ipcc.ch/ipccreports /ar4-syr.htm

[2] Dasgupta, S., Laplante, B., Murray, S., & Wheeler, D, Sea-Level Rise and Storm Surges: A Comparative Analysis of Impacts in Developing Countries. World Bank Policy Research Working Paper 4901, 2009.

[3] El-Raey, M., Impact of climate change on Egypt, Environmental Software and Services. www.ess.co.at/GAIA/CASES/EGY/impact.html

[4] Inman, D.L., Elwany M.H.S., Khafagy A.A., & Golik A., Nile Delta profiles and migrating sand blankets. *Proceedings of 23th International Coastal Engineering Conference*, ed. Billy Edge, Venice, Italy, ASCE, New York, pp 3273–3284, 1992.

[5] Fanos, A.M., The impact of human activities on the erosion and accretion of the Nile Delta coast. *J. Coastal Research*, **11**, pp 821–833, 1995.

[6] Stanley, J.D., & Warne A.G., Nile Delta in its destructive phase. *J. Coastal Research*, **14**, pp. 794–825, 1998.

[7] Intergovernmental Panel on Climate Change (IPCC), Climate Change 2001: The Scientific Basis. www.ipcc.ch.

[8] Dasgupta, S., Laplante, B., Meisner, C., Wheeler, D., & Yan, J, The Impact of Sea Level Rise on Developing Countries; A Comparative Analysis. World Bank Policy Research Working Paper 4136, 2007.

[9] National Research Council (NRC), Responding to changes in sea level: engineering implications. National Academy Press, Washington DC, 1987.

[10] Nicholls, R.J, & Leatherman, S.P., Adapting to sea-level rise: relative sea-level trends to 2100 for the United States. *Coast Manage*, **24**, pp.301–324, 1996.

[11] El-Nahry, A.H. & Doluschitz, R., Climate change and its impacts on the coastal zone of the Nile delta, Egypt. *Environ. Earth Science*, **59**, pp. 1497–1506, 2010.

[12] El-Raey, M., Nasr, S., Frihy, O., Desouki, S. & Dewidar, Kh., Potential impacts of accelerated sea level rise on Alexandria governorate, Egypt. *J. Coastal Research*, special issue **14**, 1995.

[13] EL-Raey, M., The cost of coastal vulnerability to climate change. *Conference on climate change and coastal cities, beaches and the Delta*, Cairo, Egypt, 2009.

[14] Rekacewicz, P., & Simonett, O., Impact of sea level rise on the Nile delta. www.grida.no/publications/vg/water2/page/3292.aspx

[15] El-Raey, M., Dewidar, Kh, & El Hattab, M., Adaptation to the impacts of sea level rise in Egypt. *Climate Research*, **12**, pp. 117–128, 1999.

[16] Intergovernmental Panel on Climate Change (IPCC), First Assessment Report: Climate Change 1990. Response Strategies Working Group, www.ipcc.ch.

[17] El-Shinnawy, I.A., Vulnerability assessment and adaptation policies for CC Impacts on the Nile delta coastal zones. www.arab-hdr.org /publications/ahdr/ /2009/alexandria/pres

Yachting and pleasure crafts in relation to local development and expansion: Marina di Stabia case study

C. Bizzarri[1] & D. La Foresta[2]
[1]*Department of Economics and Law, Sapienza University of Rome, Italy*
[2]*Department of DADAT, Federico II University of Naples, Italy*

Abstract

Tourism is increasingly seen as a significant component in the social and economic fabric worldwide. The tourist industry generates important employment opportunities and income within local economies but along with the developments it generates it can also be the cause of decline in the quality of life for residents living in tourist locations.

In the particular case of private yachting and sailing crafts and the harbours built to accommodate them, many social and environmental benefits have been noted. There is for example the increased cash flow that visitors bring and the possibility of using the various terminus points as venues for a series of associated activities and cultural events. However, nautical tourism is not immune from problems caused by increased congestion and pollution as well as damage to places of historic heritage. Moreover, the anticipated economic benefits are uncertain and vulnerable to external factors.

Within this broad range of possible effects, this paper considers the aspect of the relation between private sailing for pleasure and local and regional development, the spin-offs it generates in the territory, its social functions and the additional value it offers the tourist market. The focus is on a manmade harbour in Southern Italy, close to Naples: Marina di Stabia.

Marina di Stabia was born from a cooperation among a group of professional managers from the city of Castellammare di Stabia. The principal aim of their project is the requalification of the disused area to the east of Castellammare, which was once an important metallurgic centre and one of the symbols of the area's industrial supremacy in the past. The harbour was opened in June 2007, and ongoing work continues adding a constant series of improvements.

WIT Transactions on Ecology and the Environment, Vol 149, © 2011 WIT Press
www.witpress.com, ISSN 1743-3541 (on-line)
doi:10.2495/CP110051

The research is focused on the following points:

- The demand for tourist services in the port
- Marina supply system
- Marina Management models
- Issues of funding for the development of marinas and marine services
- Economic and social impacts caused by nautical tourism.

Keywords: nautical tourism, economics of marinas, environmental impacts of marinas, regional effects of nautical tourism.

1 Introduction

Sea-going private crafts spread benefits throughout the various territories wherever they dock. Nautical tourism is not limited to owning a boat and utilizing marine resources and can be listed with all those other forms of tourism which have economic, social and environmental impact on local development.

For this reason the number of berths in harbours or the number of sailing licenses issued cannot be considered an efficient indicator of the value of nautical tourism. The expenditure on land of yachtsmen and owners of private sailing crafts is a much better gauge of evaluation and analysis of economic growth.

Along with a strictly economic analysis that evaluates the possible expenditure of these sea-going tourists it is necessary to show the social and environmental impact of this tourism on the territory. In order to determine these effects, whether local or regional, it is possible to use the concept of physical capacity relative to the number of berths in the harbour and the length of coastline utilized, as well as problems arising from conflicts over resources or rather the loss of resources to other economic sectors which this form of tourism provokes by an exclusive appropriation of marine space. The construction of large ports and their infrastructure pose a limit to the use of marine resources in other sectors such as fishing (through a reduction in biodiversity caused by anchoring that damages the sea floor), bathing and all those seaside activities that require open access to the sea.

As it is not possible to have significant diversification in the use of marine resources it would be preferable to invest in projects that increase territorial and regional development. Nautical tourism produces wealth and expands as far as possible the life cycle of the tourist product

2 Nautical tourism in Italy

The study of private sailing for pleasure is quite a recent area of interest in Italy. From an analysis of data which covers only the last four years [1], there has been a notable increase in harbour berths: an increase of 13,000 units from 2007 to 2010. The most dynamic Italian regions in this sector are Liguria and Sicily, even if the Northern Tyrrhenian coast, which takes in Liguria, Tuscany and Sardinia, retain the largest number of berths with respect to other regions. As far as berths in marinas are concerned, apart from the regions already mentioned, very significant data for our research comes from Lazio, Campagnia, Basilicata, Emilio Romagna and Puglia: in 2010 an increase was registered of over 100 berths in each of these regions (see table 2.16 page 42 Report On Nautical Tourism 2010). The growth in the sector is also underlined at national level by an increase in the demand for licenses for sea going vessels: from 2003 to 2008 an additional 11,000 licenses were issued [1].

Within Europe Italy is now in third place in the classification for number of private berths coming immediately after France and Spain. In Italy however the proportion of berths to the total number of private vessels is higher than in either France or Spain and while it is true that each boat in theory has the possibility of four berths within national territory it is also true that a large number of these berths are utilised by foreign vessels, the number of which would be very difficult to estimate. Moreover, in recent years the region of Liguria enjoys an extremely high client base: almost 82% of residents, within the region and on its borders possess their own mooring in local ports.

3 Economic analysis of yachting and pleasure sailing in Italy

The phenomenon of pleasure crafts is very interesting from an economic point of view. In comparison to other harbour users most yachtsmen have higher spending power and are mostly educated qualified professionals, often self-employed (see sampling correlated by Report On Nautical Tourism).

From research sampling carried out by the National Observer On Nautical Tourism and published in the Report On Nautical Tourism it can be seen that expenditure noted by a sample of yachtsmen for port services and refuelling is very high and constitute 80% of their daily expenditure within the port. The rest of their expenditure (about 20% of the daily expenses) is very segmented: transport, restaurants, bars and cafés, shopping, entertainment and cultural activities. It is also noteworthy that the cost of transport over land is very low (approximately 10 euro per person) [1]. The spending patterns of yachtsmen is proportionate to a lack of attractions in the outlying territory. If, on the other hand, the principle motive for their journey was to visit the areas beyond the port transport expenditure and in particular private transport such as car rentals or use of taxis would constitute a consistent and very significant expenditure.

However, the average yachtsman directs most of his unessential expenditure on shopping and restaurants and cafés compared to either entertainment or cultural activity. No substantial difference in average expenditure has been noted between yachtsmen whether in transit or local but spending is however much greater than for other forms of nautical tourism. For this reason the yachtsmen are obviously preferable to other sea-going visitors who spend little whether on services strictly attaining to the port or for those of the hinterland. Marina construction is therefore more desirable than multipurpose ports in that the demand is fixed and the private sailor's expenditure is spread around connected business and services and has a higher turnover than for any other type of nautical tourism is maintained.

From a geographical point of view the Upper Tyrrhenian coast enjoys the highest turnover of all, estimated as being 50% of the national turnover for all linked business and services.

This advantage however is not of great significance over the gross national or local product because of the high seasonality of the sector. This continues to be a difficult problem for future development.

4 Analysis of the environmental impact

From the point of view of the environment the Italian coast line constitutes a very fragile ecosystem and the negative impact of large ports cannot be ignored. If the economic effects of large ports is predictable in the short and medium term, the impact on the environment is gradual and very difficult to demonstrate due to the impossibility of putting an economic evaluation on natural surroundings.

In Italy at the moment multipurpose ports constitute approximately 63% of all ports. This type of port has a strongly negative environmental impact because of its very large infrastructure and due to the size of the ships and systems of anchoring that cause a significant loss in marine biodiversity. Marinas however being built for smaller vessels put far less pressure on resources and, from an economical point of view, as we have been able to ascertain, enjoy a fixed demand and makes the whole sector more profitable.

The pressure on marine resources and on our coasts moreover is demonstrated by the proportion of the number of berths for every kilometre of cost which is estimated at 20.5 and on the average distance between ports which works out as a port every 14 kilometres. If the spread of ports on one hand represents a possibility of unblocking roads as well as openings for large scale economies, on the other hand there is notably strong pressure on resources and consequent increased loads on capacity [2]. Such a negative impact is amplified by the element of seasonality in nautical tourism and pleasure sailing.

Another problem is born from the conflict between use of resources seeing that the construction of large ports inhibit bathing and fishing for residents along with the practise of some sports such as sub aqua and canoeing. A consequent element of conflict is the protection of natural surroundings. With the formation of ports and marinas these are radically and permanently transformed. It is exactly this characteristic of non-modification which makes the construction of marina or port particularly unsustainable. Should their economic life cycle come to an end due to financial crisis or deficiency in services, the large infrastructure of ports would be very difficult to dismantle and a return the original state of the environment almost impossible.

5 Nautical tourism in the region of Campania, Italy

The Campania Region has been characterised for the last ten years by valuable reorganisation in the sector of yachting and pleasure crafts and connected activities. Notwithstanding a deceleration due to the recent financial crisis the refurbishing and requalification of present structures has resulted in an increase in the number of vessels along the coast of the region (15,588 units registered in 2009, representing 15.3% of total number), and an increase in the number of berths available (14,405 berths in 2009 equal to 10.5% of the national total with an increase from 1997 to 2009 by 58%) [3].

However, the positive quantifying data does not adequately describe the actual situation in terms of what this sector can in fact offer nor can the increased registrations of vessels compensate for the highly seasonal peaks in international or national demand, which do not permit high positive economic or occupational returns but only a state of fluctuation

Notwithstanding the efforts of local administration the availability of berths is not adequate to the demand (29 berths per kilometre of coast line as against a national average of 19 kilometres and an average of 101 vessels for every 100 berths compared to a national average of 69.5 for every 100). [4] There is also a widespread lack of organization throughout the region along with a fragmented offer of services in the tourist harbours: a disorderly arrangement of market offers, in terms of services and equipment and the level of the services offered by the infrastructure for yachts and pleasure craft are often unsatisfactory to the needs of the customs. There are an inadequate number of available berths, and many berths designed moreover for coastal vessels of only average dimension. These aspects are indicative of an explicit difficulty in the development of nautical tourism within this region and place the frame of the offer below the national average discouraging any significant flux of foreign tourists who instead crowd into better equipped and decidedly cheaper Turkish, French and Greek marinas. Many world famous locations in this Italian region have a reduced capacity to receive sailing craft and furthermore are almost always characterised by inadequate standards of quality.

To be noted in particular is the casual proliferation of seasonal moorings organised outside the port structures (4600 vessels moored to buoys) or along urban coastlines or along strips of coast intended for bathing, and other methods of inadequate mooring inside ports in terms of security and comfort for users. This phenomenon goes hand in hand with the overcrowding of many organised harbour structures and the minimal use of other port structures, in particular those which were designed for non-tourist uses but are no longer in use having lost their original role and not yet adequately reconverted for private sailing needs.

In some ports a mix of functions can be seen as a resource for private sailing and yachting (for example the cohabitation of dockyards, fishing vessels and pleasure craft) but in other situations is often seen as detrimental and forces decisions on designation (this is especially the case in the presence of both sailing crafts and shellfish farming or in the case of bathing and mooring). Such decisions are further prompted by indicators of overcrowding around closely packed port structures, which cause local disruption and a limit to a stable territorial development.

From these circumstances it is clear to see how in the first place the impossibility of structures to accommodate essential requirements and high seasonal peaks has caused a progressive appropriation of parts of the coast by a number of interests incompatible with the private sailing sector and secondly how the spread of market segments often becomes a grey area difficult to control.

However, notwithstanding the scarcity of tourist harbours and low numbers of available berths, strong demand has been registered for moorings, not only by a local customs (permanent moorings) but also by a customs from other Italian regions (transit craft, chartered boats and yachts of large dimension). The private sailing demand is also satisfied by many unauthorised structures, the result of small business improvisation, offering berths or more often moorings attached to buoys in locations of the highest attraction from every point of view.

The number of pleasure craft in transit is however limited by the very high percentage of utilization of infrastructure by permanent customs. This means during the high season regional tourist harbours cannot be relied upon by those private pleasure sailing crafts wishing to organise cruises along the Tyrrhenian coast.

6 The marina: a useful resource for revitalizing the coastline or a manifestation of environmental impoverishment?

Without reducing the importance of either the inherent difficulties or consequent criticisms of ports for tourist crafts, it seems evident, however, that there are considerable potential rewards connected with the sector, taking into account both the capacity for expanding the available services and expanding the demand.

In this context, marinas certainly represent a strong opportunity for economic growth for the territory and the local community being able to influence significantly the gross internal product. Possibilities for proliferation of production and employment in this sector are considerable and without doubt higher than in any other sector. In occupational terms it is sufficient to bear in mind that along with the 12,000 working units directly employed in the nautical industry there are linked business concerns with over 80,000 skilled workers [5]. The construction of a sailing vessel represents the first link in a chain which permits the creation of a series of additional activities: commerce, rents, hiring, institutes for professional formation, specialised press, financial services, professional services, conferences and industrial fairs. Port structures themselves require areas for maintenance and repair, security personnel, reception areas and restaurants and bar facilities and represent the most significant factor in this chain of economic growth for the territory. In recent years, in line with the above, tourist harbours have undergone restructuration with the aim of satisfying the increasing demand for new berths while the Italian political authorities have felt the necessity to promote the construction of new shipping ports.

Given the situation an increase in marinas can only be considered as an indispensable condition for consolidating the growth of private pleasure craft enterprise but only in the presence of adequate essential infrastructures to sustain the efforts made by these tourist harbours to attract and welcome yachtsmen and their vessels. The integration of tourist harbours with the hinterland and urban areas offering high standard services, including those focusing on environmental control, are fundamental and distinguishing elements for the recognition of the primary role of the "portus", which came into being to do act as exactly that: "the door to the inland territory beyond", offering the opportunity to all the various localities to promote their historic richness, their cultural and gastronomic specialities. In order to succeed in this challenge the coastal territories must be assured of more ports and for existing ports new and improved quays, jetties and berths to be characterised by a management that has an eye for innovation, but also for the requalification of those little utilized or totally unused areas, as well as easy access to all port facilities, the implementation and improvement of services and lastly a guarantee of the defence of the environment and an assurance of the least possible impact on its coastline.

In this sense marinas can represent an important safeguard for the environment in a process of a requalification already in act for some time, along with the continued improvement in services, safety and information.

7 Marina di Stabia

The interest shown by various institutions in private pleasure sailing and the linked economic effects is demonstrated in the notable increase in financing for construction or the reconversion of port structures for private sailing. These constructions are often based on overly conventional planning criteria which are in some cases obsolete, but there is however an earnest attempt to find solutions to the growing demand and to take advantage of geographic opportunities and the possibilities the market offers.

To be included in this general trend is the project for the construction of Marina di Castellamare di Stabia. This has been financed by a large public contribution (euro 2,500, 000 from P.O.R budget resources 2000-20006) and also sustained by the participation of a private business consortium. The Marina di Stabia is in the centre of the gulf of Naples and is a unique tourist harbour for both its dimensions and for its technological design which assures a continuous circulation of water and safety for docking. The structure extends over an area of 450,000 square metres, with 1300 berths, of which 161 have protective covering; fuel bunkering for each berth; control towers; Port Authority and docks for repair with ship yard, crane facilities, service slides, reserved parking and land services.

Nine hundred berths are able to receive ships of up to 80 metres in a landscape of incomparable beauty, overlooked by Vesuvius and by Mount Faito, extending from the Sorrento peninsula to Capri and Ischia making this port ideal for those who wish to have a tourist experience with a wide range of choice: diversity in time, in space and substance.

The vicinity to such tourist destinations of excellence as Capri, Amalfi, Positano, Sorrento and to important archaeological sites such as Pompeii, Ercolano and Stabia etc, as well as the possibility to reach easily the Pontine and the Eolie islands make the port a truly rich tourist proposition. Hotels and restaurants, cinemas, commercial centres, sports centres and beauty salons and a new railway station to be constructed near to the marina will all contribute to the unique experience.

The sum of all these features make Marina di Stabia the most important tourist port of the new generation in Campania, functioning as a hub for surrounding harbours with inferior moorings and services that exist along the coast but in need of substantial re-equipment.

The superb natural beauty and the high value to the territory of the physical infrastructure of the port are all within the framework of a broad project for recuperation of this length of coast, contributing to the requalification of an area that had an important industrial past, situated where the old Italian metallurgic plants stood on the northern outskirts of Castellammare di Stabia close to the border with Torre Annunziata and near the estuary of the Sarno river. This area characterised in past decades by high density industrial plants underwent a serious crisis during the eighties when many factories were closed and this transformed the zone into a series of crumbling unused buildings and vacant lots. These events, linked to the general impoverishment of the national and regional economic situation, initiated a process of considerable decline for the city, characterised by a high level of unemployment, medium to low levels of education, wide spread petty crime, high urban density, a progressive increase of population and an absence of social services.

Notwithstanding preliminary confidence and expectation the Stabia port, even though a mark of excellence has not yet had the positive bearing on the territory that could be expected from this type of infrastructure.

In the first place the City of Castellammare appears to be totally distanced and physically cut off from the marina. The urban disorder that the construction of the new port imposed on adjacent areas, were not balanced by any immediate attempts to integrate the urban system with that of the port. The zone cushioned between the port and the city is still occupied by industrial plants, some of which are unused, contrasting sharply with the concept of the port. In fact, lines of buildings of considerable length impede access to the sea and this is an obvious obstacle to the aim of re-launching the area with tourism.

There is a concrete risk that this experience is not a portent of economic benefit for both the city and territory but only constitutes an additional aggression in a situation of unstable equilibrium whether for the hydro- geologic organisation of the territory or for the richness if its archaeological sites.

8 Conclusions

Notwithstanding the international financial crisis and the numerous difficulties Italy confronts, the Italian marinas continue to grow and represent a potential key factor in processes of revitalising coastal areas. However, while it is now accepted that private sailing for pleasure generates economic benefits equal to 10% on initial outlay, opinions diverge in the real benefits to the territories involved. In fact, the required complex infrastructure, often too large and designed with the reasoning applied to shopping centres or tourist villages, needs long term heavy investment and often produce constructions that risk rapid obsolescence, contributing even more to the degradation of coastal areas without stimulating any favourable development in the neighbouring hinterland. A complete revision of the politics of the sector seems necessary, which should be directed towards a greater flexibility and marinas of smaller dimensions strongly integrated with the entire system of regional transport. The tourist product offered can therefore be further enriched with integrated activities linked to the local identity. These local activities can be centred in those areas of reclaimed coast free of port infrastructure and local communities could be important actors in the process of territorial regeneration. With the right awareness and participation and through the recuperation and maintenance of local uses and activities communities could regain possession of these free zones and contribute to both the control of the environment and the process of revitalisation, requalification and a diversification of territory.

Acknowledgements

This study is the result of the research of both the authors, but the introduction and the paragraphs 1-2-3-4 are ascribed to C. Bizzarri, while paragraphs 5-6-7 and conclusion are ascribed to D. La Foresta.

References

[1] Osservatorio Nautico Nazionale, 2010, *Rapporto sul turismo nautico*, numero 2, Genova

[2] Bennett W., Bristow R. S., 1994, Boating Opportunities a Geographical Analysis of Travel Patterns and Motivations, in: Mahoney E. M., Stynes D. J., 1996 , Recreational Boating Carring Capacity: a Framework for Managing, Inland Lakes, Proceedings of a Workshop, Michingan State University

[3] Osservatorio del turismo della Regione Campania, 2008, *I prodotti turistici in Campania: il turismo Nautico*, Napoli

[4] Osservatorio del Turismo della Regione Campania, 2008, *Il Charter nautico in Campania,* Napoli

[5] Ministero delle Infrastrutture e dei Trasporti, 2008, *Il diporto nautico in Italia*, Roma

Section 2
Coastal processes and GIS

Cartography and remote sensing for coastal erosion analysis

M. Basile Giannini, P. Maglione, C. Parente & R. Santamaria
University of Naples "Parthenope", Department of Applied Sciences, Directional Centre of Naples, Italy

Abstract

Italian coasts are subjected to morphological modifications that in the last decades have, in many cases, been the cause of considerable coast-line withdrawal. A detailed cognizance on dynamics and relative consequences on territory and environment is necessary to plan actions for limiting these events and their impacts. Reconstruction of temporal shoreline changes can be realized using historical and recent maps, remote sensed images and topographic survey results. Because of the heterogeneity of these data, their different levels of accuracy must be considered. In this paper one of the most emblematic Italian shorelines affected by erosion, the Domitian shoreline located in Northern Campania Region (South Italy), is considered. GIS environment is used to compare ancient and recent maps as well as remote sensed images. Cartographic layers include IGMI (Istituto Geografico Militare Italiano) products from 1876 to 1984, on a scale 1:25.000 or 1:50.000. Others are an aerial photo (1:10.000) produced in 1998 and an IKONOS panchromatic image acquired in 2005. Taking into account the heterogeneity of these documentations, to realise comparisons of their contents, all layers are related to the same cartographic datum (Gauss-Boaga Roma40), so a reconstruction of the coastal evolution from 1876 to 2005 is obtained.

Keywords: coastal erosion, GIS, historical maps, high resolution remote sensed images, accuracy.

1 Introduction

In Europe, every year, an area of about 15 km^2 is lost or seriously impacted by erosion. In fact from 1999 to 2002 over 250 houses were abandoned because of

WIT Transactions on Ecology and the Environment, Vol 149, © 2011 WIT Press
www.witpress.com, ISSN 1743-3541 (on-line)
doi:10.2495/CP110061

coastal erosion risk, while the value of another 3,000 houses is decreased by at least 10%. The undermining of coastal dunes and sea defences has augmented the risks of coastal flooding, so several thousands of square kilometres and millions of people are threatened. Erosion phenomena are induced by a combination of natural as well as human factors: the first include wind, near shore currents, sea level rise, vertical land movement; the second include coastal engineering, gas mining and water extraction, dredging, vegetation clearing. The combination of the coastal erosion, infrastructure development and the erection of defences to protect them have decreased the extensions of many coastal zones. (Doody *et al.* [1]).

The erosion process for sandy beaches includes three phases: dissolution, transport and deposition of sediments. The interactions are closely related to the dynamics of the water column on which waves, tides, coastal currents, geodynamic and climatic phenomena are active. Until the middle of last century the effects caused primarily land advancement; later, this trend was reversed because of anthropogenic pressures and, more recently, changes in sea level related to the greenhouse effect. The most severe effects in the medium term are related to the work of man.

To repair the damage and to mitigate the future impacts, phenomena models must be prepared using all available data to reconstruct evolution of shoreline, such as ancient and recent maps, air photo and satellite images.

2 Study area

The area considered concerns Campania coastal zone named Domitian that is located between the following latitudes (referred to WGS84 datum):

$$\phi_1 = 40° \, 57' \, 12'' \, \text{N}$$
$$\phi_2 = 40° \, 56' \, 19'' \, \text{N}$$

Regarding the situation resulting at 2005, the extreme longitudes of the area referred to Greenwich are:

$$\lambda_1 = 13° \, 59' \, 57'' \, \text{E}$$
$$\lambda_2 = 14° \, 00' \, 33'' \, \text{E}$$

The period extends from 1876 to 2005.

The area is included in the territory of Castelvolturno and it is shown in the following figure, fig. 1. The Domitian coast is the edge of the Campanian Plain. It is limited by Aurunci (North-West), Caserta (North), Avella (North-East) and Lattari Mountains (South). The Plio-Pleistocene coastal graben is filled by powerful layers of alluvial materials of the Volturno and Garigliano Rivers and pyroclastic deposits of the Roccamonfina and Campi Flegrei volcanic complexes. The physiography of the place is described by the tangential directions to Cape Point and Miseno Banner – Gaeta. The prevailing wave (70% frequency) comes from the west northwest and it is associated net transport towards the south-east coastal. The area study has a temperate Mediterranean climate. Rainfall is concentrated

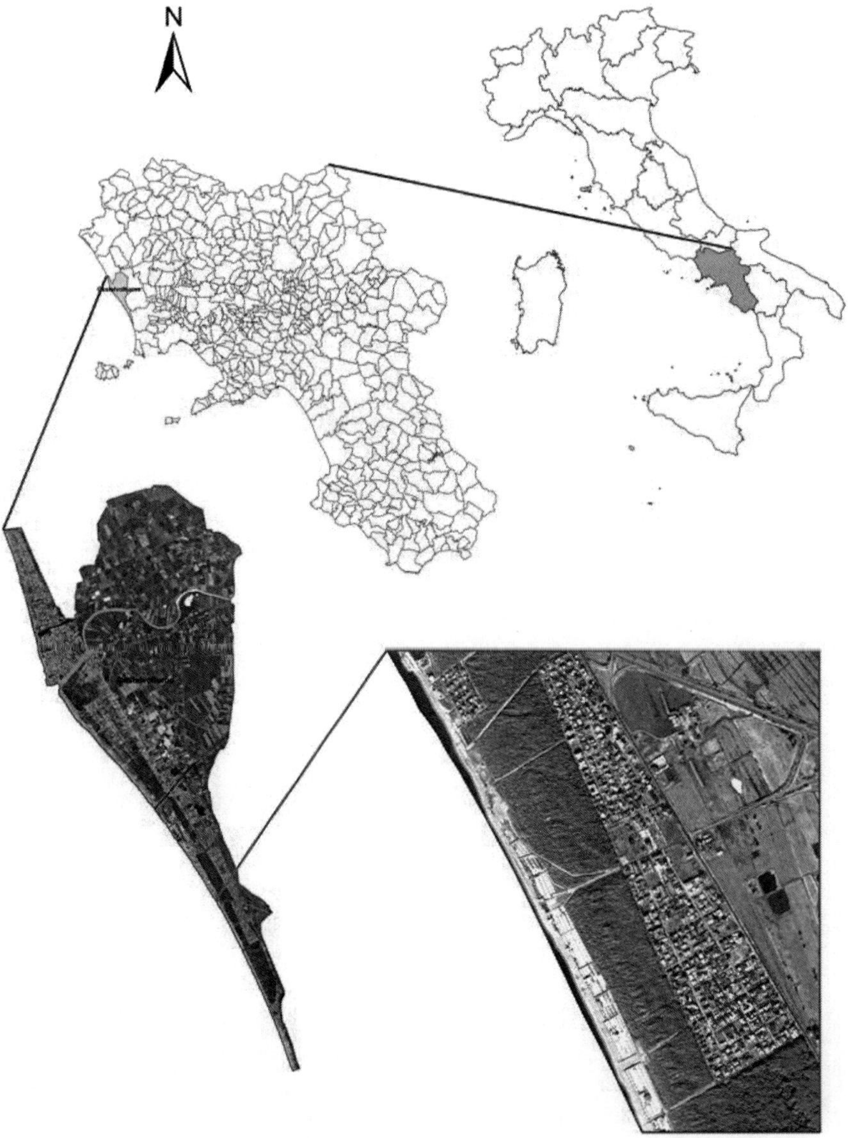

Figure 1: Castelvolturno and study area territorial framework.

during autumn and winter and usually has its maximum in the months of October and December. The minimum rainfall occurs during the period of summer (from May to September) with minimum in July. The prevailing winds on the coast are those coming from the west and southwest (Gatto [2], Cocco *et al.* [3, 4]).

3 Data set and early elaboration

In this paper six information sources concerning Domitian coast are considered. The first two are paper maps referred to Bessel ellipsoid:

- IGM map named "Marano di Napoli" (1876), in scale 1:50.000;
- IGM map named "Lago Patria" (1905), in scale 1:25.000.

In Bessel's datum, the geodetic coordinates of the origin are: ϕ_0 = 41° 55′ 25″.42 N and λ_0 = 12° 27′ 14″.00 East of Greenwich, (for consequence λ_0 = 00° 00' 00" because considered the national Prime Meridian), and with an azimuth to Monte Gennaro: α_0 = 62° 38′ 20″.03. These geodetic coordinates were derived from the astronomic observatory of Capodimonte near Naples. The projection of these maps is the Natural or Samson-Flamsteed projection (Surace [5], Clifford *et al.* [6]).

The third and the fourth paper maps are referred to Hayford ellipsoid and they are:

- IGM map named "Lago di Patria" (1954), in scale 1:25000;
- IGM map named "Quagliano" (1984), in scale 1:25.000.

For both maps Hayford ellipsoid is oriented on Potsdam in Germany (ED50 datum). The projection is the Universal Transverse Mercator (UTM) (Bezoari *et al.* [7]).

The fifth information source is a digital aerial ortho-photo produced by Campania Region in 1998, in scale 1:10.000, and referred to Gauss-Boaga Roma 40 datum.

The last one is an IKONOS panchromatic (digital) image. It is acquired on 11.04.2005 and referred to WGS84 datum.

Obtained raster files of the four maps, georeference operations have been conducted with GIS software.

For georeferencing maps and remote sensed image polynomial functions (PFs) are usually applied (Errico *et al* [8]):

$$P_n(X,Y) = \sum_{i=0}^{m_1} \sum_{j=0}^{m_2} a_{ij} X^i Y^j \tag{1}$$

where, if n is the level of polynomial function, the following relations must be satisfied:

$$0 \le m_1 \le n;$$
$$0 \le m_2 \le n;$$
$$m_1 + m_2 \le n.$$

The coefficients a_{ij} of the polynomials are computed using points named Ground Control Points (GCPs) of which coordinates (X, Y) in the original image as well as (X', Y') in reference to cartographic datum are known. The minimum number of points required for the computation of the coefficients is given by the formula:

$$N = \frac{(n+1)\cdot(n+2)}{2} \tag{2}$$

Additional data are used to get the optimal transformation with the smallest overall positional error in the selected points. These errors are originated by poor positioning of the mouse pointer on the image and by inaccurate measurement of coordinates on the map. The accuracy of the transformation is indicated by the average of the errors in the GCPs as well as in other additional points named Check Points (CPs).

In this application IGMI maps in Samson-Flamsteed projection have been georeferenced with 2th order polynomial functions while for the other ones 1st order PFs have been sufficient. Statistical characterizations of the results are shown in tables from 1 to 4.

Table 1: Residuals (in meters) obtained for the 8 Ground Control Points used to georeference the IGMI map named "Marano di Napoli" (1876).

	Δxy medium (m)	Δxy minimum (m)	Δxy maximum (m)	σ_{xy} (m)
Polynomial Function 2nd order	**5.39**	**1.67**	**12.71**	**4.24**

Table 2: Residuals (in meters) obtained for the 12 Ground Control Points used to georeference the IGMI map named "Lago Patria" (1905).

	Δxy medium (m)	Δxy minimum (m)	Δxy maximum (m)	σ_{xy} (m)
Polynomial Function 2nd order	**2.97**	**1.26**	**5.26**	**1.25**

Table 3: Residuals (in meters) obtained for the 17 Ground Control Points used to georeference the IGMI map named "Lago di Patria" (1954).

	Δxy medium (m)	Δxy minimum (m)	Δxy maximum (m)	σ_{xy} (m)
Polynomial Function 1st order	**0.92**	**0.14**	**1.84**	**0.47**

Table 4: Residuals (in meters) obtained for the 18 Ground Control Points used to georeference the IGMI map named "Quagliano" (1984).

	Δxy medium (m)	Δxy minimum (m)	Δxy maximum (m)	σ_{xy} (m)
Polynomial Function 1st order	**2.69**	**1.18**	**5.28**	**1.12**

IKONOS panchromatic image (pixel size: 1 m x 1 m; spectral band: 0,45 μm – 0,90 μm) is a product named GEO with accuracy of 25 meters (Space Imaging [9]). To ensure better accuracy, a clip of this image concerning only the study area has been considered and re-georeferencing operation has been conducted using polynomial functions, with 30 GCPs; also 15 Check Points (CPs) have been introduced for better evaluation of the accuracy, fig. 2. Plane coordinates in Gauss-Boaga Roma40 of GCPs as well as CPs have been derived from the above mentioned aerial ortho-photo. Satisfactory results have been obtained with 5th order transformations.

Figure 2: IKONOS Pan image with Ground Control Points and Check Points used for georeferencing.

Statistical characterizations of the results are shown in table 5 for GCPs and in table 6 for CPs.

To obtain a correct overlay, also the four IGM maps have been reported in Gauss-Boaga Roma 40.

Table 5: Residuals (in meters) obtained for the 30 Ground Control Points to georeference the IKONOS image.

	Δxy medium (m)	Δxy minimum (m)	Δxy maximum (m)	σ_{xy} (m)
Polynomial Function 5th order	**0.62**	**0.04**	**1.94**	**0.49**

Table 6: Residuals (in meters) obtained for the 15 Control Points to georeference the IKONOS image.

	Δxy medium (m)	Δxy minimum (m)	Δxy maximum (m)	σ_{xy} (m)
Polynomial Function 5th order	**0.97**	**0.33**	**1.72**	**0.41**

4 Shoreline evolution

Every single coastline considered has been obtained by vectorizing the coast profile from geo-referenced maps or remote sensed images concerning different periods, fig. 3.

The study of the shoreline evolution has been conducted considering pairs of temporally consecutive coastlines and evaluating, for each pair, the surface of the beach eroded or added, and its metric accuracy, fig. 4–5.

Particularly graphical errors (0.2 mm referred to scale) for maps, spatial resolution for remote sensed images, georeferencing and datum transformation errors have been considered, taking into account propagation of variance (Cina [10]).

Propagation of variance for a linear case (such as the combination of graphical and georeferencing errors) is supplied by:

$$\sigma_y^2 = a_1^2 \cdot \sigma_1^2 + a_2^2 \cdot \sigma_2^2 + \ldots + a_n^2 \cdot \sigma_n^2 \tag{3}$$

for $y = a_1 * x_1 + a_2 * x_2 + \ldots + a_n * x_n$.

Propagation of variance for a nonlinear case (necessary to calculate eroded and added area) is supplied by

$$\sigma_y^2 = \left(\frac{\delta f}{\delta x_1}\right)_{(0)}^2 \cdot \sigma_1^2 + \left(\frac{\delta f}{\delta x_2}\right)_{(0)}^2 \cdot \sigma_2^2 + \ldots + \left(\frac{\delta f}{\delta x_n}\right)_{(0)}^2 \cdot \sigma_n^2 \tag{4}$$

for $y = f(x_1, x_2, \ldots, x_n)$.

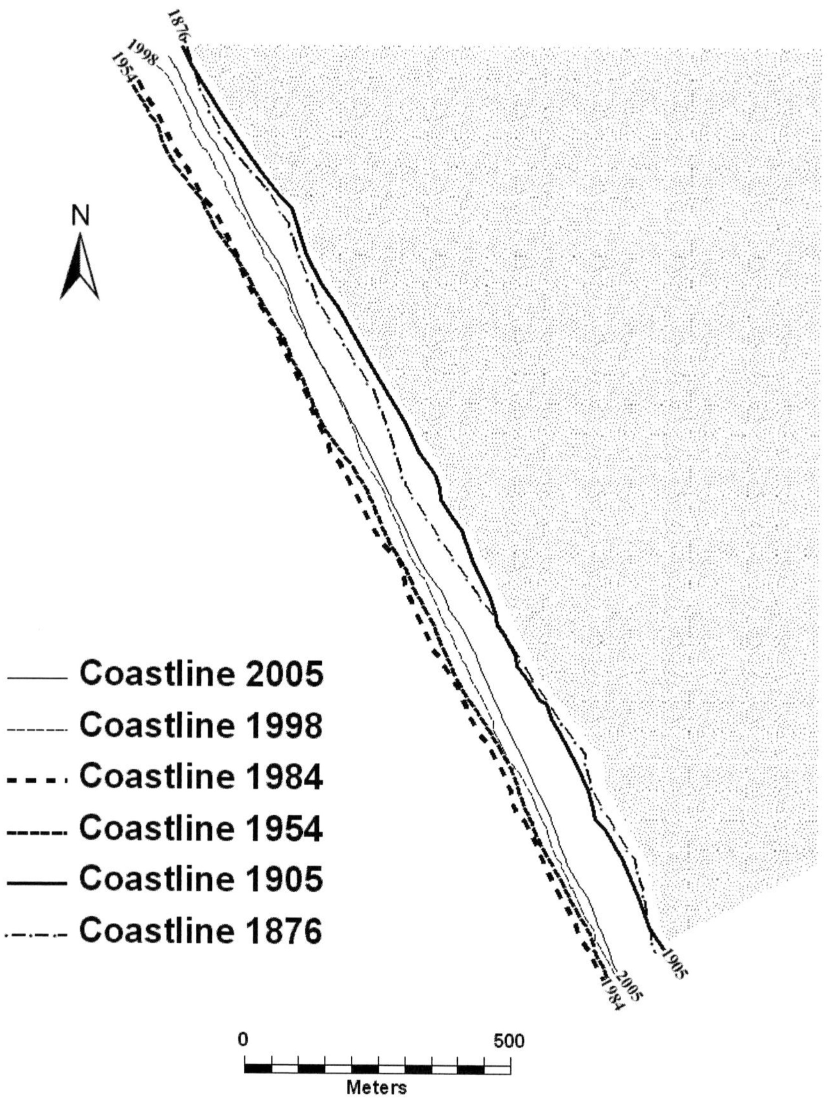

Figure 3: Coastline evolution from 1876 to 2005.

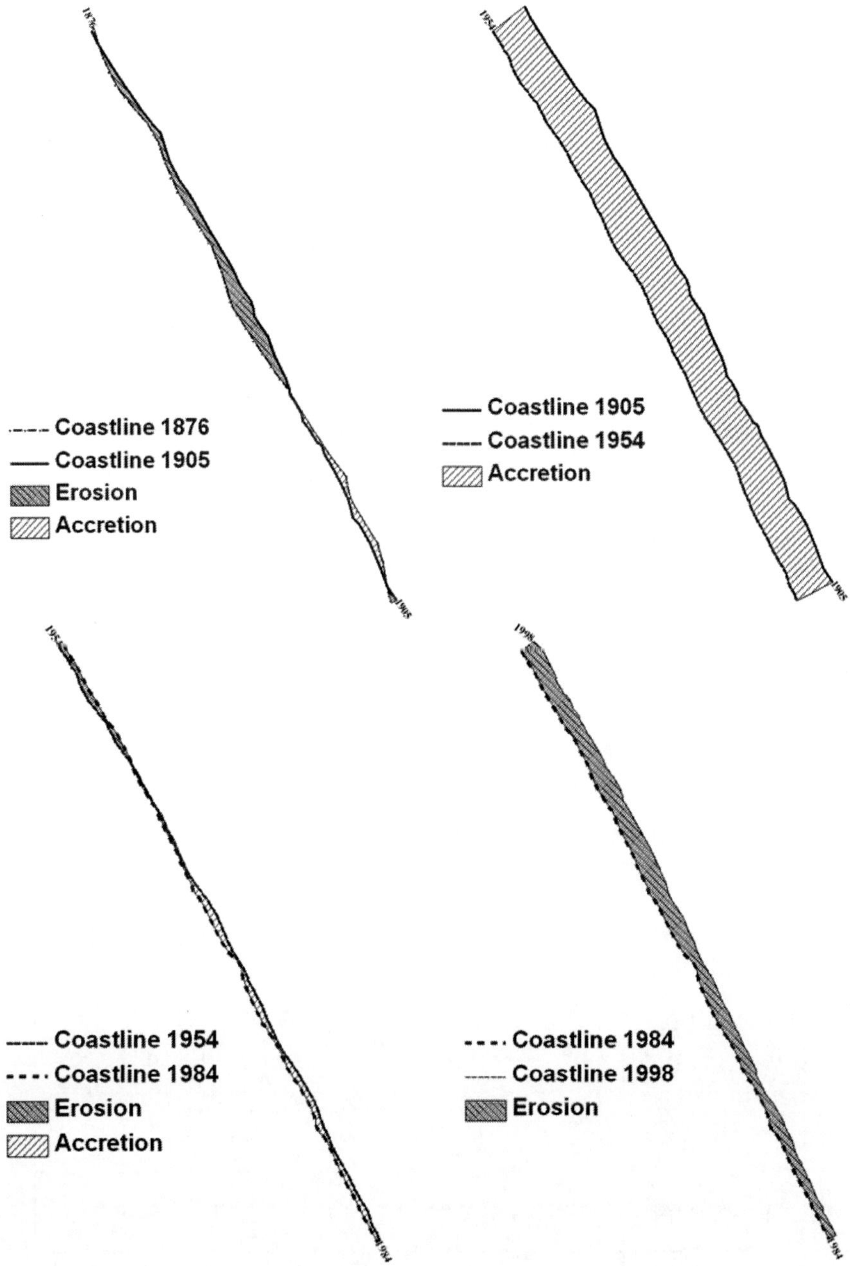

Figure 4: Shoreline evolution for each pair of temporally consecutive coastlines from 1876 to 1998.

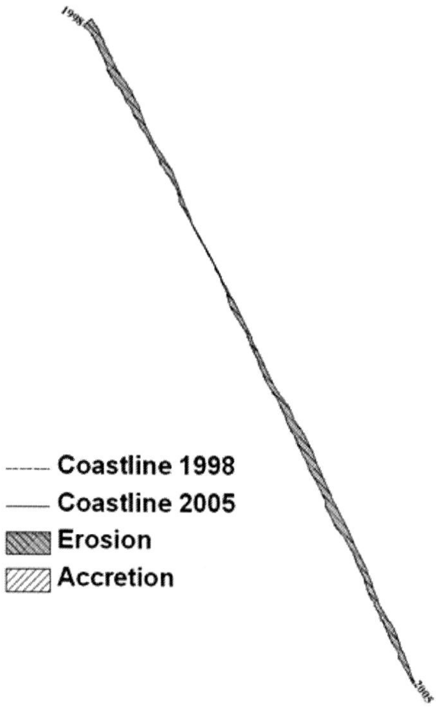

Figure 5: Shoreline evolution for each pair of temporally consecutive coastlines from 1998 to 2005.

The results are shown in the following table (Table 7).

Table 7: Eroded and added area for each time period.

Time period	Eroded Area (m^2)	Added Area (m^2)
1876 – 1905	**28356± 2586**	**8276 ± 1397**
1905 – 1954	**0**	**227118 ± 3471**
1954 – 1984	**4469 ± 484**	**17783 ± 966**
1984 – 1998	**75026 ± 1983**	**0**
1998 - 2005	**26541 ± 256**	**287 ± 27**

5 Conclusions

The conclusions to be drawn from this work are of two basic types, some about the methodological aspects, others the evolutionary process of the considered coast that is emblematic of the erosion processes that affects a considerable part of the Italian coasts.

Regarding methodological aspects, it should be noted that the availability of maps and high-resolution remote sensed images can provide useful data for the reconstruction of coastal dynamics. Considering the accuracy of different information sources and the law of propagation of variance, a quantitative analysis of the erosion magnitude and the reconstruction of the evolutionary framework are possible. The working environment consists of the most useful GIS. A fundamental role is played by the geo-referencing of information layers that often, because they refer to different historical periods, have different cartographic datum.

The overlap is possible using geometric transformations based on Ground Control Points whose coordinates can be obtained either directly from maps or, in the case of remote sensed images, from topographic survey or by comparison with other already georeferenced sources in the same datum taken as a reference. Regarding evolutionary aspects of the Domitian coast, it is interested, from 1876 to 1905, by coastal dynamics that determine the contemporary presence of eroded parts and added ones. From 1905 to 1954 only natural beach nourishment growing up and again from 1954 to 1984 eroded and added areas are present. From 1984 to 1998 there is only erosion that continues, although with less tight rhythms, until 2005, deadline for this study.

References

[1] Doody, P., Ferreira, M., Lombardo, S., Lucius, I., Misdorp, R., Niesing, H., Salman, A., Smallegange, M., *Living with coastal erosion in Europe – results from the eurosion study*, Office for Official Publications of the European Communities: Netherlands, 2004.

[2] Gatto, A., www.wwfaversa.it/DossierIschitella.htm.

[3] Cocco, E., De Magistris, M.A., De Pippo, T., Perna, A., Dinamica ed evoluzione del litorale campano-laziale: 3. Il complesso di foce del fiume Volturno. *Atti del VI Congresso A.I.O.L.*, Livorno, 1984 .

[4] Cocco, E., De Magistris, M.A., Iacono Y., Modificazioni dell'ambiente costiero in Campania (Litorale Domizio, Golfo di Gaeta) in conseguenza delle opere umane. *Il Quaterario,* **7(1b)**, pp. 409-414, 1984.

[5] Surace, L., La georeferenziazione dell'informazione territoriale. *Bollettino di Geodesia e Scienze Affini*, **2(98)**, 1998.

[6] Clifford, J. Mugnier, C.P., C.M.S., Italia Republic. *Photogrammetric Engineering & Remote Sensing*, **8(2005)**, pp. 889-890, 2005.

[7] Bezoari G., Monti C., Selvini, A., *Fondamenti di rilevamento generale*, Ulrico Hoepli Editore: Milano, pp. 668-674, 1989.

[8] Errico, A., Maglione, P., Parente, C., Santamaria, R., Impiego di immagini telerilevate da satellite per la produzione e l'aggiornamento di cartografia. *Atti della 13° Conferenza Nazionale ASITA*, Bari, pp. 1012-1013, 2009.
[9] Space Imaging www.spaceimaging.com/products/Ikonos/geo_techspec.htm
[10] Cina, A., *Trattamento delle misure topografiche*, Celid: Torino, pp. 28-34, 2002.

Changes in land use and their impact on the coastal zone in northern Sinaloa (Gulf of California, Mexico)

P. Muñoz[1], L. Menanteau[2] & D. Escobedo[3]
[1]*CIIEMAD-IPN, Mexico*
[2]*CNRS-Université de Nantes, France*
[3]*CIIDIR-Sinaloa-IPN, Mexico*

Abstract

Sinaloa State is the second largest producer of shrimp in Mexico, after Sonora, (in 2008 more than 35,000 tones, and 300 shrimp farms covering approximately 28 000 ha); with most of the production being exported to the United States. Sinaloa is one of the 31 states of Mexico, located in the northwestern part of the country. The state covers an area of 57,377 square kilometers and has 2.2 million inhabitants. The coastal plain is a narrow strip of land that stretches along the length of the state; this is located between the ocean and the mountain range "Sierra Madre Occidental", which dominates the eastern part of the state. Sinaloa is traversed by many rivers, which carve broad valleys. The largest of these rivers are the "Culiacan", "El Fuerte" and "Sinaloa". The main productive activities are agriculture, fishing, livestock breeding, shrimp aquaculture, commerce and industry. Apart from being the second largest shrimp producer in the country, Sinaloa is the most prominent state in terms of agriculture and has the second largest fishing fleet. As a result of these economic activities, the northern region of the State has experienced major changes in land use, mainly in the river deltas, being mostly devoted to aquaculture and agriculture. In terms of aquaculture, the river delta of "El Fuerte", one of the three largest rivers in the State, has undergone significant changes with a vast increase in surface for the cultivation of shrimp, from 364 ha in 1990 to 7,194 ha for year 2008 (ISA, 2007), with the consequent impact on the coastal lagoons of the region, including mangroves (LANDSAT and SPOT images).In the Topolobampo lagoon system there are evident impacts on the quality of water from adjacent coastal ecosystems due to the concentration of two nutrients: nitrates and phosphates,

WIT Transactions on Ecology and the Environment, Vol 149, © 2011 WIT Press
www.witpress.com, ISSN 1743-3541 (on-line)
doi:10.2495/CP110071

from the aquaculture activities in 2007 (11 farms) with values of 2,98 tons/year of total phosphorus and 30,02 tons/year of total nitrogen in its different forms (DIN, SRP, PO4), contributing significantly to the process of coastal eutrophication.
Keywords: land use, aquaculture, mangrove forests, coastal eutrophication.

1 Introduction

Globally diverse human activities impact the environment and ecosystems more quickly than the population is aware of in its effects. Aquatic ecosystems include extremely complex environments and for decades have been intensively used. In particular, the impact on the subject of this study, where ecosystems have been located mainly on coastal water bodies, and where much of the exploitable marine resources are concentrated.

On the Mexican Coast, coastal systems support various and significant fisheries, are reservoirs of biological diversity and provide unparalleled recreational opportunities. Due to such characteristics and to the growing development of human settlements in these environments, the demand on marine resources increased considerably which has led to the deterioration of the ecosystems with the subsequent impact on productive activities by the disappearance of habitats, creating an environmental problem, and increasingly complex economic and social problems.

Sinaloa State is located in Northwest Mexico, limited by the States of Sonora, Chihuahua, Durango, Nayarit and California Gulf. Territorial surface is of 58,092 km, including 656 km of coast with abundant estuarine systems. Agriculture, livestock and aquaculture are the main economic activities in the State, and on a smaller scale mining (Ag, Au, Zn and Cu), which is one of the largest producers in the Northwest region of the country. Agriculture activities are favored by the plentiful irrigation (Figure 1), only in the northern portion of the State are located the basins of the rivers El Fuerte and Sinaloa districts of irrigation 075 strong River 076 Valley of the Carrizo and 063 River Sinaloa, supplying an area of approximately 600,000 hectares. For this reason the use of various types of fertilizers and pesticides are undoubtedly one of the main concerns about the possible impact on ecosystems (freshwater and marine), and the quality of the products (agricultural, aquaculture and fisheries) which are intended for human consumption. Another problem presented is wastewater discharge in the bed of the rivers, through drains, creeks, lagoons or bays as result of inefficiency or lack of treatment plants in some of major cities such as Los Mochis, Guasave and Guamúchil. This situation creates a serious problem for the coastal north of the region, if it is considered that there are about 700 km^2 systems lagoon where they develop different species of commercial shrimp, oyster, clam, crab, axe callus and a great diversity of species of scale, all serving the ecosystems.

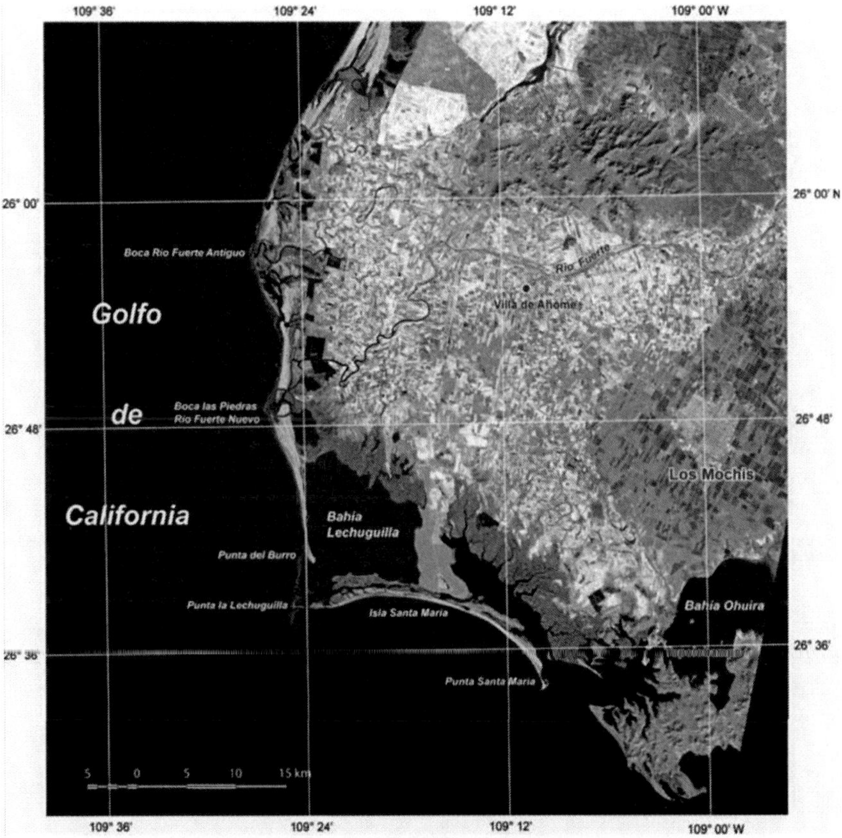

Figure 1: Delta of river "El Fuerte".

At least two of the most important activities in the region, agriculture and aquaculture occur in areas where the change in land use has been constantly growing.

2 Change in land use

Agricultural expansion is a special case of the frequent changes brought about by human land use, or the type of utilization of terrestrial ecosystems. Today, such a change is an important part of the so-called global change, along with climatic alterations and changes in atmospheric composition. Its consequences exceed local or regional level (Carabias [1]). Agricultural expansion influences on climate cycles of water, carbon and nitrogen in the biosphere, the emission of gases causing the greenhouse effect and biodiversity. At the same time, to an increasing demand for food and fiber, the increase in agricultural land appears as an inevitable process Elvira [3]. Social and environmental consequences also underline the importance of carefully planning the expansion of the areas intended for cultivation worldwide, as well as those aimed at aquaculture in growing expansion. Mexico land suffers from the same threats as the rest of the soils of the world. Our losses in this category are: by water erosion, 56%; by wind erosion, 28%; by chemical degradation 12 per cent and physical degradation 4 per cent. Human activities that cause soil degradation in our country - the same in the world - are: deforestation, with 29 per cent of the total damage; inadequate agricultural practices with 28%; overgrazing with 35% M; extraction of firewood, with 7%, and industry and urbanization, with 1%. The causes of deforestation in Mexico include: agricultural cuttings with 82% of the total; damage illegal logging with 8% wildfires with 3%; pests and diseases with 3%; authorized changes with 2%, and others, with 2%. Change of land use in our country is regarded as the second largest source of CO_2 generation (Paruelo [3]).

In the State of Sinaloa agriculture is significant, however this study analyses primarily land occupation by another important activity of the State as it is the case of the shrimp culture, Sinaloa State is the second largest producer of shrimp in Mexico, after Sonora (in 2008 more than 35 000 tones, 300 shrimp farms, covering approximately 28 000 ha; Figures 2 and 3).

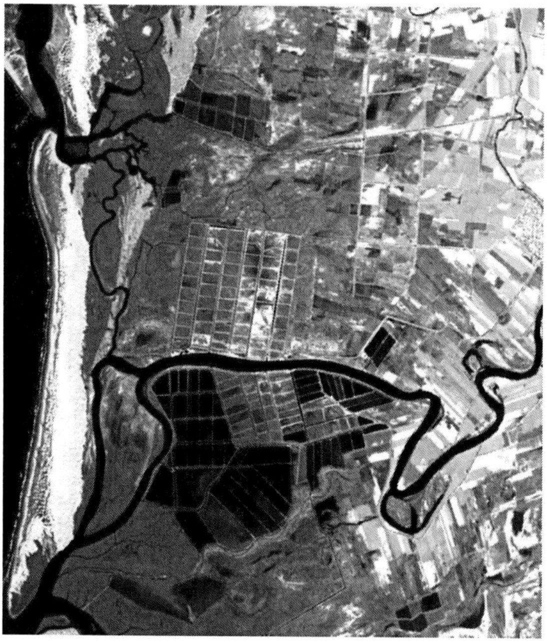

Figure 2: Shrimp farms, in the ancient delta of river "El Fuerte".

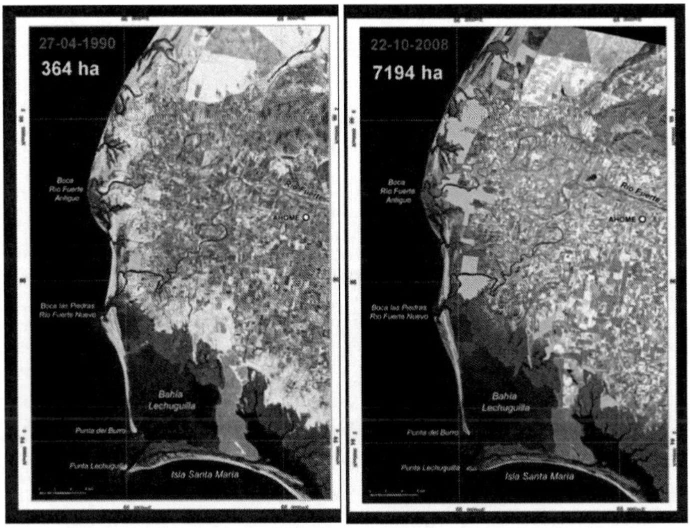

Figure 3: Use land change by development of shrimp farms (1990-2008) in northern Sinaloa, Mexico.

2.1 Aquaculture activity in Sinaloa

On a global scale, aquaculture is one of the food cultures where production is very fast with a 10% annual growth (Biao and Zhuhong [4]). In Latin America, prawn production passed 172 million tonnes in 1984, 921 million tonnes in 1994 Tobey et al. [5]. In Mexico, the production of prawns and crab almost increased four-fold from 13 million tonnes in 1996 to 47 million tonnes in 2001 respectively. In the high seas alone the prawns captured was 38 million tonnes during 2001 (Flores-Nava and Euán-Avila [6]).

However, the intensity and increase in prawn production activities has caused considerable alterations in the ecosystems. The mangroves ecosystem very close to the coastal region is the main area which is affected and nearly 5% of these areas were affected during the last decade (Phillips and Beveridge [7]). It is very well established that nearly 25% of the mangroves are destroyed in the world due to this activity (Clay [8]); also they modify the hydrodynamic patterns and the quality of water of many coastal ecosystems (Tobey et al [5]; Páez-Osuna [9]; Páez-Osuna [10]; Flores-Nava and Euán-Avila [6]). The activity is also responsible for the introduction of exotic pathogens microorganisms with several consequences in many countries NRC [11].

In the Sinaloa State, the prawn cultivation started experimentally during 1971 in a big manner in the Lagoon system of Caimanero. It attained great significance and they were commercialized during 1983, after the increase in production from the regions of River Fuerte in the Northern part of Sinaloa State (Lyle and Romero [12]). The prawn culture activity was affected seriously during 1994 to 1997 due to a virus which is well known as "Yellow Head" and "White Spot". The diseases decreased the production level by nearly half a million tonne/Ha^{-1} on average in 1996. Actually, the overall average is between 0.8 to 1.2 tonnes per hectare for non intensive culture (Anonymous [13]; Lyle and Romero [14]).

3 Methodology

The intensity of aquaculture development in the northern part of Sinaloa State has increased the effluent input into the coastal region and at the present time there is no direct effluent treatment way to reduce their effect in the receiving aquatic regions. In the lagoon system of Topobampo we found 10 shrimp farms covering a total area of 1,257 has CONANP [15]. However, the contribution of effluents due to the aquaculture activity was never monitored, for that reason a sampling was done every fifteen days in the area during March to September 2007, in the entry points (canal region) and in outlet points (drainage points), as it is in the reservoirs. We used for each station a multi-parameter probe (Horiba U-10) to register the temperature, salinity, pH and dissolved oxygen. Samples were also taken to identify nutrients (NH_4^+), dissolved inorganic phosphate (DIP), silicate (Si), total nitrogen (TN) and total phosphorous (TP), total suspended solids (TSS), particulate organic matter (POM) and chlorophyll (Cl) in all the regions. For the last years in the study area the white spot syndrome virus (WSSV) has reoccurred, for that reason was not possible to get environmental data to avoid the use of probes that could produce a contagion effect between different farms.

The samples collected every fifteen days were filtered using a GF/F glass filter (pore size 0.7 μm) for the analysis of inorganic dissolved nutrients (NO_3, NO_2, NH_4^+, PO_4^{3-}, SiO_3^{2-}). The analysis of ammonia was done instantly, whereas a part of the samples were filtered for subsequent analysis. The preserved filtered samples were used for the analysis of total nitrogen and total phosphorus. The samples were also analyzed for chloride and the calculation was done based on the equations of Jeffrey and Humphrey.

The determination of inorganic nutrients was done using spectrophotometer following the method of Strickland and Parsons (1972). The analysis of total nitrogen (TN) and total phosphorus (TP) was done using the method of Valderrama [16] and Strickland and Parsons [17].The determination of TSS and POM was done on the glass filtered samples using Whatman GF/C, 1.2 μm, which was dried at 550°C. Likewise, the filters were dried sufficiently to measure TSS and POM using total ash technique as described in APHA [18]. The calculated activity of nutrients was done using a multi monitoring equipment which was based one day before harvesting to identify the contribution by aquaculture activities. Moreover, the infection due to virus, "White Spot" was also observed in many areas. The average concentration of DIN, TN, DIP and TP (Kg Ha^{-1}) for one cycle is a simple multiplication for the extensive activity of N and P contribution to the ecosystem, in tonnes per cycle^{-1} , and per year this can be multiplied by a factor of 1.68 depending on the number of cycles during the activity (Lyle-Fritch and Romero [14]).

In other terms, the differences in determination of each site vs each entry points often vary from time to time. Hence, the ANOVA variance analysis or the Kruskall-Wallis non-parametric analysis is used when the analyzed variables did not show a normal distribution (Zar [19]).

4 Results

The calculated contributions of nutrients due to aqua culture activity on an overall average were 0.88 and 0.54 Kg Ha^{-1} per year of dissolved inorganic nitrogen and dissolved inorganic phosphorus (Table 1). Likewise, it was 25.2

Table 1: Values of nutrients concentrations, SST y MOP, in the shrimp culture pools located in the north of Sinaloa (2007–2009).

Variables		Concentration ($mg\ L^{-1}$)
	Max	0.29
NID	Min	0.01
	Average	0.05
	Max	0.65
PID	Min	0.0005
	Average	0.03
	Max	2.92
NT	Min	0.24
	Average	1.50
	Max	0.65
FT	Min	0.02
	Average	0.15
	Max	348.00
SST	Min	8.67
	Average	104.72
	Max	73.64
MOP	Min	3.72
	Average	24.93

and 2.49 Kg Ha^{-1} per year of total nitrogen and total phosphorus respectively. Miranda and Voltolina [20] calculated an annual average of 122 Kg Ha^{-1} of total nitrogen and 14 Kg Ha^{-1} per year of total phosphorous from the aqua culture region. These estimations were generated from the present study. The contribution from the aquaculture activity measured for other regions around the World by Lemonnier and Faninoz [21] and Jackson and Preston [22] reported a total nitrogen values between 55.1 to 121.2 Kg Ha-1 per year. The big difference between contributions from aqua culture is attributed to other reasons, where sampling technique also plays a role (Páez-Osuna and Guerrero-Galván [23]), but the main problem is the lack of analysis of the entry and outlet points, particularly for the different
forms of dissolved nitrogen that can vary, even among reservoirs from the same farm. On the other side in our work we encountered very low values which are attributed due to the low contribution of nutrients. These are mainly due to the low contribution of the nutrients measured in the water column after the aquaculture cultivation. The low quantity of organic material and particles in the sediments with nutrients also acts as a natural medium for the nutrient transport. Additionally, most of the work on other sites was based on blank calculation of masses with no direct measurement in the field, which could underestimate the contributions. The data from the present study are almost similar to the other authors, for that reason we conclude that is necessary to homogenize methodologies for the calculation of farms contribution; only in that way the results can be compared.

However, when the total nutrient values measured in the reservoirs were very high and the inorganic nutrients relatively low, inclusive lower than average contribution reported for the adjacent lagoon system (Escobedo Urías and Hernández-Real [24]; Magaña-Alvarez, [25]; Poot-Delgado [26]). This situation wrongly suggests that the reservoirs act as a nutrient filters in many cases and thus effluent treatment is not necessary.

As we can observe, for the present study the major part of nitrogen and phosphorous generated in the reservoirs is often present in organic form (dissolved or particulate) which is not measured routinely during the cultivation cycle. According with this concept the result is a bad practice related to the reservoir fertilization and a lack of treatment for the waters going out directly for the system to the coastal lagoons, many of these bodies with a strong environmental impact.

References

[1] Facultad de Ciencias, UNAM, Cambio de uso de suelo y servicios ambientales en el contexto de cambio climático; Carabias, Julia: México. Online:http://www.cambioclimaticoyseguridadnacional.org/documentos/Julia%20Carabias%20Cambio%20de%20Uso%20de%20Suelo.pdf

[2] Procuraduría Federal de Protección al Ambiente, El cambio de uso de suelo y sus repercusiones en la atmósfera, Elvira-Quesada, Juan Rafael: México. Online: http://www2.ine.gob.mx/publicaciones/libros/508/cambiouso.pdf

[3] Facultad de Agronomía. UBA, Expansión agrícola y cambios en el uso de suelo. Paruelo, J.M., Argentina, Online http://www.cienciahoy.org.ar /ln/hoy87/expansion.htm

[4] Biao, X. &Zhuhong, D. Impact of the intensive shrimp farming on the water quality of the adjacent coastal creeks from Eastern China. *Marine Pollution Bulletin*, 48(5-6), pp. 543-553. 2004

[5] Tobey, J., Clay, J. &Vergne. P. *The economic, environmental and social impacts of shrimp farming in Latin America*. Coastal Management Report #2202, Coastal Resource Center, University of Rhode Island: USA. 1998.

[6] Flores-Nava, A. & Euán-Avila, J. La Acuacultura en el marco del Manejo Integral de la Zona Costera: Reflexiones Generales, *El manejo costero en México, ed.* E.G. Rivera, EPOMEX-CETYS-Universidad de Quintana Roo-SEMARNAT: México, pp. 551-560, 2006.

[7] Phillips, M.C. &Beveridge, M. Shrimp culture and the environment: lessons from the world's most rapidly expanding warm water aquaculture sector. Environment and Aquaculture in Developing Countries, Ed. M.C. Pullin, ICLARM: Bangkok, pp. 171-197, 1993.

[8] Clay, J. Markets Potential for Readdressing the Environmental Impact of Wild Captured and Pond Produced Shrimp, Manuscript. World Wildlife Fund: Washington, 188 p. 1996.

[9] Páez-Osuna, F. *Camaronicultura y medio ambiente,* ICMYL-UNAM, Colegio de Sinaloa: México, 451 p, 2001

[10] Páez-Osuna, F. Retos y perspectivas de la camaronicultura en la zona costera. *Revista Latinoamericana de Recursos Naturales,* 1, pp. 21-31, 2005

[11] NRC (National Research Council). *Marine Aquaculture: Opportunities for Growth,* Nat. Acad. Press: Washington, p. 118, 1992.

[12] Lyle, L. & Romero, E. Desarrollo y características de las granjas camaronícolas en Sinaloa. *Camaronicultura y medio ambiente, ed. F.* Páez-Osuna, Instituto de Ciencia del Mar y Limnología. Programa Universitario de Alimentos, El Colegio de Sinaloa: México, Pp. 351-372.

[13] Anonymous. *Salud, ambiente y acuicultura en la costa Pacífica de México,* PACRC/UHH, UAS, Ecocostas, A.C./Ecuador, CIAD, Centro de Recursos Costeros / Universidad de Rhode Island, Conservation International México A. C., CESASIN: México, 149 p, 2006.

[14] Lyle-Fritch, L. P. & Romero-Beltrán, E. A survey on use of the chemical and biological products for shrimp farming in Sinaloa (NW Mexico). *Aquacultural Engineering*, 35(3), pp. 135–146. 2006.

[15] CONANP. Sistema Lagunar San Ignacio-Navachiste-Macapule. Ficha Informativa de los Humedales de Ramsar (FIR), 24 p. 2007.

[16] Valderrama, J. C. The simultaneous analysis of total nitrogen and total phosphorus in natural waters. Marine Chemistry 10(2), pp. 109-122. 1981.

[17] Strickland, J. D. & Parsons. *A practical handbook of the sea-water analysis.* J. Fisheries Research Board of Canada. Bull: Ottawa, 167 p. 1972.

[18] APHA. *Standard methods for the examination of water and wastewater*. 17 Edition. Washington, D.C., 1989. Folio variado.

[19] Zar, J.H. *Biostatistical analysis*. 2nd ed., Prentice-Hall: New York, 718 pp, 1984.

[20] Miranda, A. &Voltolina, D. Budget and discharges of nutrients to the Gulf of California of a semi-intensive shrimp farm (NW Mexico). *Hidrobiológica,* 19(1), pp. 43-48, 2009.

[21] Lemonnier, H. &Faninoz, S. Effect of water exchange on effluent and sediment characteristics and on partial nitrogen budget in semi-intensive shrimp ponds in New Caledonia. *Aquaculture research*, 37(9), pp. 938-948, 2006.

[22] Jackson, C. & Preston, N. Intake and discharge nutrient loads at three intensive shrimp farms. *Aquaculture research,* 35( ), pp. 1053-1061, 2004.

[23] Páez-Osuna, F. & Guerrero-Galván, S. R., et al. Fluxes and mass balances of nutrients in a semi-intensive shrimp farm in north-western Mexico. Marine Pollution Bulletin 34(5), pp. 290-297, 1997.

[24] Escobedo-Urías, D. & Hernández-Real, M. T. et al. Calidad bacteriológica del Sistema Lagunar de San Ignacio Navachiste, Sinaloa. *Ciencia y Mar*, III (9), pp. 17-27, 1999.

[25] Magaña-Álvarez, M. Distribución de nutrientes y su efecto en el nivel trófico de la laguna Macapule, Sinaloa. Tesis de Maestría. IPN-CIIDIR Sinaloa: México, 115 p. 2004.

[26] Poot-Delgado, C. Estructura de la comunidad fitoplanctónica con énfasis en las especies tóxicas y/o nocivas de la laguna de Macapule, Sinaloa, tesis de Maestría, IPN-CICIMAR: La Paz, B.C.S., México, 142 p. 2006.

Intervisibility analysis of an offshore wind farm using GIS tools

A. Errico, F. Guastaferro, C. Oliviero & R. Santamaria
Department of Applied Sciences, Faculty of Sciences and Technologies, University of Naples "Parthenope", Naples, Italy

Abstract

One of Kyoto's protocol targets is to limit greenhouse gas emissions produced by human activities through renewable energy and the conversion of wind energy into a useful form of energy (for example electricity) which is a kind of renewable source undergoing strong development.

In this work, the study is focused on offshore aeolic parks. The term "offshore aeolic" indicates wind turbines installed a few miles from the coasts of seas or lakes, exploiting the exposure of strong currents. The offshore aeolic farms have some disadvantages including the impact of these generators on the landscape, but they produce a minimum visual disturbance compared to onshore aeolic farms. Environmental Impact Assessment (known as the EIA), introduced in Europe by the 377/85/EEC Directive and amended three times, in 1997 (Directive 97/11/EC), in 2003 (Directive 2003/35/EC) and in 2009 (Directive 2009/31/EC), ensures that natural resources, usability and identity of landscapes are not compromised by engineering operations. Planning applications must be accompanied by an Environmental Statement (ES), with the purpose of giving an idea about the effect that the operations will have on the local environment. One of the main aims of the ES is the understanding of the likely visual intrusion of the proposed wind farm, which necessitates the use of visual techniques. A visual technique commonly used is the intervisibility map, normally represented by a two-dimensional map, centered on the location of the wind farm. Digital Terrain Models (DTMs) are used to achieve Thematic Maps obtained with GIS software. In this paper, GIS tools are used to identify the effects of wind farms on the landscape; in particular, attention is focused on Map Algebra functions and the study of the analysis of the intervisibility of an offshore aeolic farm is

WIT Transactions on Ecology and the Environment, Vol 149, © 2011 WIT Press
www.witpress.com, ISSN 1743-3541 (on-line)
doi:10.2495/CP110081

tested in a zone in Italy, localized in the Adriatic sea, along the coast of the Apulia and Molise Regions.
Keywords: offshore aeolic farm, intervisibility, viewshed, GIS, DTM.

1 Introduction

In Italy aeolic resources are less exploited than abroad, and one of the causes is the absence of a law about wind energy, unlike solar energy (http://it.wikipedia.org/wiki/Energia_eolica [1]). In recent times the interest in alternatives energy resources to traditional ones, such as offshore wind energy, has been increasing.

In Italy there are not offshore aeolic farms but there are many plans of parks located in islands and in southern areas of Italy such as in the Campania and Apulia Regions where there are the best conditions for wind resources available. In particular, the Adriatic side of Apulia is one of the Mediterranean areas with good potential for exploitation of offshore wind power.

Recently a collection of Guidelines ("Linee Guida per l'inserimento paesaggistico degli interventi di trasformazione territoriale" published by the Environmental Ministry [2]) has been realized. One of these concerns is the correct integration of large works into the environment, especially the insertion of wind power parks in the landscape. The aim of these Guidelines is to give an input to all persons who want to plan or to assess the great engineering works.

The offshore aeolic parks, in fact, have some disadvantages including the impact of these generators on the landscape [3]. The construction of an outsize engineering operation has a significant impact both on the site of insertion, called by Guidelines Local Impact Area (A.I.L.), and on the surrounding area, Potential Impact Area (A.I.P.); the second one represents the geographical space where the impact on the landscape is likely to occur and it is a circular area where the radius R depends on the number of wind turbines and their height.

At distances greater than the radius calculated, however, the impact is considered negligible.

2 Intervisibility map

The minimum distance of visibility is a basic concept to approach the intervisibility. It represents the distance, expressed in nautical miles, at which an obvious beacon is visible from a boat diametrically opposite it on the horizon.

Baldwin et al. [4] defines viewshed as: "two locations in a landscape are said to be in view from each other if they enjoy an uninterrupted view of each other. When one location is specified as the viewing location, and the visibility of all other locations in the study area is analysed, the resulting map of the study area is known as the viewshed or the visible area".

Simulating the insertion of a structure, a map of intervisibility is generally represented by a two-dimensional map whose purpose is to show areas in which the engineering operation will be visible. The aim of these "simulations" shows how the aeolic farm will insert inside the landscape and how it will appear from different points, allowing an assessment of environmental impact (Sparkes and Kidner [6]).

Using high-resolution elevation data, standard GIS software can make the intervisibility analysis for landscaping and environmental impact assessment of works characterized by a strong visual impact. The maps, in fact, parameterize the visibility impact of a work on the surrounding areas using different scales or parameters of incidence.

Depending on the type of work analyzed, three kinds of maps will be produced:

1) Map of theoretical visibility, that is independent of distance or size of objects that produce impact;
2) Map of cumulative exposure, that is ideal for maps of wind power farms with large areal extension;
3) Map of the incidence of weighed visibility distance: the visual impact of a wind turbine of 90 m in height is very different when viewed at a distance of 1 or 10 km, even if the generator is always fully visible.

The map produced for this paper covers both the second and the third case. It is a cumulative map because more aeolic generators are considered and the distance from these turbines is important; it is a weighted maps because the wind park is at some kilometers distance from the coast and so the intervisibility analysis will have more weight for greater radius.

Maps should be themed to make the analysis as well as scientifically accurate, clear and understandable to the layman.

3 Intervisibility method in ArcGIS 9.x (ESRI)

To realize an intervisibility map, the software ArcGIS 9.x (ESRI) processes three-dimensional data of a particular area, or rather a Digital Terrain Model (DTM) that shows the height of obstacles in the landscape, with the target represented by a points file.

It calculates if there is visibility between a generic point of observation (observer or spot) and a point to be observed (target). Intervisibility is achieved in the GIS environment by the tool "viewshed": it is a map algebra function that classifies the area around the target into two classes, visible and not visible areas, so that it is possible to develop thematic maps (fig. 1).

The function "viewshed" of ArcGIS 9.2 (ESRI) builds a raster, overlapping on the territory investigated, where for each cell a value corresponds indicating from how many points of view, previously set by the user, the cell is visible. If the observation point is one, the value assigned to the pixel is equal to 1 or 0 depending on whether or not one can see the enclosed area from it.

When the target is not only one but the visibility needs to be evaluated from multiple targets, intervisibility function records the number of targets visible by the observer and the value assigned to the pixel is equal to 0 (not visible) or different to 0 (visible).

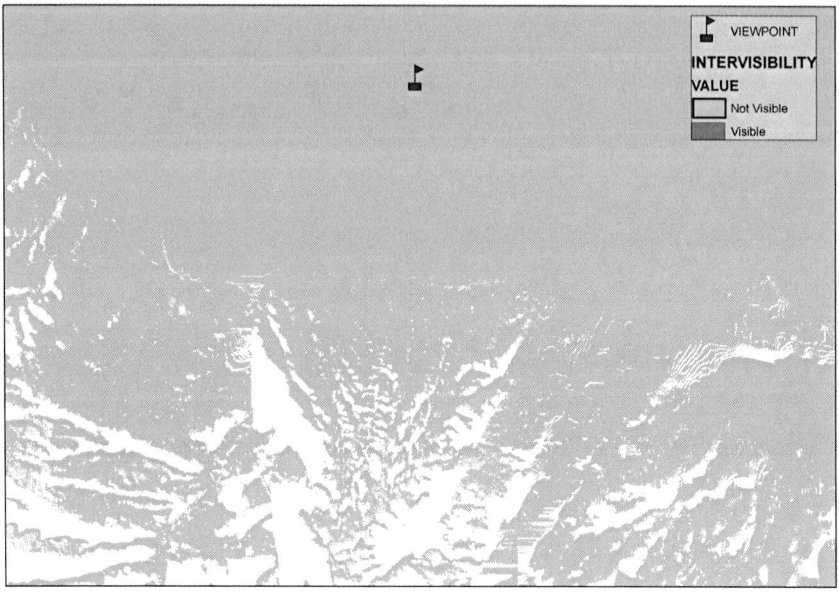

Figure 1: Example of an intervisibility map delineating visible and non-visible areas from one viewpoint.

Generally, visibility is evaluated considering more observers localized in different points of the area. Then, using a DTM, visibility from each observer it is possible to verify.

Visibility analysis is reciprocal; it can be made from both directions. In this paper viewshed tool is used to determine from which pixel can see a target located at a certain position.

The result, obtained by more observers and more targets, is a map where two colors indicate visible or not visible.

The user achieves an intervisibility map with ArcGIS, providing to the software some features such as:

- the DTM size cells;
- the height of the observation point;
- the increase to add to the surface elevation of the viewpoint (Offset A);
- the increase to add to the target surface cells (Offset B);
- the beginning and end of the angle of horizontal view (Azimuth1 and Azimuth2).

Values are in degrees from 0 to 360, with 0 oriented to the North;

- the upper and lower vertical angle of view (Vert1 and Vert2). Values are from -90° to +90°;
- the internal and external radius, Radius 1 and Radius 2 (between the rays visibility area from the point of view is delimited).

Further, the software considers the Earth's curvature and refraction, and makes corrections on the odds provided by the DTM using the following formula:

$$Z_a = Z_s - 0{,}87F\left(\frac{D^2}{2R}\right) \tag{1}$$

where:
Z_a is the correct value of the height (the actual height);
Z_s is the initial value of the height (the surface height);
D is the planimetric distance between the observer and the target;
R is the radius of Earth, considering equal to 6375 meters.

The simplest viewshed calculations suppose that light moves in straight lines in a Euclidean space (the Earth is not curved and no refraction occurs). This is a good approximation for distances of several kilometers. Over large distances it is necessary to consider the Earth's curvature and the transparency of the atmosphere (Burrough and McDonnell [5]), and so these parameters are taken into account for the paper, setting standard parameter provided by the software.

4 Applications for an offshore wind farm

The purpose of this study is to assess the visibility of an offshore wind power park still at planning. This area is localized in the Adriatic sea, along the coast of Apulia and Molise Regions, close to the town of Marina di Chieuti.

Environmental Impact Assessment project, conducted by Tecnogaia Srl behalf of the Company TREVI Energy SpA, is considered; this plan involves the installation of 50 large size turbines of the power of 3.0 MW each, for a total capacity of 150 MW [7].

The 50 turbines will then be placed into 8 parallel rows, oriented between 50° North and 60° East-North East, spaced with a step of 900 m and the first turbine will be placed in a row distance of 5 km from the coast. Each row will consist of 6–7 turbines spaced with a step of about 700 m.

The rotor ray is about 60 m, the height of the center of the rotor from the mean sea level is 90 m, the submerged part of the tower is between 17 m and 24 m. The territorial framework of the study is presented in Figure 2.

The potential site, identified for the construction of offshore wind farm, is selected for several factors such as distance from the coast, the nature and the depth of the seabed, the possibility of connection to the national electricity grid through power lines located in coastal areas and the presence of any non-technical constraints (military zones, marine protected areas, etc.). The visible elements of the project are, therefore, all the wind turbines.

To determine the level of visual disturbance caused by the installation it is necessary to realize a map of intervisibility that underlines the major points of visual exposure.

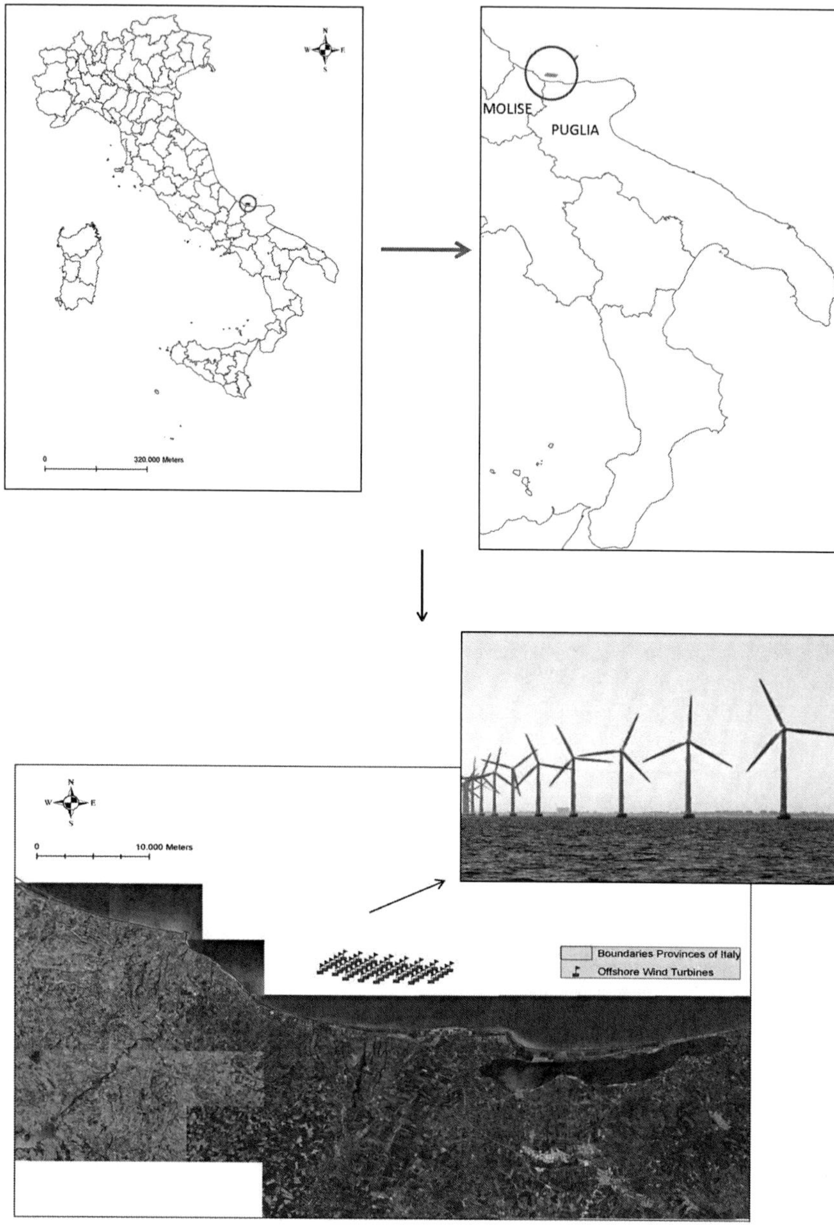

Figure 2: Territorial framework of the study area.

Data used in this work are the following:

1. Cartography (*.shp format) of the administrative boundaries of municipalities of the Puglia and Molise regions, referred in the UTM-WGS84 system;

2. DTM of Puglia and Molise regions, produced by IGM (Istituto Geografico Militare) in the UTM-WGS84 system and with a grid cells size of 20 m x 20 m;
3. Orthophoto (*.tif format) of the area affected by the analysis of intervisibility (flight 1998), scale 1:10.000, in the UTM-WGS84 system;
4. Points representing wind turbines in the Gauss-Boaga Roma40 System.

The aeolic generators are located in the GIS environment considering the coordinates of the turbines reported in the plan of Environmental Impact Assessment, framed in the Gauss-Boaga Roma40 System. To overlap the different layers, the UTM-WGS84 has been chosen as a reference system. So, Gauss-Boaga Roma40 data are converted in the new reference system using coordinate transformation software, Traspunto v.2.52 produced by Environmental Ministry.

Intervisibility analysis is conducted using the tool "viewshed" by software ArcGIS 9.2 (ESRI).

Because visibility analysis is reciprocal, for this study the location of each wind turbine is a point of observation, a target and it is defined from which pixel it is possible see a turbine located at a certain position (Errico et al. [8]). Intervisibility analysis allows us to study all 50 points together.

The intervisibility map produced for this study is conducted from elevation data with a 20 m step; the DTM used is a mosaic of Molise and Puglia, realized by PCI Geomatica v. 9.1.

Figure 3: Photo of study area.

The guidelines [2] for landscape assessment of wind farms, prepared by the Ministry of Cultural Heritage and for the assessment of visual impact, suggest that the theoretical limit of visibility should be considered equal to 20 km. The resolving power of the human eye at a distance of 20 km (corresponding to a period of 1 minute or 1/60 of a degree) is about 5.8 m, which means that, at this distance, you can see objects with a size higher than about 6 m.

For this reason a visibility ray R= 20 Km is considered in this analysis.
An "offset A" equal to 150 m is taken into account because the effective height is represented by the ray of the rotor added to the height of the hub above sea level (60 m + 90 m).

The Offset B is zero because is taken into account only surface elevations derived from DTM.

Azimuth1 and Azimuth2 are the parameters set by the software because for this study is necessary have a complete visibility (from 0° to 360°). The same is true for Vert1 and Vert 2 (from -90° to 90°).

5 Conclusions

Intervisibility analysis is a useful study that may help to clarify some potential effect of new buildings in a landscape.

GIS software provides tools for quick and simple application. The studies proposed in the literature are mostly based on the identification of viewpoints and the development of an intervisibility map: in the case study in question, taking into account the bidirectionality which can be considered the phenomenon, the approach has been completely reversed.

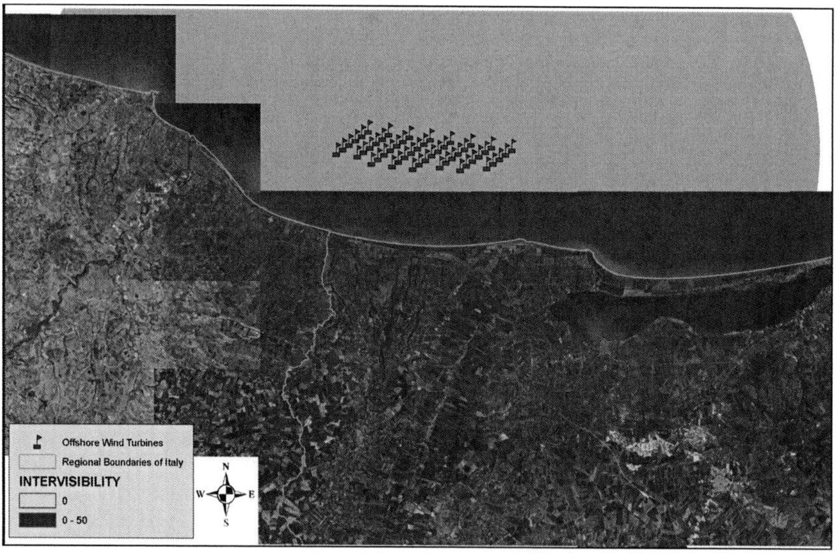

Figure 4: Viewshed analysis of study area.

With the intervisibility map it is possible to see the potential visual effect of an aeolic farm in the landscape (figure 4) but this analysis is not an end in itself. In order to get closer to a good understanding of actual visibility, the results must be tested on the field. An appropriate ground check will establish if areas of potential visibility are areas of actual visibility or not.

References

[1] *http://it.wikipedia.org/wiki/Energia_eolica*
[2] Ministero dell'Ambiente, 2006. Linee Guida per l'inserimento paesaggistico degli interventi di trasformazione territoriale. Gangemi Editore, Roma, pp 34. Available at *http://www.parc.beniculturali.it/ita/paesaggio/linee_guida / inserimentopaes /min_beni_imp_eolici.pdf.*
[3] *http://www.parc.beniculturali.it/ita/paesaggio/VIA/energiaeolica.htm*
[4] Baldwin, J., Fisher, P., Wood, J. and Langford, M. Modelling Environmental Cognition of the View with GIS. 1996 (Available at *http://www.ncgia.ucsb.edu/conf/SANTA_FE_CD-ROM/sf_papers/fisher_ peter/baldwin.html)*
[5] Burrough A. and McDonnell R.A. Principles of Geographical Information Systems, Oxford U Press ISBN 0-19-823365-5, 2000.
[6] Sparkes A. and Kidner D. A GIS for the Environmental Impact Assessment of Wind Farms. In Proceedings of the ESRI European user conference, 1996 Available at: *http://proceedings.esri.com/library/userconf/europroc96/ PAPERS /PN26/PN26F.HTM*
[7] Progetto Valutazione Impatto Ambientale Centrale Eolica di Chieuti: Available at:http://www.dsa.minambiente.it/via/DettaglioProgetto.aspx? ID_Progetto=240
[8] Errico, A., Maglione, P., Parente, C. DTM e GIS per l'analisi di intervisibilità nello studio degli impatti sul paesaggio di opere di ingegneria, Convegno Nazionale Sifet 2009, Mantova 24-26 Giugno 2009, pp. 249-258, ISBN 88-901939-7-2.

Section 3
Coastal geomorphology

Wave climate and coastal processes in the Osmussaar–Neugrund region, Baltic Sea

Ü. Suursaar[1], R. Szava-Kovats[2] & H. Tõnisson[3]
[1]*Estonian Marine Institute, University of Tartu, Estonia*
[2]*Institute of Ecology and Earth Sciences, University of Tartu, Estonia*
[3]*Institute of Ecology, Tallinn University, Estonia*

Abstract

The aim of the paper is to investigate hydrodynamic conditions in the Neugrund Bank area from ADCP measurements in 2009–2010, to present a local long-term wave hindcast and to study coastal formations around the Neugrund. Coastal geomorphic surveys have been carried out since 2004 and analysis has been performed of aerial photographs and old charts dating back to 1900. Both the *in situ* measurements of waves and currents, as well as the semi-empirical wave hindcast are focused on the area known as the Neugrund submarine impact structure, a 535 million-years-old meteorite crater. This crater is located in the Gulf of Finland, about 10 km from the shore of Estonia. The depth above the central plateau of the structure is 1–15 m, whereas the adjacent sea depth is 20–40 m shoreward and about 60 m to the north. Limestone scarp and accumulative pebble-shingle shores dominate Osmussaar Island and Pakri Peninsula and are separated by the sandy beaches of Nõva and Keibu Bays. The Osmussaar scarp is slowly retreating on the north-western and northern side of the island, whereas the coastline is migrating seaward in the south as owing to formation of accumulative spits. The most rapid changes have occurred either during exceptionally strong single storms or in periods of increased cyclonic and wave activity, the last high phase of which occurred in 1980s–1990s.
Keywords: wave hindcast, coastal dynamics, impact crater, climate change.

1 Introduction

The Neugrund Bank is an intriguing locale at the entrance of the Gulf of Finland. In the 1990s, the nearly circular-shape shoal was recognised as a meteorite

WIT Transactions on Ecology and the Environment, Vol 149, © 2011 WIT Press
www.witpress.com, ISSN 1743-3541 (on-line)
doi:10.2495/CP110091

impact structure [1, 2]. Since this discovery, the area has drawn much attention from geophysicists, marine archaeologists and biologists. The shoal has been proposed as a possible location for a wind farm. In order to study long-term variations in coastal processes and to incorporate the almost tideless Estonian coastal sea with wave hindcasts, the shoal was chosen as the third location after Vilsandi-Harilaid [3] and Letipea-Kunda [4] regions (Fig. 1).

The coasts encircling the shoals are very diverse, ranging from the limestone cliffs of the Pakri Islands and Peninsula to the "singing sands" of Keibu and Nõva beaches (Fig. 2). Despite having seen far less study than others, e.g. the coasts of West Estonia, these coasts are nonetheless experiencing increasing pressure, both climate change-related, as well as anthropogenic. For example, the Pakri Islands have been proposed as a possible location for the first Estonian nuclear power plant.

The aim of this study is to investigate hydrodynamic conditions in the Neugrund Bank area on the basis of wave and current measurements during five months in winter 2009/10, to present a local wave hindcast for the period 1966–2009 and to study coastal formations and their development with respect to temporal variations in climatological and hydrodynamical forcing conditions.

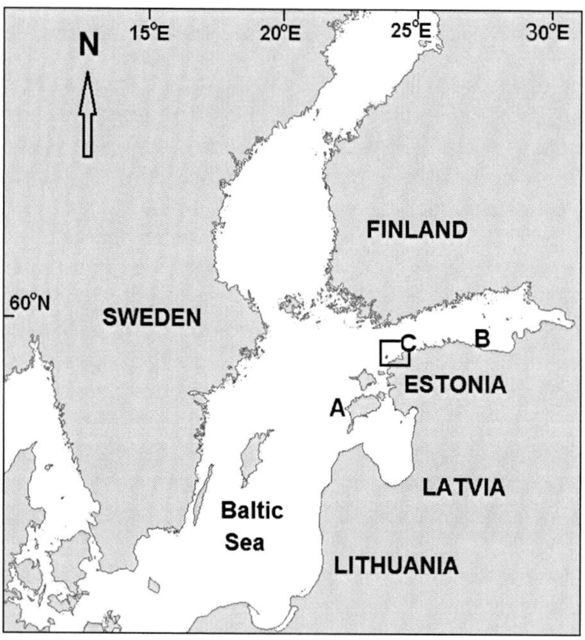

Figure 1: Map of the study area. The rectangle shows the Neugrund study area (Fig. 2). The locations of meteorological stations and respective wave hindcasts: Harilaid-Vilsandi (A), Letipea-Kunda (B), and Pakri Peninsula (C).

2 Study area

2.1 The Neugrund impact structure

Both the *in situ* measurements of waves and currents, as well as the semi-empirical wave hindcast are focused on the area known as Neugrund submarine impact structure, a ca. 535 million-year-old meteorite crater with a rim-to-rim diameter of 7–8 km [1]. The buried but partially re-exposed (during Neogene erosion) structure is located in the Gulf of Finland, about 10 km off the Estonian coast. The water depth at the central plateau is 1–15 m, whereas the adjacent sea depth is 20–40 m shoreward and 60–70 m in the circular canyon-like depression to the north (Fig. 2). The great number of shipwrecks lying on the seabed on and around the shoal testifies to the unusually rugged bottom topography in this region. More than a hundred isolated erratic boulders consisting of brecciated Precambrium metamorphic rock are situated on the seabed and coasts around the structure. Mineralogic and microscope evidence of the fine-grained glassy gneiss breccias indicate shock metamorphism. In the late 1990s, this evidence – together with detailed geophysical mapping, seismoacoustic and magnetic profiling – served to demonstrate that the Neugrund Bank is a meteorite crater with an initial structural altitude of about 500 m [2]. The current structural elements consist of a rimwall attaining 50 m, the crater proper, ejecta, and a 20 km outer ring fault (Fig. 2) [1, 2].

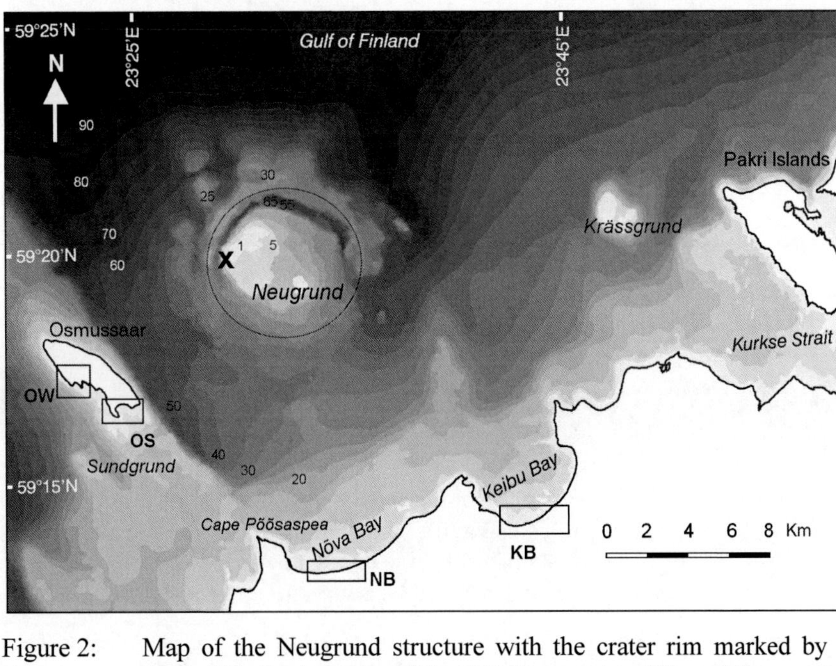

Figure 2: Map of the Neugrund structure with the crater rim marked by a circle [1]; the location of the RDCP mooring (X) and the coastal study sites at Osmussaar Island (OW, OS), Nõva (NB) and Keibu (KB) Bays.

2.2 The coasts of Osmussaar, Nõva-Keibu and Pakri

The sandy beaches of Nõva and Keibu bays separate the limestone scarp and accumulative pebble-shingle shores that dominate Osmussaar Island in the west and Pakri Peninsula to the east (Figs. 2, 3) [5–7]. The study area is experiencing post-glacial land uplift with a current rate of 2.8 mm/yr [8].

Osmussaar is a 5 km long relict island of the Baltic Clint, overlain by a thin layer of Quaternary deposits, which emerged from the Baltic Sea about 3000 years ago [7]. The highest elevation is approximately 8 meters above mean sea level. An Ordovician limestone cliff up to 6 m high is the most characteristic feature on the northern part of the Island, whereas accumulative gravel-pebble beach ridges 2–3 meters high cover the southern part of the island [5, 7].

Figure 3: Northern coast of Osmussaar Island (a), the 20 m high limestone cliff at Pakri Peninsula (b), coastal destruction after the 2001 winter storm at Pakri Peninsula (c), and sandy coast at Keibu Bay (d). (Photos by H. Tõnisson and Ü. Suursaar.)

Increments of the spits often rejoin the main island and therefore several coastal lagoons or lakes can be found behind the spits.

The coastal sea is less than 2 meters deep above a 200–300 m wide limestone bench on the western side of the island. Further offshore, the depth increases rapidly. The main shore processes are erosion from the cliff, which recedes approximately 9 cm/yr, longshore southward transport of gravel-pebble and accumulation of material as beach ridges at the southern end of the island [9].

Thc Pakri Clint Peninsula is a plateau bounded by steep escarpments – the Baltic Clint. The elevation of the Pakri cliff attains ca. 25 meters, which decreases from north-west to south-east. Pakri cliff rose from the sea about 7000 years ago, upon which it has been subject to wave erosion [5,6]. Erosion of the cliff by wave activity and weathering is relatively slow. The abraded sediments are transported southward along the shore.

Keibu and Nõva Bays are located off the Estonian mainland, approximately 20–25 km west of Pakri Peninsula and 10–15 km south-east of Osmussaar (Fig. 2). The coastal sea in both bays is relatively shallow and coastal slope is very gentle. Ancient coastal formations and dunes are located at different heights landwards with older formations being farther from and higher than the contemporary shoreline. The present shores themselves are sandy with a grain size varying between 0.15–0.25 mm. Glacial sediments are abraded from the sides of the peninsulas bordering the bays, and the material is transported towards the heads of the accumulative bays [5, 9].

3 Material and methods

3.1 In situ measurements of waves and currents

The hydrodynamic study is based on measurements of waves and currents using the Doppler-effect based current profiler (RDCP-600 by AADI Aanderaa) in the Neugrund Bank area of the Gulf of Finland (59°19′47″N; 23°29′33″E, Figs. 1, 2) between 20 November 2009–24 April 2010. The instrument was installed on the seabed at a depth of ca. 15 m. Over 155 days the instrument recorded hourly data on currents at seven depth layers (without cell overlap), wave parameters, sea level variations and water column properties, such as water temperature, salinity and turbidity. The wave data included significant wave height (Hs), as well as maximum wave heights and wave periods.

3.2 Meteorological data

Meteorological data were obtained from the Pakri meteorological station (59° 23′22″N; 24°02′24″E), located about 30 km east of the mooring site and operated by the Estonian Meteorological and Hydrological Institute (EMHI). The Pakri station has been operational since 1865, but the exact position of the station on the Pakri Peninsula has slightly changed, most recently in September 2003. Digitized wind records are available since 1966. Winds have been measured by "weathercocks" (wind vanes of Wild's design) during 1966–1976, automatic anemorhumbographs during 1976–2003, and MILOS-520 automatic weather stations since September 2003.

3.3 Wave hindcast

In order to assess long-term variations in forcing conditions for the coasts encircling the Neugrund, a wave hindcast was performed using a semi-empirical SMB-type wave model. The SMB-model, also known as the significant wave method, is based on the fetch-dependent shallow-water equations by Sverdrup, Munk, and Bretschneider [10]. Based on wind data, the model calculates the significant wave height, wave period and wavelength for a chosen location. As the role of remotely generated waves (swell) is small and the memory time of the wave fields in the Baltic Sea is relatively short [11], this relatively simple method can deliver reasonably good and fast results for this study area [3, 4].

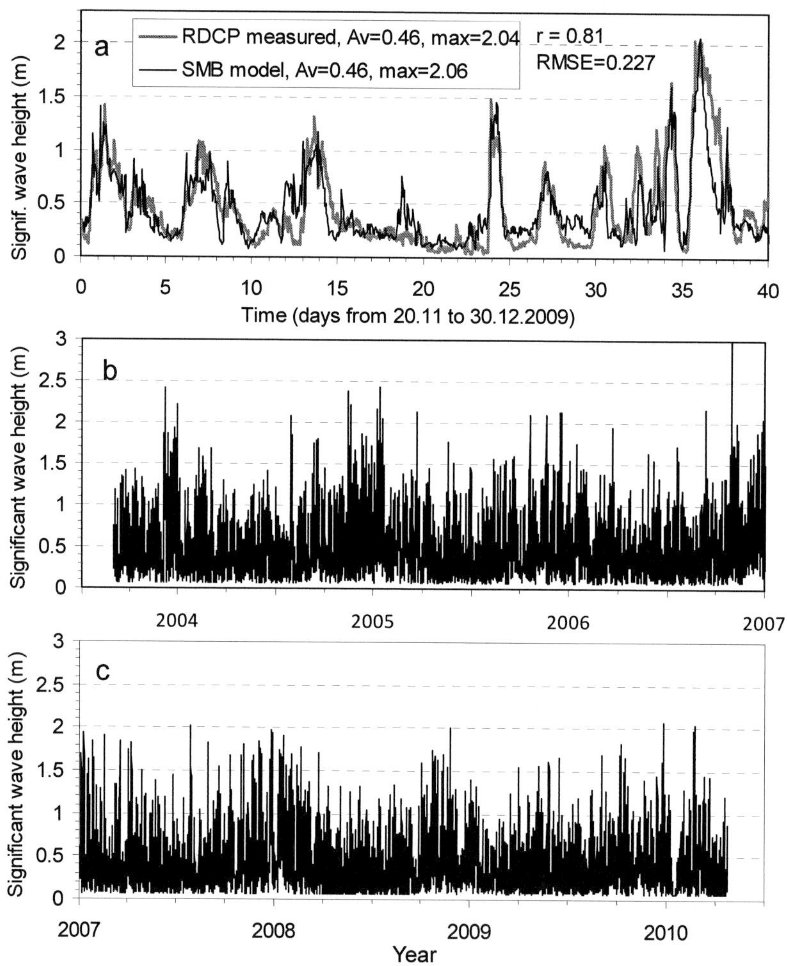

Figure 4: Wave measurements and the model calibration (a), wave hindcast at the Neugrund from September 2003 to April 2010 (b, c).

Our earlier comparison of hindcasts between the SMB model and 3rd generation WAM model [12] showed that SMB-type models can be used for site-specific long-term hindcasts when the gridded geostrophic wind data required for the 3rd generation wave models are not available in the desired resolution or length.

In this study, the wind data from Pakri station were used to run the wave model, and the wave data obtained by the RDCP served to calibrate the model. The calibration was done by prescribing fetch distances for different wind directions with a step of 10° and searching for appropriate depth parameter. The fetch is usually measured as the headwind distance from the nearest shore for that wind direction, and a procedure for taking into account basin properties in a wider wind sector is applied [13]. In our case, the calibration was done iteratively, in an attempt to keep the maximum and average wave heights equal in the modelled and reference series, to maximize the correlation coefficient and to minimize the RMS error. In fact, the procedure modified slightly the "real" geographical fetches in order to compensate for the wind speed impediment of certain wind directions at the Pakri station. For the hourly-based 40-day calibration period (Fig. 4a), we obtained a correlation coefficient 0.81 and an RMSE of 0.227 m (relative RMSE=11%). The site-dependently calibrated model was then used in multi-year wave hindcast at 1h or 3h interval, as it would have been measured by the same RDCP at the same location.

The calibration quality in this study was somewhat lower than in our previous efforts near Harilaid-Vilsandi (r=0.88, RMSE=0.23 m, 7.4%) [3] and Kunda-Letipea (r=0.92, RMSE=0.22 m, 7.8%) [4], due largely to the greater distance (about 30 km) between the wind measurement site at Pakri and the wave hindcast site at Neugrund. The corresponding distances at Harilaid-Vilsandi and at Kunda-Letipea were only 7 and 10 km, respectively. Calibration using Ristna wind data (90 km away) and Vilsandi station data (140 km away) yielded r=0.61 and r=0.51, respectively. In addition, the correlation was impaired by the rugged bottom topography at the Neugrund site. The modelling quality is lower also due to some islands (e.g. Osmussaar to SW) and shallow banks (to NE and SE), as the simple wind-wave fields may be occasionally "contaminated" by refraction, breaking and other shallow water effects [13].

3.4 Coastal studies

The dynamics of the seashores (Figs. 2, 3) were assessed by comparing areal changes in shore formations and changes in shoreline positions in space and time. Topographic maps in a 1:25 000 scale from 1947, 1961 and 1981, as well as orthophotos (from 1998, 2005 and 2008, pixel sizes between 25 and 100 cm) were compared to analyze the geomorphology, changes in shoreline position and character of shore processes in the study sites. Topographic maps from 1900 (1:42 000), 1935 and 1939 (1:50 000) were used to analyse the general tendencies in shore development.

GPS-measurements were taken during 2004–2008 to ascertain short-term changes in shoreline position, contours of beach ridges and locations of their crests. The boundaries of different shore types were mapped using Garmin 12 and Garmin 60CS GPS instruments, both of which are accurate to within 3 m. Data acquired in the additional field surveys in 2004, 2005 and 2009 were used to determine sediment origin and its direction of transport. These data were processed using MapInfo software, which was also used to calculate reductions in the erosion areas and increases in accumulative areas (see [14] for more details).

4 Results and discussion

4.1 Measured hydrodynamic conditions in 2009–2010

As a result of meteorologically-induced sea level variations, the depth of the instrument varied between 13.9 and 15.2 m over the 155 days. The maximum wave height attained 3.1 m (Hs=2.1 m) on 20 February 2010 (Fig. 4a) with an sustained wind speed of 12.6 m/s. In fact, the average wind speed did not exceed 13.2 m/s (with 21 m/s gusts) throughout the measuring period. The average measured sub-surface (1–3 m) current speed module was 9.8 cm/s, and the maximum attained 36 cm/s in this location. Ordinary waves and sea level variations during occasional (up to 2–3 m) storm surges (e.g. in 1967, 2001, 2005) are evidently the main hydrodynamic factors for the coastal changes in the almost tideless Estonian coastal sea [14, 15]. The site was covered by ice during 40 days in February and March – a rare event in the mild winters of the last decades [16]. The water temperature varied between -0.2 and 8°C, and salinity between 5.9 and 7.2. Variations in near-bottom water turbidity were influenced by wave generated resuspension from nearby shoals.

4.2 Hindcast wave conditions during 2003–2010

The maximum wave height (Hs) during the most reliable hindcast period from September 2003 to April 2010 attained 3.2 m (Fig. 4). Wave statistics suggest the maximum wave height would have been about 5 m. Owing to the rugged bottom topography and fetch limitations, the waves at the Neugrund Bank are not particularly high compared to our previous study sites [3, 4]. Both the average wave heights and higher percentiles showed clear seasonal variations (Fig. 5a) confirming that rough seas in the Baltic Sea prevail in autumn and winter, and calm conditions are more likely in spring and summer. The overall decreasing trends (Fig. 5a) are similar to those at Harilaid and Letipea for the same period [4]. However, it is impossible to extrapolate changes in wave conditions for longer periods. Although the hindcast was constructed for the entire period covered by the digitized wind data at the Pakri station (Fig. 5b), both the unnatural structure of the series as well as information about the exact changes in the location of the Pakri station indicate to significant inhomogeneity. As the model calibration is based only for the present location of the wind instrument, the periods 1966–1992 and 1993–2003 should be considered questionable. However, lying geographically between Harilaid and Kunda, the coastal developments can be viewed on the basis of these results (Fig. 5c).

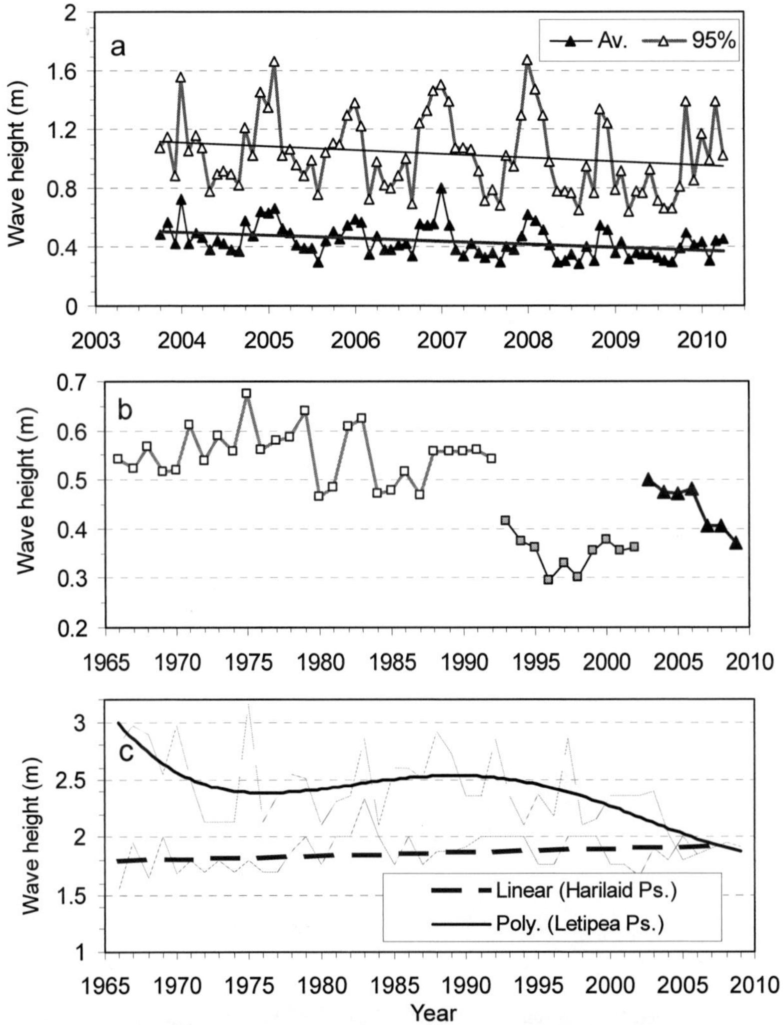

Figure 5: Average and 95% wave heights (in terms of Hs) computed on monthly basis from September 2003 to April 2010 (a); averages of annual wave hindcast data (1966–2010) (b); different markings refer to inhomogeneity in wind forcing data series. Long-term tendencies in high wave events (99-percentiles) according to data obtained at Harilaid [3] and Letipea [4] (c).

4.3 Coastal developments in relation to long-term changes in wave climate

The Osmussaar scarp (Fig. 3a) is slowly retreating (9 cm/yr) on the northern and eastern coast of the island, whereas the coastline is migrating seaward in the south-west and southern part as a result of formation of accumulative beach ridges and spits (Figs. 6, 7).

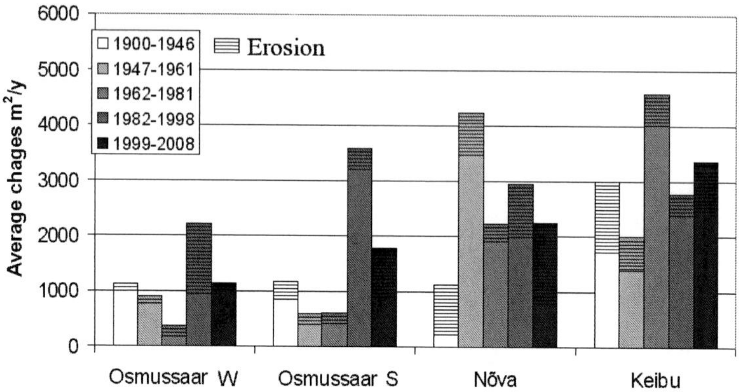

Figure 6: Average areal changes in the study sites during different time periods. Horizontal stripes (top part of the bar) indicate the share of erosion.

The spit in the southern part of the island has grown nearly 800 m south-eastward in the last century (Fig. 7). The former small bay in the southern part of the island became a lake when another spit-like formation closed the entrance into the bay between 1900 and 1935. In addition, the southern coast of the island has migrated over 120 meters southward towards the small bay. The most rapid

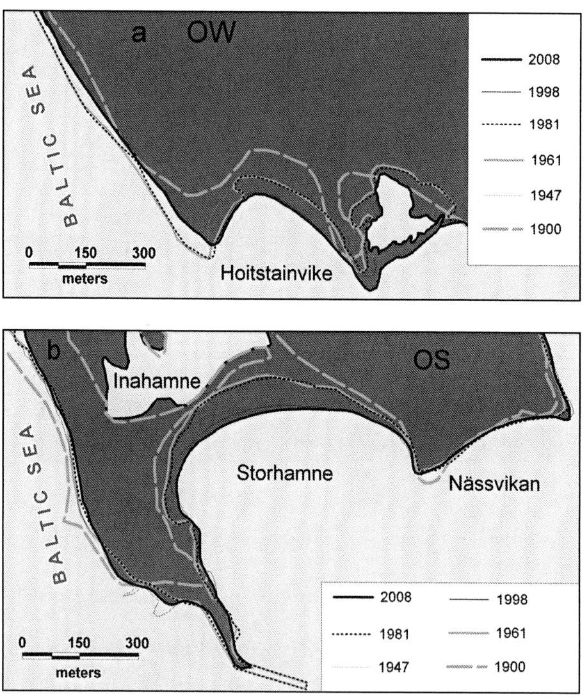

Figure 7: Changes in coastal position at Osmussaar OW and OS study sites.

changes in Osmussaar have occurred either during exceptionally strong storms or in periods of increased cyclonic and wave activity, the last high phase of which occurred in 1980s–1990s (Fig. 5) [3, 15]. Map analyses indicate that the processes were twice as rapid during 1982–1998 than in other periods (Figs. 6, 7). Erosion in the proximal part of the spit or former transport zone has increased rapidly since the beginning of the 1980s. Similar tendencies have been observed in the west Estonian study sites [14].

Nõva Bay (Fig. 3d) is a good example of the impact of the increased storminess on the sandy beaches. The most rapid changes in Nõva beaches occurred during 1947–1961 (Fig. 6). The rate of the shore processes has decreased slightly but the degree of erosion has increased significantly since this period. Erosion has even exceeded accumulation during the last decade. Therefore, we can conclude that formerly accumulative sandy shore at the head of the Nõva Bay has started to retread as a result of increased storminess.

Keibu Bay is the least exposed study site and therefore the impact of the increased west-storms is not clearly evident. The rate of the shore processes in Keibu Bay was the most rapid during 1962–1981 (Fig. 6). However, the share of erosion never exceeded a third of the total changes during the analysed periods. Due to sea level rise and increased storminess [11, 14, 15], very few sandy beaches in the world are still accumulative, and we can conclude that exposure to the prevailing storm winds allows the sandy shores in Keibu Bay to belong among those few accumulation dominated shores.

5 Conclusions

Based on hydrodynamic measurements and the semi-empirical wave hindcast at the Neugrund submarine impact structure, it was found that the main hydrodynamic forces acting on seashores in the studied locations are waves and occasional strong storm surges, which may elevate the sea level up to 2–3 m above the mean sea level. Owing to rugged bottom topography of the Neugrund bank, the wave conditions are rather site dependent. The coastal processes on the westerly exposed coasts of Osmussaar are continuously active, whereas in the northerly exposed coastal sections and Keibu Bay accumulation prevails as wave activity is declining. The shores in Nõva Bay have been accumulative until the last decade, but subsequent storms have caused them to retreat. Although the Osmussaar scarp is slowly retreating on the north-western and northern side of the island, the coastline is migrating seaward in the south as a result of the formation of accumulative beach ridges and spits. The most rapid changes have occurred either during strong storms or in periods of increased cyclonic and wave activity, the last high phase of which occurred in 1980s–1990s.

Acknowledgements

The study was supported by Estonian Science Foundation grants No.7609 and 7564, as well as by target financed themes SF0180104s08 and SF0280009s07.

References

[1] Suuroja, K. & Suuroja, S., Neugrund structure – a submarine meteorite crater at the entrance to the Gulf of Finland. *Estonia Maritima*, 4, pp. 161–189, 1999.

[2] Suuroja, K. & Suuroja, S., Neugrund structure – the newly discovered submarine early Cambrian impact crater. *Lecture Notes in Earth Sciences*, 91, pp. 389–416, 2000.

[3] Suursaar, Ü. & Kullas, T., Decadal variations in wave heights off Cape Kelba, Saaremaa Island, and their relationships with changes in wind climate. *Oceanologia*, 51, pp. 39–61, 2009.

[4] Suursaar, Ü., Waves, currents and sea level variations along the Letipea – Sillamäe coastal section of the southern Gulf of Finland. *Oceanologia*, 52 pp. 391–416, 2010.

[5] Orviku, K., *Estonian seacoasts*. Valgus: Tallinn, 112 pp., 1974.

[6] Kink, H. & Miidel A., (Eds.). *The Pakri Peninsula – Nature and Human Activity*. Estonian Academy Publishers: Tallinn, 94 pp., 1996.

[7] Mäesalu, M., *Landscape structure and development of Osmussaar Island. Bachelor thesis in Tallinn Pedagogical University*. 63 pp., 2005. [manuscript in Estonian with English summary].

[8] Vallner, L., Sildvee, H. & Torim, A., Recent crustal movements in Estonia. J. *Geodyn*., 9, pp. 215–223, 1988.

[9] Orviku, K., Some aspects of shallow water sedimentation along the Estonian coast. *The Baltic Sea*, ed. B. Winterhalter, Geological Survey of Finland: Special paper, 6, pp. 73–77, 1988.

[10] Seymour, R.J., Estimating wave generation in restricted fetches. *J. ASME WW2, May 1977*, pp. 251–263, 1977.

[11] Soomere, T., Myrberg, K., Leppäranta, M. & Nekrasov, A., The progress in knowledge of physical oceanography of the Gulf of Finland: a review for 1997–2007. *Oceanologia*, 50, pp. 287–362, 2008.

[12] Räämet, A., Suursaar, Ü., Kullas, T. & Soomere, T., Reconsidering uncertainties of wave conditions in the coastal areas of the northern Baltic Sea. *Journal of Coastal Research*, SI 56, pp. 257–261, 2009.

[13] Massel, S.R., Surface waves in deep and shallow waters. *Oceanologia*, 52, pp. 5–52, 2010.

[14] Tõnisson, H., Orviku, K., Jaagus, J., Suursaar, Ü., Kont, A. & Rivis, R., Coastal Damages on Saaremaa Island, Estonia, Caused by the Extreme Storm and Flooding on January 9, 2005. *Journal of Coastal Research*, 24, pp. 602–614, 2008.

[15] Suursaar, Ü., Kullas, T. & Szava-Kovats, R., Wind- and wave storms, storm surges and sea level rise along the Estonian coast of the Baltic Sea. *Ravage of the Planet II*. Book Series: *WIT Transactions on Ecology and Environment*, 127, WIT Press: Southampton, Boston: pp. 149–160, 2010.

[16] Sooäär, J. & Jaagus, J., Long-term variability and changes in the sea ice regime in the Baltic Sea near the Estonian coast. *Proc. Estonian Acad. Sci. Eng*., 13, pp. 189–200, 2007.

Radar and optical hybrid images to improve geoenvironmental mapping in Rio Grande do Norte state, Brazil

B. R. De Barros Pereira, V. E. Amaro & D. R. Valentim da Silva
PPGG - Programa de Pesquisa e Pós-Graduação em Geodinâmica e Geofísica, Universidade Federal do Rio Grande do Norte, Brazil

Abstract

The northern littoral of the Rio Grande do Norte state is a region under a strong dynamic influence concerning high energy coastal processes. These processes are increased by antropic actions joining with natural processes that change the natural landscape constantly. This area has lots of natural resources and it is very important economically, therefore it has been a target of studies of the main changes in the environmental aspects. In this context the geoenvironmental mapping have been defined and are included according to their physical and biological characteristics of the area and the territorial units that are linked to them. The territorial units follow the geological, vegetation, land-use/land-cover parameters. The geomorphology was the main parameter to define the geoenvironmental units. The geoenvironmental units were mapped on 1:50.000 scale based on the remote sensing images interpretation, using the fusion of multi sensor images (radar and optical). The images used were Radarsat-1, Landsat 5 TM (both from September 2006) and the SRTM from February 2000. The purpose of this methodology was to attend new resources to mapping geoenvironmental and their territorial units; previously each data has individual characteristics (an optical image has a spectral response while radar image provides textural and topographic data). The SRTM image was applied to emphasize the altimetry and morphology of the area. The digital image processing tools to integrate these data involved the precise registration, geometric correction and fusion through RGB and HSI color. The hybrid images created by these procedures was integrated and interpreted with the DEM (Digital Elevation Model) to merge four parameters: texture, morphology, spectral information and elevation in a single data. Then interpretation of the

WIT Transactions on Ecology and the Environment, Vol 149, © 2011 WIT Press
www.witpress.com, ISSN 1743-3541 (on-line)
doi:10.2495/CP110101

hybrid images allowed mapping and accurate description of the geoenvironmental units.
Keywords: remote sensing, hybrid images, coastal dynamic, geoenvironmental and digital image process.

1 Introduction

The observation of environmental changes through multi-temporal analysis of optical images became a very important methodology in the case of coastal zones. These areas suffer changes constantly in a short period of time caused by a strong natural dynamic due to their geographical position. In the work area during the last ten years this methodology was applied in a lot of works and obtained high success, showing its importance. It can be found in Grigio *et al.* [1], Amaro and Araújo [2], De Barros Pereira *et al.* [3], De Barros Pereira *et al.* [4] and Souto *et al.* [5] for example.

In this paper, further than the use of traditional optical images, fusion with radar images was made. The main propose of this technique has shown a new methodology of digital image process to map geoenvironmental units in the coastal zone. The product of the radar and optical image fusion shows in only one image the capacity to analyze many aspects as: texture, morphology, relief and altimetry, dragged out from radar data, and spectral information, extracted from optical image. It was applied the optical image LANDSAT 5 TM and the radar image Radarsat-1, both data from September 2006. To extract the altimetry data and create the DEM was used the STRM image (from February 2000).

The interpretation of the hybrid image and the DEM allowed the mapping of units of land use, vegetation, geology and geomorphology. Joining these four parameters it was possible to define the geoenvironmental units. All the other units are included in a geomorphology unit so it was possible to use the geomorphology as the main parameter to define a geoenvironmental unit.

1.1 Work area

The work area is located in the northern coast of Rio Grande do Norte state, including the cities of Macau and Porto do Mangue in northeast Brazil. It is approximately 1.100 km^2. The region is characterized with a high economic potential which is inserted into the oil industry, salt and shrimp farms and tourism.

The main access is made by the BR 406 road, around 190 km from Natal, the capital of the state (Figure 1).

This area is located in the equatorial zone that is characterized by high temperatures, high evaporation rates and low rainfall which the seasons are totally irregular presenting a rain period (April to July) and a dry period (the rest of the year). According to Nimer [6], the clime is semi arid, with annual average temperature of 27°C, 600 mm of rainfall and 70% of relative humidity (e.g. EMPARN [7]). The Piranhas-Assu River is the main one on the area, crossing it in a N-S trend until the Atlantic Ocean.

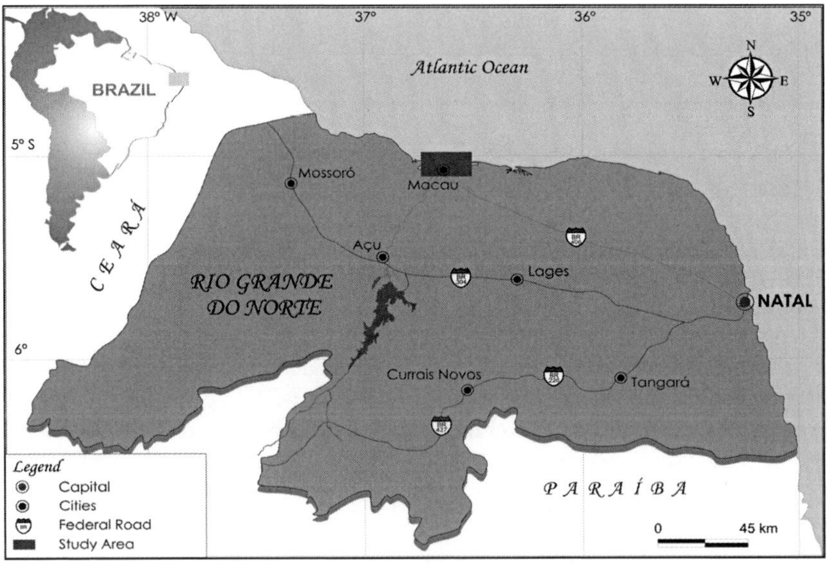

Figure 1: Located map of the study area.

The main coastal process agents that affect the area are: wind, marine currents, waves and tides. According to Caldas [8], there are two main wind trends: NE and ENE. The NE trend occurs among March and July and its velocity average is 4.8 m/s while the ENE trend is frequent in the rest of year having velocity average of 7.7 m/s. The middle tidal is predominant which reaches 3.3 m in the highest level. The wave average height is 56 cm.

2 Methodology

The methodology applied in this work had intention to extract the maximum information from each image individually using a sequence of process for each data. All the images passed for pre-processing, processing and then, was made the fusion of them creating the hybrid image with spectral, textural, morphology and relief characteristics.

The images used were: Landsat 5 TM, point/orbit 215/064, with seven spectral bands (six of them with 30m of spatial resolution and one thermal band with 120 m), captured on September 2nd, 2006; Radarsat-1 Mode standard has 12.5 m of spatial resolution and was obtained on September 26th, 2006; and SRTM from February 2000.

2.1 Pre-processing

Firstly, all the data were pre-processed. This step consists in the georeferencing of the images which was done in the Landsat and Radarsat-1 by the addiction of 55 points collected in situ with a geodesic GPS. It was used the UTM projection,

Zone 24S and SAD-1969 Datum. In SRTM image the Datum was modified from WGS-1984 to SAD-1969.

2.2 Processing

The processing steps seek to extract the major differences in the surface units disposed in an image. Sought, do all these steps for each data separately respecting the single characteristics of each image. All of them were made in ArcGis 9.2 and Er-Mapper 7.1 software.

2.2.1 Landsat image

The processes applied in Landsat 5 TM were: RGB compositions (using different spectral bands or band ratio in each channel); image enhancement; and creation of Principal Components (PC).

It was used two methods to choose the bests RGB composition. 1) It is based in the statistical parameters as standard deviation and variance of spectral bands correlation. The analysis of this method provides the OIF (Optimum Index Factor) that represent the best combination of three spectral bands. As major the OIF value major will be the superficial differences among the units in the image (e.g. Cháves *et al.* [9]). The best combination involved the bands 1, 4 and 5, and it was chose RGB541 as better to this work; 2) It is based in the knowledge according to the spectral response. The best compositions were RGB753, RGB531, RGB432, RGB453, RGB345, RGB7/1-4/1-3/1, RGB7/3-5/3-4/3 and RGB7/4-5/3-4/2. After choosing the RGB compositions, all of them were enhanced to equalize the histogram, aiming to show a better contrast in the image. At last, the Principal Components were created. PC involves all the dimensional components in an image which each band represent one dimensional space. Through this technique is possible to extract a major amount of information from images (e.g. Medrano *et al.* [10]). The analysis of PC shows the major difference among superficial units in the composition R(PC1) G(PC2) B(PC3) from the 6 bands of Landsat image.

2.2.2 Radarsat image

The processes on Radarsat image consisted in the rescheduling of image, application of filters and enhancement. Firstly, the rescheduling was made due

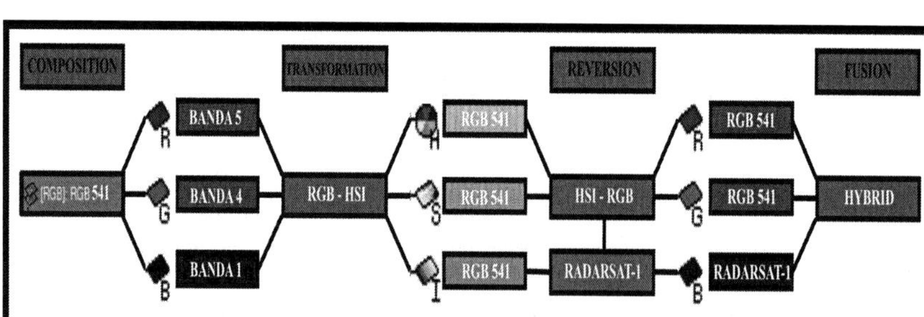

Figure 2: Flowchart showing the strategy of processing used in the fusion transformation RGB-HSI-RGBI.

the original image had 16 bits. It was rescheduling to 8 bits to facilitate the processing of image on the software. Later, *frost* 3x3 filter was applied to remove the *speckle* noise. Following, the *kernel* N-S directional filter was applied together with the change of solar zenith angle to 70°Az/30°. These applied tools improved the texture, topography and relief of superficial feature in the shadow direction and across it. Finally, the image was enhanced, *SRTM IMAGE.*

2.2.3 SRTM image

The SRTM passed by the change of Datum, previously WGS-1984 to SAD-1969 to join with the hybrid image. Posteriorly, shaded images were created through the change of solar zenith angle to 355°Az/20°, extraction of topography contour lines and insertion of vertical exaggeration of 45 times to create the DEM.

2.3 Fusion of data

The fusion of optical and radar image consist in a combination of both data to create a hybrid image which can be analyzed in the aspects of texture, morphology, topographic and spectral (color). To combine the data was applied the transformation of color system RGB-HSI-RGBI (Figure 2). In this method, one composition of three bands is inserted RGB channels, then the bands passed by transformation to HSI system. The second step is to reverse the color system HSI to RGB and finally, the 'I' (intensity) channel, is added in the RGB system to integrate the radar image, forming the new system RGBI.

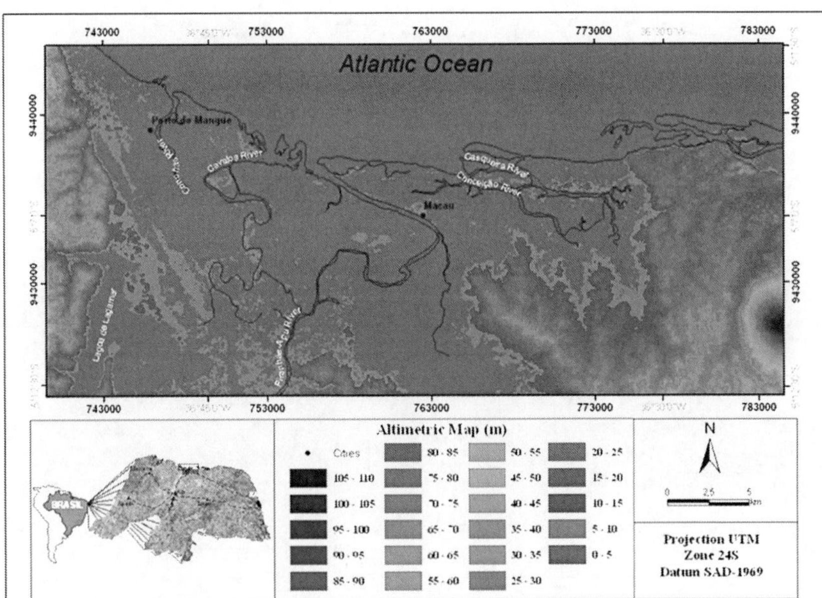

Figure 3: Altimetry map of the work area extracted from SRTM image.

The DEM was created by the extraction of the altimetry from the SRTM (Figure 3). This data was integrated with the hybrid image and then it passed for many different view tests to see the best point of view to define all the topographic surfaces in the work area.

3 Geoenvironmental units

The characterization of geoenvironmental units became possible through analysis of the field aspects, hybrid image and DEM.

The field work analysis was made in situ which all the units referent to geology, vegetation, pedology and land use were mapped; the hybrid image analysis combined visual investigation and interpretation in the perception of the author. In the hybrid image the topographic and textural pattern were increased by the properties of color (hue, saturation and intensity) from the RGB-HSI transformation and also from the position of Radarsat-1 image into I (Intensity) channel. The DEM analysis consisted in the interpretation of textural, morphological color and altimetry parameters of superficial features. The altimetry in the work area varies from 0 to 110 m (Figure 3).

Four geoenvironmental units were mapped joining altimetry with all the superficial units which obey to geology, vegetation, pedology and land aspects showed in the hybrid images: planation surface, fluvial-estuarine/tidal flat, interdune flat and interdune depression (Figure 4).

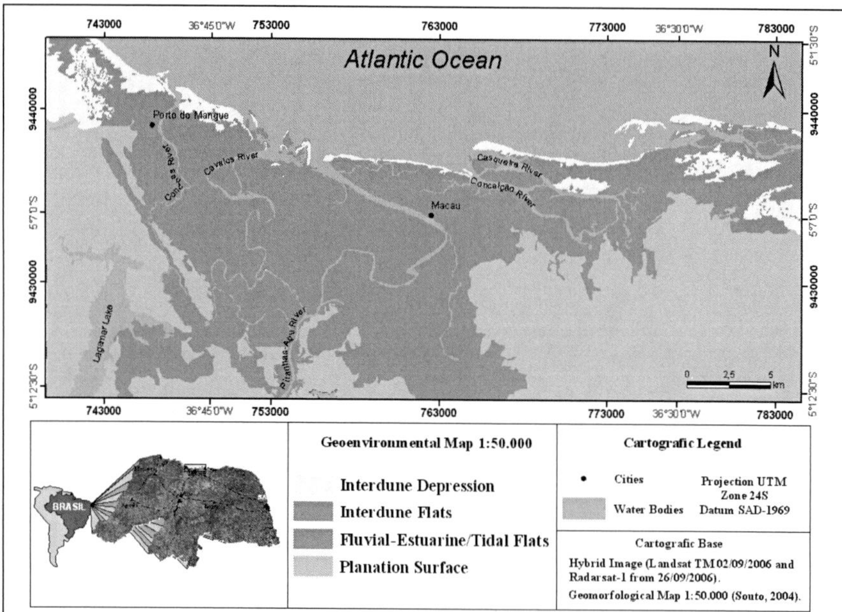

Figure 4: Geoenvironmental units map on the study area, at 1:50.000.

3.1 Planation surface

This unit shows a flat morphology, sometimes a little wavy in some sectors and presents the highest elevation in the work area, reaching about 110 m. The highest point is found in Mangue Seco Dome (Figures 5 and 6).

The planation surface was characterized by the geological units composed by calcarenites and calcilulites (Jandaíra Fm.); medium and fine grained sandstones and calcarenites (Tibau Fm.); volcanic rocks (Macau Fm.), conglomerates and coarse grained sandstones (Barreiras Fm.) and Quaternary alluvial deposits. Predominantly, this unit is cover by the caatinga vegetation and its soil is classified as soil association 1 (PLA) that is typical in caatinga's area. This soil is composed by quartz coarse grained sand, red and yellow coarse grained soil (e.g. Souto *et al.* [5]). On the NW portion of the study area the limit between the highs of planation surface and the low estuarine flats is marked by the morphological expression of a tectonic fault system, known as Afonso Bezerra fault system (Figure 5).

In the hybrid image were included in this unit: caatinga vegetation in RGBI541(RADARSAT-1), showing light green; the exposed soil and the cultivation area in red tones (Figure 6).

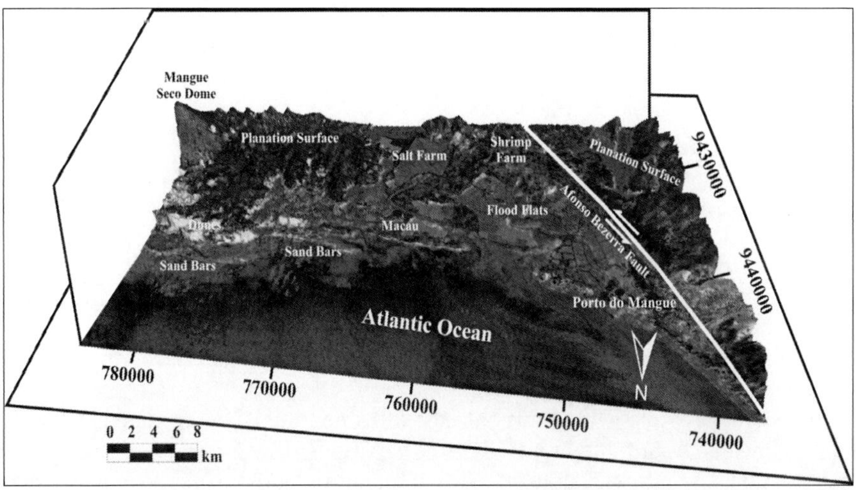

Figure 5: DEM combining the hybrid image R(PC1)G(PC2)B(PC3)I (RADARSAT-1) with SRTM. It shows the planation surface in purple tone; dunes fields in yellow tones; the flood area in yellow to light green tones; the Mangue Seco dome is the highest point while the flood flat is the lowest altimetry zone in the work area; and the Afonso Bezerra fault. (Colour in online version.)

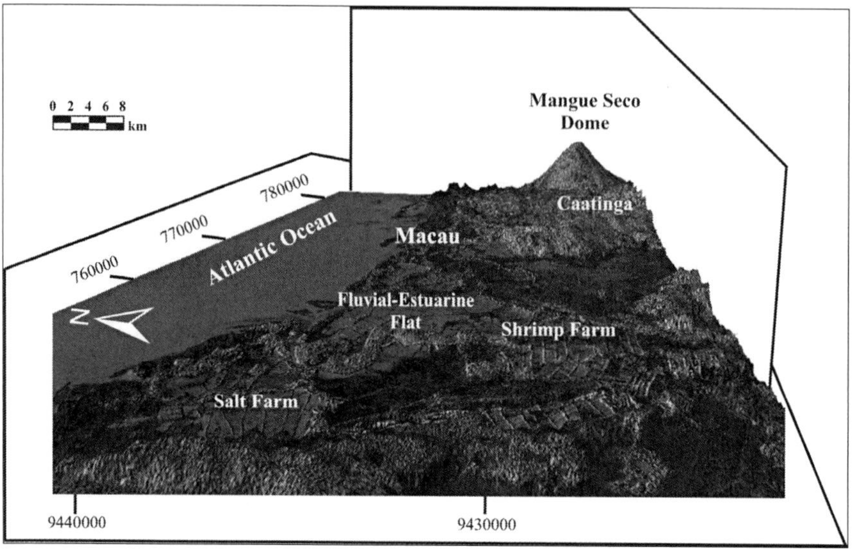

Figure 6: DEM combining the hybrid image RGBI541(RADARSAT-1) with SRTM. It shows the Caatinga in green tones; dunes fields in dark blue to purple tones; the square geometry of salt and shrimp farms. (Colour in online version.)

3.2 Fluvial-estuarine/tidal flats

This unit is located in the center of area, presenting the lowest altimetry values (0 to 5 m). It is linked with tidal flats geological unit, having the flood and ebb tidal deposits.

In this area, the hydrodynamic processes have an important configuration in the changes of natural landscape. The main hydrodynamic agents are waves, tidal and fluvial channels that modified the coastal zone reworking it and creating erosion and depositional areas.

This unit presents a flat morphology with fine grained sediments as silt and mud, typically from flood areas. The soil is classified by Souto *et al.* [5] as Soil Association 2 that is characterized by mangrove mud soil. The mangrove vegetation is typical in this unit. In some local, the vegetation is substituted by salt and shrimp farms.

In the hybrid image, RGBI753(RADARSAT-1), the mangrove appears in light green tones, while in R(PC1G(PC2)B(PC3)I(RADARSAT-1) the tones are dark purple; the flood area was emphasized in it hybrid image in green tones; this image also represent the shrimp farm in beige tone and the salt farm in light green tones (Figure 7).

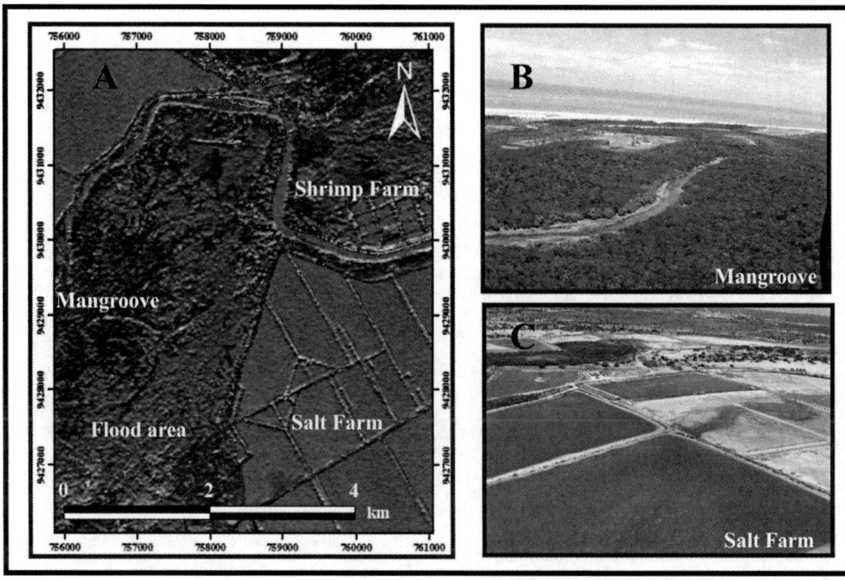

Figure 7: A Hybrid image R(PC1)G(PC2)B(PC3)I(RADARSAT-1) presenting salt farms in light green tone, flood area in dark green tone, mangrove in purple tone and shrimp farms in beige tone; B Mangrove; C Salt farms. (Colour in online version.)

3.3 Interdune flat

This unit is overlain by fixed dunes, situated around the deflation zone until the intertidal zone, with elevations until 20 m. This unit is characterized by typical vegetation that protects the migration of dunes. This fixation following the standard called "sheet" dunes which they follow the Wind trend. (ENE and NE). The typical soil find in dunes is classified as (DQA) quartz sand, well selected, very round and frost brightness (e.g. Souto *et al.* [5]).

In the hybrid image, RGB753(RADARSAT-1), the fixed dunes were mapped in dark green tone (Figure 8).

3.3.1 Interdune depression

This unit is located in the beach zone and incorporates the field of non fixed dunes. It is characterized by well selected sand grains, composed by round quartz, very frost brightness, white to gray tones, deposited by the wind and waves action, showing the wavy pattern of morphology. In this region, the migration of dunes shows the "half moon standard" called *barchan* dunes. The altimetry of this unit reaches in the top of coastal zone around 10 m and 20 m in the far part of coastal line.

In the hybrid image, RGB753(RADARSAT-1), the field of dunes was characterized red and purple tones and its textural aspect (Figure 8).

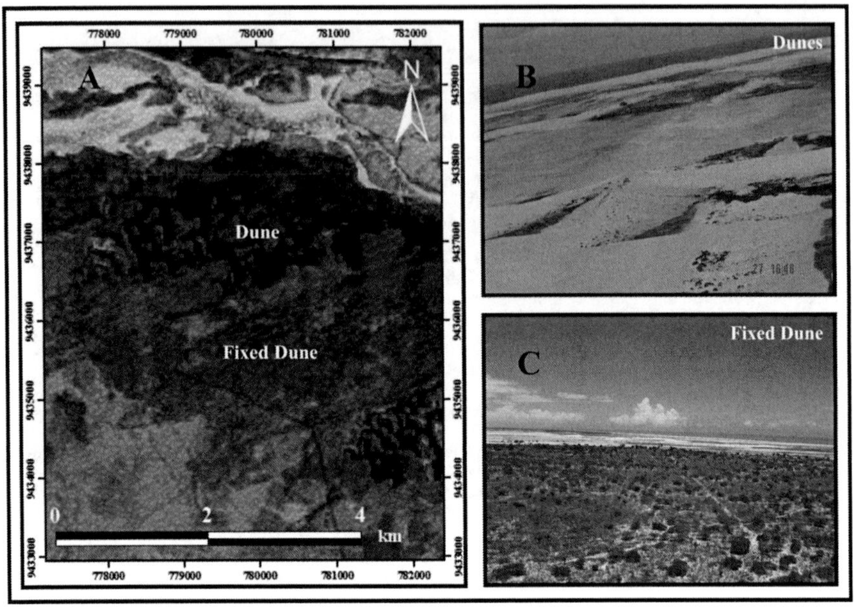

Figure 8: A Hybrid image RGBI753(RADARSAT-1) showing the dunes body in red and purple tones, and fixed dunes in dark green tone; B Field of dunes; C Typical vegetation fixing de body of dunes. (Colour in online version.)

4 Conclusions

The methodology to integrate multi-sensor images was satisfactory because all the objectives were fulfilled. The analysis of hybrid images gives a new data with textural, morphology, topographic and color (spectral) potential in only one data. Add to those parameters the altimetry is essential to determinate the geoenvironmental unit conclusions.

The bests hybrid images were: RGBI541(RADARSAT-1), RGBI753(RADARSAT-1) and R(PC1)G(PC2)B(PC3)I(RADARSAT-1). They gave a good response to define the superficial units, enhancing their differences with contrasting color tones and textural aspects.

The analysis of DEM shows the variation of altimetry in the work area from 0 to 110 m presenting as the highest point the Mangue Seco Dome.

The integrated analysis of field works, interpretation of hybrid images and DEM made possible to map four geoenvironmental units: planation surface (until 110m), fluvial-estuarine/tidal flat (0–5m), interdune flat (until 20m) and interdune depression (until 20m).

Acknowledgements

To the geoprocessing laboratory of UFRN (GEOPRO) and all people that work hard to make it a good work place. To my supervisor Prof. Dr. Venerando Eustáquio Amaro and Dalton Rosemberg Valentim da Silva. To Rede Cooperativa de Pesquisa em Monitoramento Ambiental de Áreas sob influência da Indústria Petrolífera, through PETRORISCO and HIDROSEMA projects.

References

[1] Grigio, A.M.; Amaro, V.E.; Diodato, M.A., Dinâmica espaço-temporal do uso e ocupação do solo no período de 1988-2004, do baixo curso do rio Piranhas-assu (RN): sugestões de acompanhamento integrado das atividades socioeconômicas impactantes em área costeira. Geografia (Rio Claro), v. 34, pp. 141-161, 2009.

[2] Amaro, V.E.; Araújo, A.B., Análise multitemporal da morfodinâmica da região costeira setentrional do Nordeste do Brasil entre os municípios de Grossos e Tibau, estado do Rio Grande do Norte. In: Revista de Gestão Costeira Integrada, v. 8(2), pp.77-100, 2008.

[3] De Barros Pereira, B. R.; Amaro, V. E.; Souza, A. S.; Aquino, A. G.; Valentim da Silva, D. R., Análise de dados orbitais multitemporais para avaliação de linha de costa nas áreas adjacentes ao município de Areia Branca/RN, com enfoque na região próxima ao estuário do Rio Apodí-Mossoró. In: *Anais of XIII Remote Sensing Brazrazilian Simposium*: Florianópolis, Brasil, pp. 2993-2998, 2007.

[4] De Barros Pereira, B. R.; Amaro, V. E.; Souza, A. S., Avaliação da Dinâmica Costeira da Foz do Rio Apodí-Mossoró e Áreas Adjacentes, Através de Produtos de Sensoriamento Remoto. *Anais of 43° Brazilian Geology Congress*,: Aracajú/SE, Brazil, pp. 219, 2006.

[5] Souto, M.V.S., Castro, A.F., Grigio, A.M., Amaro, V.E. & Vital, H., Multitemporal analysis of geoenvironmental elements of the coastal dynamics of the region of the Ponta do Tubarão, City of Macau/RN, on the basis of remote sensing products and integration in GIS. *Journal of Coastal Research*, v.39, pp. 1618-1621, 2006.

[6] Nimer, E., Climatologia da Região Nordeste do Brasil. Introdução à Climatologia Dinâmica. Revista Brasileira de Geografia, v. 34 (2), pp. 3-51, 1972.

[7] EMPARN, Empresa de Pesquisa Agropecuária do Rio Grande do Norte, http://www.emparn.rn.gov.br/links/meteorologia/meteorologia.asp

[8] Caldas, L. H. O., Late Quaternary coastal evolution of the northern Rio Grande do Norte coast, NE Brazil. PhD thesis, University of Kiel, Germany, pp.92, 2002.

[9] Cháves, P.S.J.; Berlin, G.L.; Bowel, J.A. Statistical method for selecting Landsat MSS Ratios. *Journal of Applied Photogrammetry and Engineering*, v. 8 (1), pp. 23-30, 1982.

[10] Medrano, M.C.N.; Corona, A.H.; Barajas, A.M., Caracterización litológica de regiones desérticas mediante técnicas de percepción remota: Un ejemplo en la franja costera central de Baja California, Mexico. Revista Mexicana de Ciências Geológicas, v. 22 (2), pp.229-245, 2005.

Far-infragravity and infragravity "pulses" in a rip current

B. Greenwood[1], R. W. Brander[2] & E. Joseph[2]
[1]*Department of Physical & Environmental Sciences, University of Toronto Scarborough, Canada*
[2]*School of Biological, Earth & Environmental Sciences, University of New South Wales, Australia*

Abstract

Rip currents are narrow, jet-like flows directed offshore through the breaker zone, often fed by shore-parallel feeder currents and forced by variable alongshore set-up induced by breaking gravity waves. Current speeds are not "*quasi-steady*", but "*pulse*" at a range of frequencies, some much lower than that of the incident waves. Relatively few field measurements identify these low frequency motions and even fewer record their sediment transport potential. Cross-shore velocities and sediment concentrations measured in a rip neck are used to: (a) identify the frequencies of these *pulses* and (b) assess the role of low frequencies in the transport of suspended sediment. Velocities at z = 0.13 and 0.39 m and concentrations at z = 0.13, 0.26 and 0.39 m were recorded continuously at 4 Hz over several hours in a rip neck at Bennett's Beach, NSW, Australia. Spectral and cross-spectral analyses of ~33-minute records were used to identify low frequency modulations of the velocity and concentration fields, as well as the suspended sediment flux. A lack of spectral variance around 0.005 Hz marked the partition frequency between infragravity (IG) and far-infragravity (FIG) motions. At high tide when the mean cross-shore current at z = 0.13 m was at its maximum (~0.38 m s^{-1}): (a) the dominant frequency was a FIG *pulse* at 0.0026-0.0028 Hz (~5-6 minutes). (b) IG motions also occurred with a 0.0199 Hz (~50 s) *pulse* dominating. (c) near the bed, concentrations were strongly modulated at FIG frequencies, but at higher elevations, the spectra were distinctly "*red*". (d) co-spectra revealed an offshore sediment flux dominated by a FIG *pulse* centered at 0.00244 Hz (~6-7 minutes). IG *pulses* between 0.0129

WIT Transactions on Ecology and the Environment, Vol 149, © 2011 WIT Press
www.witpress.com, ISSN 1743-3541 (on-line)
doi:10.2495/CP110111

and 0.0132 Hz (76-78 s) were less dominant, although both FIG and IG frequencies produced a net offshore transport of suspended sediment.

Keywords: rip currents, sediment concentrations, far-infragravity and infragravity pulses, sediment flux.

1 Introduction

On barred coasts, rip currents are narrow, topographically-constrained, "*quasi-steady*" flows directed offshore through the breaker zone as "jet-like" flows reaching speeds up to ~2 m s^{-1} and forced by variable alongshore set-up induced by incident gravity waves breaking on a segmented nearshore bar. It is clear from the literature, however, that rip currents are not remotely "*quasi-steady*"; they are modulated by "*pulsations*" at a range of frequencies, often much lower than those of the incident gravity waves. Frequently, "*rips*" form an integral part of a cellular water and sediment circulation in both marine and lacustrine surf zones; the classic circulation model is a net flux of water landward across the bar under the breaking waves followed by an alongshore and seaward flux through the feeder channel and rip neck respectively (Komar [1]). This flux is often modulated by the tide, even under micro-tidal conditions (Aagaard et al. [2]; Brander [3, 4]; MacMahan et al. [5, 6]; Bruneau et al. [7]). Since the early work of Shepard [8], Shepard et al. [9], McKenzie [10], Reimnitz [11], Reimnitz et al. [12], Cook [13] and Greenwood and Davidson-Arnott [14, 15] and more recently Smith and Largier [16] and Thornton et al. [17], the assumption has been that rips play an important role in maintaining both the water and sediment balance on the upper shoreface, and are a large constraint on topographic change. However, recent field and numerical experiments have seriously questioned this basic assumption, at least with respect to the flux of fluid (MacMahan et al. [18, 19]). It appears that many cellular flows in the nearshore involve closed *vortical systems* inside the breaker line and only occasionally do rip currents expel water offshore (MacMahan et al. [19]). These very low frequency "*pulses*" in the offshore flow would appear to be the main mechanism for water exchange between the surf zone and the rest of the upper shoreface. However, the nature of theses "*pulses*" and role they play in transporting sediment is virtually unknown, since measurements of sediment flux are extremely limited.

2 Low frequency motions ("*pulses*") in rip systems

Low frequency oscillations in nearshore wave and current systems have been documented for many years since the early work on "*surf beat*" by Munk [20] and Tucker [21]. More recently, Aagaard et al. [2] showed that infragravity oscillations were important to the net flux of suspended sediment in rip current systems at least at certain stages of the tide (low tide); furthermore, Greenwood et al. [22] demonstrated that infragravity oscillations do influence the net transport of suspended sediment at least in the rip neck under low to moderate energy conditions, and may in fact control the direction of the suspended sediment flux at certain stages of the tide. In this paper, we document very low frequency motions ("*pulses*") in a rip channel under swell conditions and demonstrate their potential to transport suspended sediment offshore.

3 Location and nearshore bathymetry

Bennett's Beach, NSW, Australia, is micro-tidal with a mixed semi-diurnal regime and was in an "*intermediate*" state (Wright and Short [23]) during the experiment. Samples from the medium sand beach were well-sorted, bimodal, with a mean grain size of 1.79 ø (300 μm) and a standard deviation of 0.37 ø (91 μm). The beach was subject to long period swell from the SE during the experiment, coupled with a variable wind-wave field forced by a daily sea breeze cycle. Figure 1 illustrates the study rip cell on February 19th 2004 after injection of a tracer at the head of the feeder channel. The cell was asymmetric with a dominant feeder current from the southwest and a rip which broke through the bar in a narrow channel; beyond this neck the flow diffused to mix with water outside the surf zone. The bathymetry (Fig. 2) consisted of an oblique nearshore bar welded to a shoal to the south, extending alongshore in a north-easterly direction to end at the rip neck; a trough between the bar and the beach face, varied in depth from near zero where the bar was welded to the shoal, to ~1 m before it turned offshore into the rip neck, which at high tide was ~1.75 m deep. Over three days, the bar migrated northeast and offshore, forcing the neck northward as it was constricted and deepened (Figs. 1 and 2).

Figure 1: Tracer deployed in the rip cell at Bennett's Beach, NSW, Australia. Note the breaking waves on the segmented bar, the flow in the southern feeder channel, the narrow jet in the rip neck and the diffused rip head.

4 Measurement and analytical techniques

Instruments were deployed at high tide in the centre of the rip neck (P1, Fig. 2) to measure: (a) the cross-shore and alongshore (relative to the average local shoreline orientation) horizontal currents, at elevations of z = 0.13, 0.39 and 0.50 m, using biaxial electromagnetic current meters (Marsh-McBirney, OEM 512); (b) sediment concentrations at z = 0.13, 0.26, 0.39 m, using optical

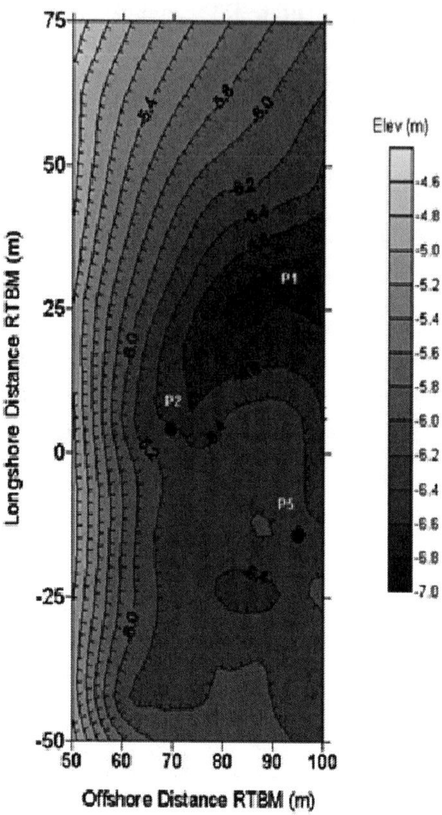

Figure 2: Nearshore bathymetry, Bennett's Beach, February 21st. P1 marks the location of the instrument pod in the rip neck and P2 and P5 marks the location of the pods in the feeder channel and on the bar crest, respectively. NB: RBTM is relative to the bench mark.

backscatter sensors (D&A Instruments, OBS-1P and OBS-3); and (c) mean water surface elevation and waves, using a strain-gauge pressure sensor (Druck-1830). Waves incident to the surf zone were recorded by a Druck pressure sensor deployed ~130 m directly offshore of the bar. For transport calculations, concentrations measured at 0.13 and 0.26 m were coupled with velocities measured at 0.13 m; concentrations measured at z = 0.39 m were coupled with velocities measured z = 0.39 m. There is the potential for error here (Austin and Masselink [24]); in all cases currents at 0.13 m were only very slightly different from those at 0.39 and 0.50 m; the latter were essentially identical. The sensors were supported on a heavy solid based support pod, which did not allow significant shifts in elevation once the support settled. The currents and sediment transports were also averaged over significant time blocks, with continuous recording at 4 Hz. Spectral analyses of "long" time series were used to identify frequency components beyond the range of incident gravity waves. This assumes that the recorded variables are "*stationary*"; a random variable or process is said to be *stationary* if its statistical parameters (e.g. mean, variance, etc.) are independent of time. In reality, only weakly stationary processes exist in nature; the assumption is that if the controlling variables in the system (e.g. incident waves, tidal level) are stable, then the underlying frequency structure of the resulting current and concentration time series can still be determined by appropriate demeaning and detrending. Blocks of 8000 points were demeaned and detrended, on the assumption that ~33 minutes is long enough to identify far-infragravity *pulses*, but short enough that the system remained quasi-stationary. Within this time frame, tidal changes did not exceed 0.11 m and the incident swell remained unchanged. Series of "*collocated*" velocities and concentrations were used to compute the time-averaged *net*, *mean*, and *oscillatory* components of the suspended sediment transport (SST) rate for a 2-dimensional, vertical section of fluid at each elevation (see Jaffe et al. [25]; Huntley and Hanes [26]; Osborne and Greenwood [27]):

$$< q_s >_{net} = \frac{1}{T}\int_0^T\int_0^h U_{(z,t)} C_{(z,t)} dzdt \qquad \text{(i)}$$

$$< q_s >_{mean} = \frac{1}{T}[\int_0^T\int_0^h U_{(z,t)}] * \frac{1}{T}[\int_0^T\int_0^h C_{(z,t)}] \qquad \text{(ii)}$$

$$< q_s >_{osc} = \frac{\Delta f}{f_c}\int_0^T\int_0^h C_{(UC)}(f) \qquad \text{(iii)}$$

where $U_{(z,t)}$ = horizontal water velocity (m s^{-1}), which can be disaggregated into cross-shore and alongshore components; $C_{(z,t)}$ = sediment concentration (kg m^{-3}); z = height above the bed (m); t = time (s); h = water depth (m); T = time of integration (s); < > = time-averaged term; $C_{(UC)}$ = co-spectrum of the horizontal velocities and concentrations; f = frequency band of the oscillation (Hz); Δf = bandwidth for spectral estimates; f_c = frequency range. The co-spectrum identifies SST due to oscillatory motions, which can be partitioned into different frequency ranges (Huntley and Hanes [26]; Davidson et al. [28]). Gravity and infragravity waves were partitioned at 0.04 Hz (25 s) and infragravity and far infragravity frequencies at 0.005 Hz (200 s), at pronounced reductions in spectral variance.

5 Results: February 21st, 2004

Measurements when the rip was at a maximum will be used to demonstrate the presence of very low frequency velocity "*pulses*" and their potential to suspend and transport sediment. Examination of ~1.5 h of record from the rip neck (P1; Fig. 2) of the near-bed, cross-shore velocity and suspended sediment concentration (Fig. 3) reveals that both signals are modulated at frequencies lower than the incident waves. Although the mean cross-shore current was ~0.38 m s^{-1} (at z = 0.13m) and directed offshore, the current fluctuated from >1.64 m s^{-1} offshore to almost -1.44 m s^{-1} onshore. Sediment concentrations revealed low frequency "*pulses*", ranging from <<2 kg m^{-3} to >30 kg m^{-3} (Fig. 3).

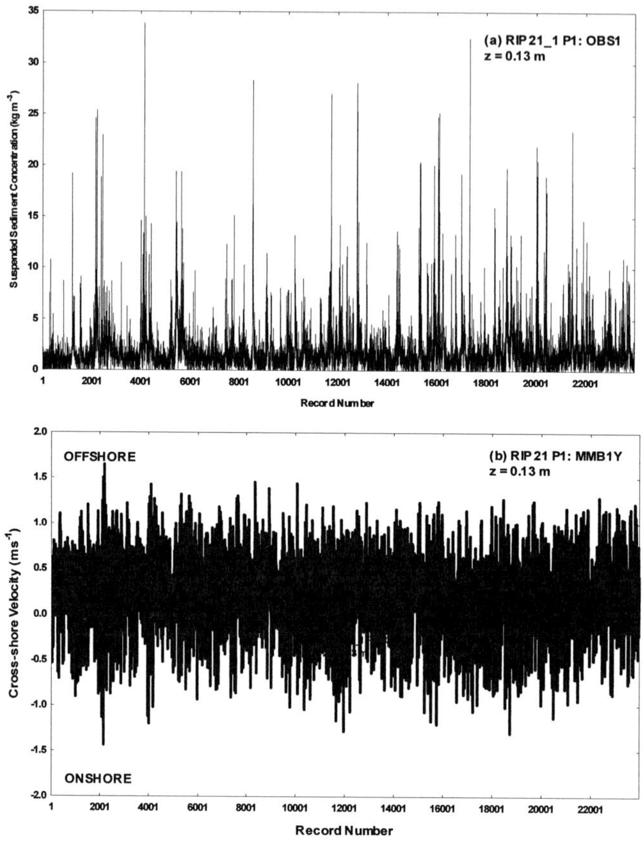

Figure 3: Time series of ~1.67 h of near-bed suspended sediment concentrations (a) and cross-shore rip current velocities (b), measured at z = 0.13 m elevation and recorded at 4 Hz in the rip neck at high tide. Note the distinct low frequency modulation of the signals.

5.1 Far-infragravity and infragravity *pulses* in the rip current

Figure 4 illustrates frequency spectra of the cross-shore currents measured at P1 (Fig. 2) at elevations of z = 0.13 and 0.26 m for ~33 minutes at high tide (1350 h, Feb. 21st), when the rip velocity was at a maximum (~0.38 m s^{-1}; see Greenwood et al. [22]). The time series were first tested for their difference from "*white noise*" using the Bartlett-Kolmogorov Smirnov distance statistic, *d* (Fuller [29]). For the ~8000-point series, the computed *d*-statistics were 0.775 for currents at z = 0.13 m and 0.786 for currents at z = 0.26 m. Clearly, the series were significantly different from white noise ($\alpha < 0.01$), and spectral densities were concentrated at both far-infragravity (FIG) and infragravity (IG) frequencies. These low frequency motions dominated the signal at this time scale. There were

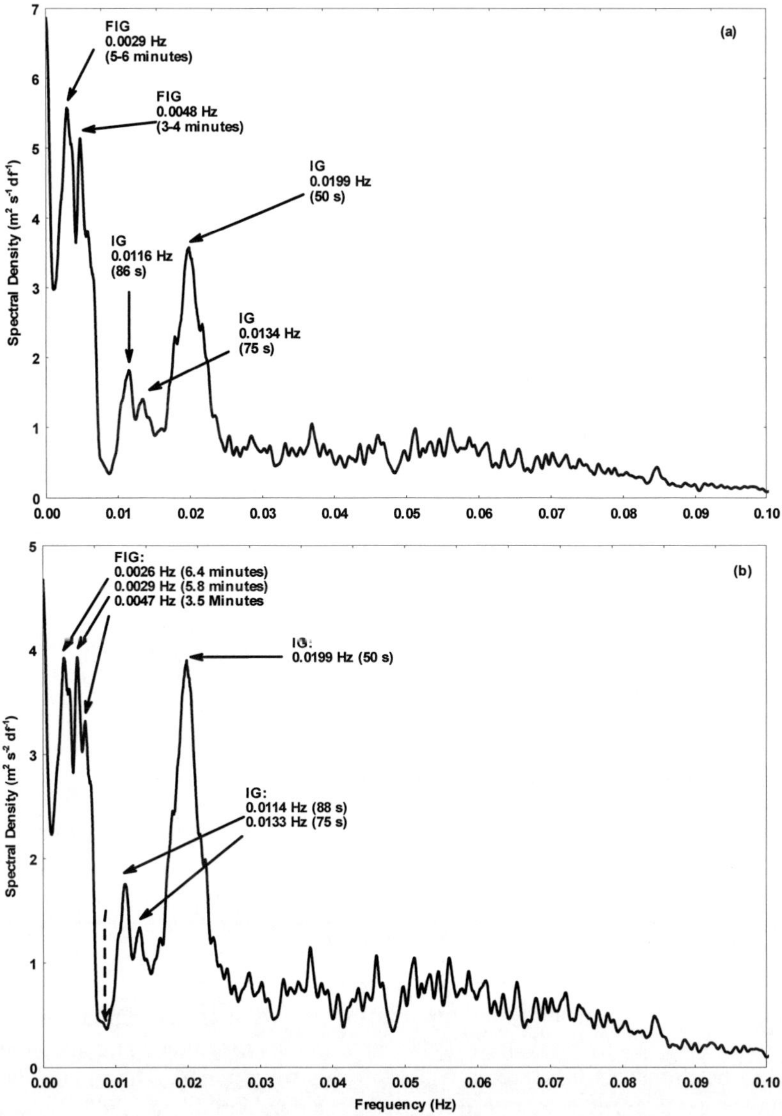

Figure 4: Frequency spectra of ~33 minutes of horizontal cross-shore currents in the rip neck at high tide at two elevations: (a) z = 0.26 m; (b) z = 0.13 m. The dashed vertical arrow in (b) represents the partition frequency between far-infragravity (FIG) and infragravity (IG) frequencies.

no significant spectral density peaks at either the incident swell frequency (0.077-0.083 Hz; ~12-13 s) or that of the wind waves (0.160-0.125 Hz; 6-8 s). FIG spectral peaks occurred at: (i) 0.0026-0.0029 Hz (~5-6 minutes); and (ii) 0.0048 Hz (3-4 minutes). The largest IG peak was at 0.0199 Hz (~50 s) almost as large as the FIG peak at z = 0.13 m; subsidiary IG peaks occurred at 0.0114-0.116 Hz (86-88 s) and 0.0133-0.0134 Hz (~75-76 s). Not all these low frequency motions were reflected in the concentration spectra. Nevertheless, the Bartlett-Kolmogorov *d*-statistic showed that all three concentration series also differed significantly from "*white noise*". The *d*-statistic values were 0.586, 0.562 and 0.458 respectively for the spectra at 0.13, 0.26 and 0.39 m. At z = 0.13 m, a large FIG peak at 0.0028 Hz (~6 minutes; Fig. 5a), matched the largest peak in the cross-shore current spectrum (Fig. 4), suggesting a direct causal link. A smaller but much less distinct peak at the same frequency was also evident in the concentrations at z = 0.26 m. However, no other FIG peaks were revealed in the concentration spectra and instead the spectra were distinctly "red", and increasingly so with increased elevation. There were small peaks in concentration at IG frequencies (0.0124 and 0.0203 Hz; ~80 and 50 s), which matched similar frequencies in the current velocities (e.g 0.0133 and 0.0198 Hz; Fig. 4). At the highest elevation (z = 0.39 m), there was some evidence for a spectral peak at 0.0198 Hz (~50 s), although the most prominent signature was that of the swell frequency (0.0889 Hz; 11-12 s). Examination of shorter time series (~17 minutes) at the beginnning of these long records revealed that the incident swell (~12-13 s) was the dominant spectral signature at this shorter time scale, with a smaller contribution from locally generated wind waves (Greenwood et al. [22]). Thus, it appears that near the bed, sediment concentrations do respond to the lower frequency water motions (FIG, IG frequencies); this linkage decreases with elevation, where the dominant incident swell, (~0.08-0.09 Hz; 11-12 s) increases in importance.

5.2 Suspended sediment flux at low frequencies

Figure 6 illustrates the co-spectra computed using the cross-shore current at z = 0.13 m and sediment concentrations recorded at z = 0.13 and 0.26 m in the rip neck for ~33 minutes at high tide. Within this time frame, the suspended sediment transported by the incident swell and wind waves is subordinate to that of the lower frequency motions. Transport of suspended load at FIG frequencies peaked at 0.00244 Hz (~6-7 minutes), with the flux directed offshore at both z = 0.13 and 0.26 m), supporting the transport by the mean rip current itself (Greenwood et al. [22]). IG transports varied in both magnitude and direction but with a net transport offshore as well over this time period. The incident wave frequencies, as expected, induced a net onshore transport of suspended sediment.

6 Conclusions

At relatively large time scales (~33 minutes), cross-shore current and sediment concentration measurements in a "stable" rip neck system at high tide when the rip was at a maximum (~0.38 m s^{-1}) lead to the following conclusions:

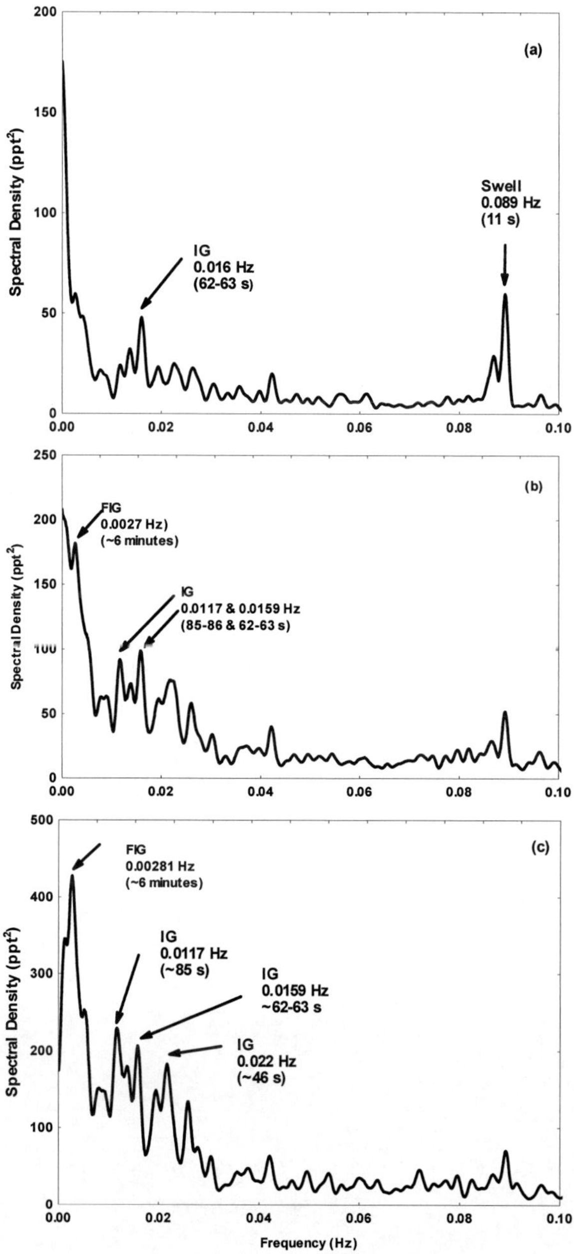

Figure 5: Frequency spectra of ~33 minutes of near-bed, sediment concentrations in the rip neck at high tide: (a) z = 0.39 m; (b) z = 0.26 m; (c) z = 0.13 m. Note the decreasing vertical scale with elevation as overall concentrations decrease.

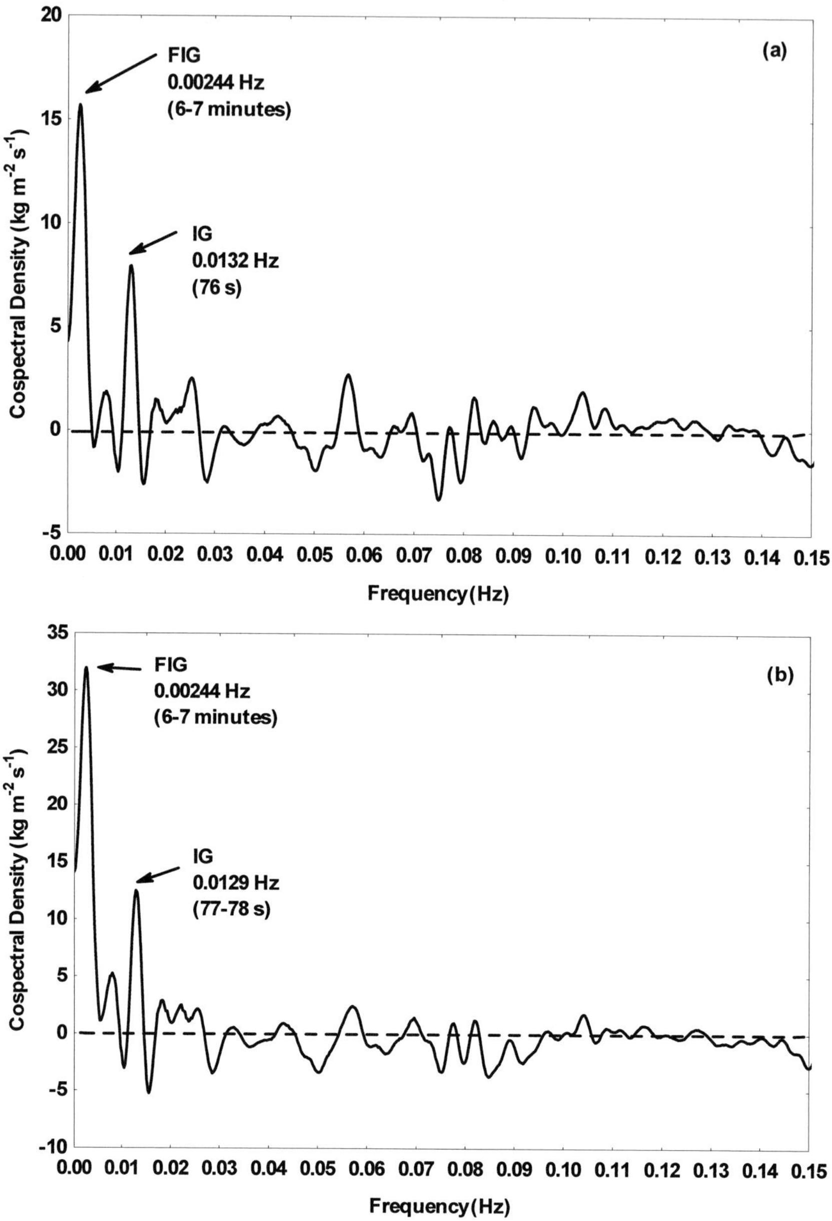

Figure 6: Co-spectra of cross-shore velocity and sediment concentration at z = 0.26 m (a) and 0.13 m (b) in the rip neck recorded over ~33 minutes. Note the difference in the vertical scales, which reflects the decreasing concentration with elevation, as there is essentially no change in the velocity field.

- current pulses at far-infragravity frequencies (>0.005 Hz; >200 s) dominate the near-bed cross-shore velocity spectrum (z = 0.13, 0.39 m). The dominant far-infragravity frequency was 0.0026-0.0028 Hz (~5-6 minutes).Further low frequency motions occurred at infragravity frequencies, with a dominant "*pulse*" at 0.0199 Hz (~50 s).
- similar very low frequency "*pulses*" were found in the near-bed (z = 0.13, 0.26 m) sediment concentration field. The strength of the signal decreased rapidly with elevation, so that at 0.39 m, the spectrum was distinctly "red" with no preferred concentration of variance at any FIG frequency. Instead, the only clear signal was that of an IG "*pulse*" at 0.0199 Hz (~50 s).
- co-spectra between currents and concentrations revealed that suspension transport was also dominated by FIG "*pulses*", at frequencies between 0.0026 and 0.0047 Hz (3.5 and 6.5 minutes), inducing a net offshore flux of sediment. IG frequencies were also present with the largest "*pulse*" (0.0129-0.0132 Hz; 76-78 s) also transporting suspended sediment offshore.
- at shorter time scales (~17 minutes) both the spectra and co-spectra were dominated by the incident swell (0.082 Hz; ~12 s) and wind waves generated by a sea breeze (0.167 Hz; ~6 s); these higher frequency waves together with the mean current were the dominant transport at short time scales (Greenwood et al. [22]).

Neither the consistency of the low frequency *pulses* through time, nor their origin(s), can be discussed in detail here, although a number of possible forcing mechanisms should be mentioned: (a) inherent modulations of the landward mass transport by Stokes drift or wave rollers over the segmented bar (Aagaard et al. [2]), which is the fundamental driver of the rip cell; (b) standing far-infragravity and infragravity waves within the rip neck itself produced by wave grouping (MacMahan et al. [6]; Bruneau et al. [7]); and (c) current shear instabilities (Smith and Largier [16]) and (d) the expulsion of vortices from the surf zone induced by wave-group forcing of large scale Lagrangian structures (MacMahan et al. [19]; Reniers et al. [30, 32]).

Acknowledgements

NSERC Canada (BG) and an FRGP grant from the UNSW (RWB) provided funding; M. Daly, M. Hughes, T. Baldock & F. Weir helped in the field. Dave Mitchell (Sydney University) is thanked for his technical and culinary skills.

References

[1] Komar, P.D. *Beach Processes and Sedimentation.* 2nd Edition, Prentice Hall, New Jersey, p. 338, 1998.

[2] Aagaard, T., Greenwood, B., Nielsen, J. Mean currents and sediment transport in a rip channel. *Marine Geol.*, 140: 25-45, 1997.

[3] Brander, R.W. Field observations on the morphodynamic evolution of low-energy rip systems. *Marine Geol.*, 157: 199-217, 1999.

[4] Brander, R.W. Morphodynamics of a large-scale rip current system at Muriwai Beach, New Zealand. *Marine Geol.*, 165: 27-39, 2000.

[5] MacMahan, J.H., Thornton, E.B., Stanton, T.P., Reniers, A.J.H.M. RIPEX: observations of a rip current system. *Marine Geol.*, 218: 113-134, 2005.

[6] MacMahan, J.H., Thornton, E.B., Reniers, A.J.H.M. Rip Current Review. *Coastal Eng.*, 53: 191-208, 2006.

[7] Bruneau, N., Castelle, B., Bonneton, P., Pedreros, R. Very low frequency motions of a rip current system: observations and modeling. *Journal of Coastal Research*, SI 56: 1731-1735, 2009.

[8] Shepard, F.P. Undertow, rip-tide or "rip current". Science, New Series, 84: 181-182.

[9] Shepard, F.P., Emery, K.O. LaFond, E.C. Rip currents: a process of geological importance. *Journal Geology,* 49: 337-369, 1941.

[10] McKenzie, P. Rip-current systems. *Journal Geology*, 66: 103-113.

[11] Reimnitz, E. Surf-beat origin for pulsating bottom currents in the Rio Balsas submarine canyon. *Geol. Soc. Am. Bull.* 82: 81-90, 1975.

[12] Reimintz, E., Toimil, L.J, Shepard, F.P., Gutierrez-Estrana, M. Possible rip current origin for bottom ripple zones to 30 m depth. *Geology* 4: 395–400, 1976.

[13] Cook, D.O., The occurrence and geologic work of rip currents off southern California. *Marine Geology*, 9: 173-186, 1970.

[14] Greenwood, B., Davidson-Arnott, R.G.D. Marine bars and nearshore sedimentary processes, Kouchibouguac Bay, New Brunswick. In: *Nearshore Sediment Dynamics and Sedimentation*, edited by Hails, J. and Carr, A., John Wiley and Sons, London, p.123-150, 1975.

[15] Greenwood, B, Davidson-Arnott, R.G.D. Sedimentation and equilibrium in wave-formed bars: a review and case study. *Canadian J. Earth Sciences*, 16: 312-332, 1979.

[16] Smith and Largier, Observations of nearshore circulation: rip currents. *Journal Geophyical Res.*, 100: 10967-10975.

[17] Thornton, E.B., MacMahan, J., Sallenger Jr., A.H., Rip currents, mega-cusps, and eroding dunes, *Marine Geology* 240: 151–167, 2007.

[18] MacMahan, J.H., Reniers, Ad J.H.M., Thornton, E.B. and Stanton, T.P. Infragravity rip current pulsations. *Journal Geophysical Res.*, 109: C01033, 2004.

[19] MacMahan, J.H., Brown, J., Brown Jenna, Thornton, E.B. Reniers, Ad, Stanton, T., Henriquez, M., Gallagher, E., Morrison, J., Austin, M.J., Scott, T.M., Senechal, N. Mean Lagrangian flow behavior on an open coast rip-channeled beach: a new perspective. Marine Geol., 268: 1-15, 2010.

[20] Munk, W. H. Surf beats. *EOS Trans. Amer. Geophyical. Union*, 30: 849-854, 1949.

[21] Tucker, M. J. Surf beats: sea waves of 1 to 5 minutes period. *Proc. Royal Soc. (A)* 202: 565–573, 1950.

[22] Greenwood, B., Brander, R.W., Joseph, R., Hughes, M.G., Baldock, T. and Aagaard, T., 2009. Sediment flux in a rip channel on a barred intermediate beach under low wave energy. In: *Coastal Processes*, edited by C.A. Brebbia, G. Benassai & G.R. Rodriguez, WIT Press, p.197-209.

[23] Wright, L.D., Short, A.D. Morphodynamic variability of surf zones and beaches: a synthesis. *Marine Geol.*, 56: 93-118, 1984.

[24] Austin, M.J., Masselink, G. The effect of bedform dynamics on computing suspended sediment fluxes using optical backscatter sensors and current meters. *Coastal Eng.* 55: 251-260, 2008.

[25] Jaffe, B.E., Sternberg, R.W., Sallenger, A.H. The role of suspended sediment in shore-normal beach profile changes. *Proceedings of the 21st Coastal Engineering Conference*, American Society Civil Engineers, p. 1725-1743, 1984.

[26] Huntley, D.A., Hanes, D.M. Direct measurement of suspended sediment transport. *Proceedings of Coastal Sediments '87*, ASCE, New York, p. 723-737, 1987.

[27] Osborne, P.D., Greenwood, B. Frequency dependent cross-shore suspended sediment transport 1: a non-barred shoreface, Queensland Beach, Nova Scotia, Canada. *Marine Geol.*, 106: 1-24, 1992.

[28] Davidson, M.A., Russell, P.E., Huntley, D.A., Hardisty, J. Tidal asymmetry on a macro-tidal intermediate beach. *Marine Geol.*, 110: 333-353, 1993.

[29] Fuller, W.A. *Introduction to Statistical Time Series*. Wiley, New York.

[30] Reniers, Ad. J. H. M., MacMahan, J. H., Thornton, E. B., Stanton, T. P., Henriquez, M., Brown, J. W., Brown, J. A., Gallagher, E. Surf zone surface retention on a rip-channeled beach. *Journal Geophysical Res.*, 114, C10010, doi:10.1029/2008JC005153, 2009.

[31] Reniers, A. J. H. M., MacMahan, J. H., Beron-Vera, F. J., Olascoaga, M. J. Rip-current pulses tied to Lagrangian coherent structures. *Geophysical Res. Lett.*, *37*, L05605, 5 p., 2010.

Section 4
Extreme events and sea level rise

A new approach to calculate extreme storm surges: analysing the interaction of storm surge components

G. Gönnert & K. Sossidi
Agency of Roads, Bridges and Waters,
Free and Hanseatic City of Hamburg, Germany

Abstract

The North Sea coast is, as well as many coastal regions, a preferred residential and industrial area. At the same time the North Sea coast is seriously threatened by storm surges. Climate change and resulting consequences will have severe effects on the safety in coastal areas. The Project XtremRisK – Extreme Storm Surges at Open Coasts and Estuarine Areas, Risk Assessment and Mitigation under Climate Change Aspects, will assist in facing this challenge. The "Source-Pathway-Receptor"-Concept will be used as a basis for risk analysis and development of new management strategies.

Within the XtremRisK project, a method to assess extreme events will be developed which takes the hydrodynamics of storm surges into account. With this method, the components of a storm surge, tide, wind surge and external surge, and their interaction will beanalysed. The study focuses on analysing (i) the highest event of each component and (ii) the interaction between tide and surge as well as the interaction between surge and external surge. This detailed analysis is needed because the components do not interact linearly. With this knowledge, an extreme event based on the observed extremes of the components can be calculated. Non-linear addition leads to a lower water level than in case of linear superposition. With this approach, an extreme event can be calculated in accordance with natural law.Subsequently, extreme events will be assessed using conditions which reflect the climate change scenarios.

Keywords: storm surges, North Sea, extreme events, non-linearity, coastal protection, design level.

1 Introduction

Climate change and the resulting consequences like sea level rise may have serious consequences on the safety of people and assets in coastal areas. To evaluate these consequences, it is necessary to first assess extreme storm surge events. In a second step scenarios of potential sea level rise are identified and investigated. These extreme scenarios are necessary for risk analyses and moreover can be used for design level.

The project XtremRisK (Extreme Storm Surges at Open Coasts and Estuarine Areas – Risk Assessment and Mitigation under Climate Change Aspects) aims to design approaches to meet the challenge of the future. The project is funded by the German Federal Ministry of Education and Research (BMBF) and was launched in October 2008.

Within the project, integrated risk analyses will be conducted and risk management policies for open coasts (pilot site Sylt Island) and estuarine areas (pilot site Hamburg) will be developed. The integrated flood risk analysis is based on the source-pathway-receptor concept (figure 1).

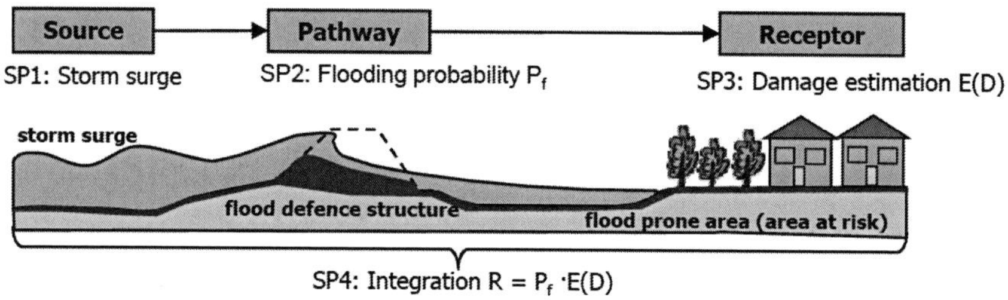

Figure 1: Source-pathway-receptor concept.

This paper illustrates the approach to calculate physically possible extreme storm surges developed in subproject 1a. With this method, the three main components of a storm surge, tide, wind surge and external surge, will be analysed and their development over time will be assessed.

The focus lies on the analysis of (i) the highest observed occurence of each component and (ii) the interaction between tide and wind surge and the interaction between storm surge and external surge. This detailed analysis is needed because the components do not interact linearly. Investigation on the hydrodynamics and physics of storm surges have shown that the components have to be added non-linearly (e.g. Tang *et al.* [1]; Horsburgh and Wilson [2]; Jones and Davies [3]), which leads to a lower water level than in case of linear superposition. With this knowledge an extreme event based on the maximum components can be calculated. The result is a realistic extreme scenario under present climate conditions.

2 Pilot sites

The analyses are conducted for one location at the open coast (Island of Sylt) and another, which is situated in an estuarine area (city of Hamburg).The chosen study sites allow for transferability of the developed methods to other coastal or estuarine areas.

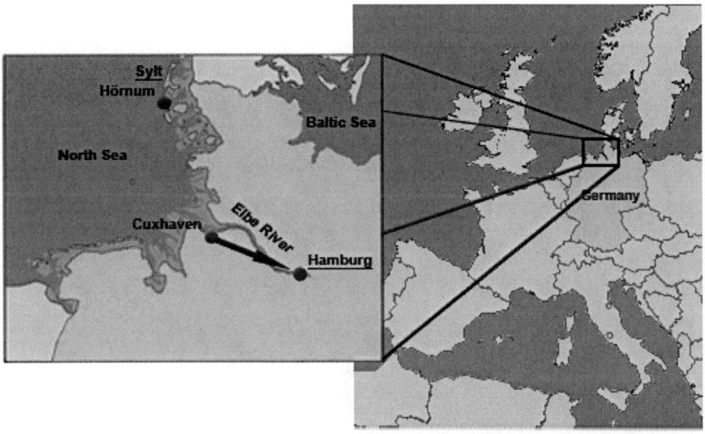

Figure 2: Location of the pilot sites Hamburg and Sylt.

The city of Hamburg is the second largest city of Germany. About 1.8 million people live in this megacity which is a centre for trade, transportation and services and an important industrial site. Furthermore, the port of Hamburg is the largest seaport in Germany.

The Island of Sylt is the biggest of the North Frisian Islands. With its 40 km of beach at the west side it is one of the most popular touristic places in Germany. Due to its exposed position in the North Sea it is vulnerable to severe erosion during storm surges.

For the characterisation of storm surges, a long consistent time series of water level observations is needed. Consequently the extreme storm surge for the city of Hamburg will be determined at tidal gauge Cuxhaven and its propagation upstream to Hamburg subsequently calculated by numerical modeling. The storm surge for Sylt will be determined at the tidal gauge Hörnum.

The approach of the analysis of extreme storm surges and the single components will be presented below. It was developed at tidal gauge Cuxhaven, because this is the best evaluated tidal gauge in Germany with a continuous time series that reaches back to 1900.

3 The methodological approach

The chosen approach is based on a deterministic procedure that takes into account the hydrodynamic interactions of the storm surge components as well as their meteorologic context. In a first step all observed storm surges are separated into their components wind surge, tide and external surge and the highest event of each component is detected (figure 3).

For the construction of extreme storm surge events it is essential to consider the highest observed occurrence of each component. Since these extreme values appeared in different storm surge events one has to combine them in a new storm surge event. To receive a result that corresponds with physical law the non-linear interaction of the storm surge components have to be taken into account. Considering this, the calculated extreme storm surge is a feasible event in fact.

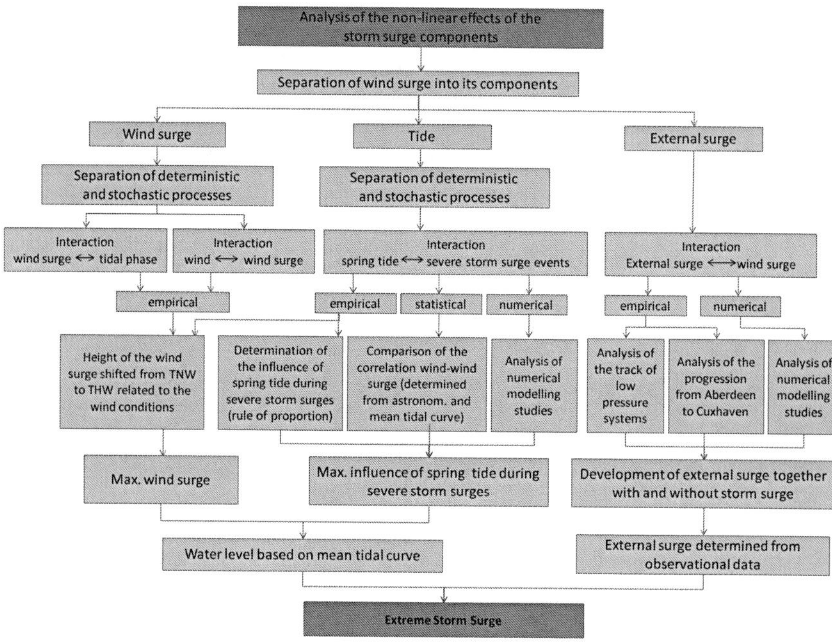

Figure 3: Methods to estimate the non-linear effects.

Before combining the components to a new storm surge, it is required to examine the physical possibility of the coincidence of the observed maximum values of the components. Therefore the possibility of a simultaneous occurrence of a high spring high tide, external surges and high wind surge at tidal gauge Cuxhaven is examined.

In the course of a comprehensive analysis the components are firstly scrutinized separately and secondly their interaction is investigated with empirical, statistical and numerical methods. The consistence of the different methods' results allows an optimal ascertainment of the physical context and a validation of the used methods. The non-linear interaction is worked out. Finally, the highest observed values of each component are added non-linearly following natural law to receive an extreme storm surge event.

3.1 The components of a storm surge

The wind is the main reason for the formation of a storm surge. It pushes the water against the coast causing higher water levels than the ordinary sea level which is influenced by the tides. The tidal range reaches its maximum at spring tide so that during spring tides high waters rise extra high. The water level at the North Sea coast can also be raised by external surges that are generated by cyclones over the Northeastern Atlantic and propagate into the North Sea (Lundbak [4]; Timmerman [5]), pushing additional water masses into the basin. The coincidence of wind surge, spring high tide and external surge leads to extreme storm surge levels along the coast (figure 4).

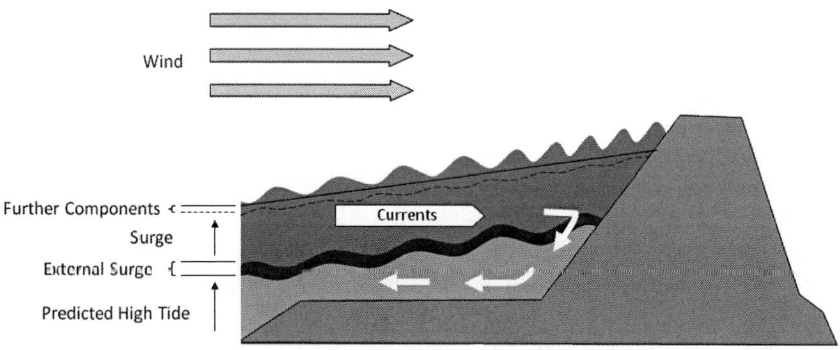

Figure 4: The components of a storm surge.

Thus a storm surge is composed of the local tides, wind surge and a possible external surge. In the project storm surges from 1901 until 2008 at tidal gauge Cuxhaven and from 1936 to 2008 were analysed and decomposed into wind surge, external surge and astronomical tide. These three components will be outlined in the following subchapters.

3.1.1 Wind surge

Wind surge is caused by wind-induced onshore surface current. The wind direction that produces the highest wind surge at tidal gauge Cuxhaven is 295° (westerly to north-westerly winds). At tidal gauge Hörnum it is 267° (westerly winds).The wind causes a rise of the water near the coast and hence a tilt of the water surface of the North Sea from the deep water area towards the coast. The height of the wind surge depends on wind speed and wind direction.

A formula to calculate wind surge as a function of wind speed and wind direction exists only for few tidal gauges at the German North Sea coast. Therefore wind surge is generally calculated by subtracting the astronomical tide (or the mean tide) from the observed tide. In case of using the astronomical tide for calculation, the wind surge also contains other factors that are not caused by the wind, like the effect of local air pressure or the external surge. Due to this circumstance it was examined which of the storm surges since 1901 occurred simultaneously with an external surge before conducting further analyses.

The analyses of the data showed that the wind surge maximum can occur to all tidal phases, but mostly occurs around tidal low water. The highest wind surge maxima also occurred around tidal low water. This confirms earlier studies that indicated that the maximum wind surge tends not to coincide with high tide (e.g. Keers [6]; Horsburg and Wilson [2]).

The highest wind surge event at tidal gauge Cuxhaven that is not influenced by an external surge consequently reached a height of 4.15 m. It occurred in January 1976 around low tide.The highest wind surge event at tidal gauge Hörnum features 3.50 m. It occurred in January 1990 around low tide.

3.1.2 Spring tide

Tide is caused by the gravitational forces of the sun and moon. At new and full moon the tidal powers of moon and sun amplify each other causing very high and very low tides which are called spring tides. At half moon these powers cancel each other partly so that the tides, that are called neap tides, are not as dramatically high and low.

The mean tidal curve is the arithmetical averaged observed tidal curve at a specific gauge of a certain time period. Thereby tidal curves of different moon phases are averaged so that the differences of spring and neap tide are indeterminable.The past analyses of storm surges were based on the mean tide because of pragmatically reasons: The mean tide is a statistical term which needs only relatively short calculation time (Siefert [7]). In the past it was mathematically difficult and a time-consuming intent to calculate the astronomical tide for any day. Nowadays there are computer programs that make it possible to calculate the astronomical tide in a comparable time frame to the mean tide. Thus it is possible to divide the astronomical effects from the wind effects and to analyse the tide-surge-interaction.

The highest values of spring high tide at the tidal gauges were detected by subtracting the mean high tide from astronomical high tide. In Cuxhaven it was 60 cm, in Hörnum it was 63 cm.

3.1.3 External surge

External surges are generated in the Northeastern Atlantic due to of air pressure variations that are generated by fast passing cyclones. They produce observable waves at the water surface as well as internal waves in the water body. The propagation of an external surge into the North Sea takes place in preference when the track of the causative cyclone leads from the sea areas between Ireland and Iceland to Mid-Norway (Schmitz and Habicht [8]). These waves propagate to the North Sea similar to a tide wave, but independent of tide phase and any periodic regularity and travel through the North Sea in counterclockwise direction.

External surges are calculated out of observed tide data at tidal gauge Cuxhaven that were recorded from the Federal Maritime and Hydrographic Agency (Gönnert [9]). Their height is calculated by subtracting the astronomical tide, wind surge and the effect of local air pressure from the observed low and high waters. The calculated values still may contain components besides the external surge. These are seiches, effects of air pressure and air pressure variations over the North Sea, effects of water temperature and temperature difference between air and water. This variety of potential causes for an increase of the water level reveals the necessity of a precise definition of external surges. So, residual surges are only defined as external surges if they can also be indentified at tidal gauge Aberdeen and Immingham at the British coast (Gönnert [9]).

The maximum height of an external surge recognised at Cuxhaven was 109 cm. This external surge reached 108 cm in Aberdeen. It did not coincide with a storm surge. The highest external surge that coincided with a storm surge reached a peak level of 1 m at Cuxhaven. Each fifth to sixth storm surge at tidal gauge Cuxhaven occured at the same time as an external surge (Gönnert [9]).

For tidal gauge Hörnum no records of external surges exist. Since external surges travel through the North Sea similarly to the tide (Banner *et al.* [10]; Pugh [11]), the external surges detected at tidal gauge Cuxhaven are shifted to Hörnum with the observed tide to estimate the time of occurrence at that tidal gauge. For the detected time periods, the height of external surges at tidal gauge Hörnumwill be calculated following the method used at tidal gauge Cuxhaven.

3.2 Investigations of the non-linear interaction

When calculating an extreme storm surge it has to be taken into account that the components of a storm surge do not superimpose linearly. Their interaction is non-linear, dependent on the water depth. The wind-induced onshore surface current generates an offshore-directed bottom return current. This return current increases with rising water level near the coast which in turn leads to a stronger decrease of the water level. Consequently, taking this physical law into account when calculating storm surge events leads to lower storm surge levels than linear superposition.

A comprehensive analysis of non-linear interactions between the storm surge components is carried out by combining empirical, statistical and numerical methods (figure 3). This analysis makes it possible to work out the effect of non-linear interaction when combining the maximum values of the components to receive the extreme storm surge's curve and peak water level at high tide. The results are presented using the example of Cuxhaven.

3.2.1 Non-linear interaction between tide and wind surge

To detect the highest observed wind surge, the wind surge curve was calculated for around 250 storm surge events. The winds surge curve was calculated by subtracting the astronomical tide from the observed tide as well as by subtracting the mean tide from the observed tide. Doing that, one can identify the effect of the astronomical dissimilarity on the wind surge.

The calculation of the wind surge curve for the whole storm surge period makes it possible to determine a maximum wind surge peak regardless of tide phase for each storm surge. The highest observed wind surge values occur around low water. Since there is no cohesion between tide phase and wind the maximum force of wind can also occur at other tide phases such as high water. The data show that the height of wind surge is less around high water, though.

To determine the ratio between wind surge height around low water and around high water the relations between wind surge, the respective wind speed and tide are analysed.Based on the wind directions that produce the highest wind surge at tidal gauge Cuxhaven (see subchapter 3.1.1) the effective wind is calculated. It describes the wind speed converted to the wind direction that produces the highest wind surge. The analysis of wind surge depending on effective wind to each tide phase showed the relation between the height of wind surge around high tide and around low tide. At tidal gauge Cuxhaven, the wind surge at high tide accounts for 82 to 94 percent of the wind surge around low tide. The amount of remaining wind surge rises with the effective wind speed.

This ratio makes it possible to detect the highest wind surge around high tide by shifting the observed wind surge events to high tide and reducing their height as appropriate. The precondition for this shifting is that the respective effective wind acts over an extended amount of time.

The highest wind surge event around high water at tidal gauge Cuxhaven that is not influenced by an external surge features 3.70 m. It occurred in January 1976 around high tide and does not have to be shifted. During this storm surge event the wind surge reached 4.15 m around previous low tide. The wind did not change during that time, so the observed wind surge values are equivalent to a reduction of 89 percent.

When wind surge coincides with spring high water it has to be investigated to which amount this astronomically induced increase of the water level affects the storm surge water level. For this investigation three different methods are used:

- Statistical analysis of the correlation between effective wind and wind surge determined from astronomical tide and mean tidal curve
- Analysis of numerical modelling studies
- Calculation of the spring tide influence on an extreme event.

The investigations on spring tide are based on the calculation of astronomical tide curves with a non-harmonically method. The cross-correlation shows the only small effect of the astronomically induced deflection of water level on the wind surge curve and peak. This result matches with the results of numerical modeling studies by Dick [12] that showed that the effect of the astronomical dissimilarity decreases as the wind surge increases.

The analysis of the relation between the height of wind surge around high tide and around low tide clarified the increasing relevance of the wind with increasing wind speed. It can be transferred to the astronomical dissimilarity. Hence the amount of reduction of wind surge due to shifting from low water to high water can be adapted to make out the reduction of effects of spring tide on storm surge water levels. Because there is a connection between overall water level and reduction, the calculation of the astronomical dissimilarity's decrease has to be made including wind surge.

For Cuxhaven, the astronomical dissimilarity in combination with the wind surge of 3.70 m reduces to 10 cm. At tidal gauge Hörnum the amount of spring tide reduces to about 44 cm when coinciding with a wind surge of 3.19 m.

3.2.2 Non-linear interaction between storm surge and external surge

The cyclone that causes an external surge in the Atlantic Ocean has negligible to substantial effects on the water level in the North Sea, depending on the cyclone's track. The investigation of the development of the external surges from Aberdeen to Cuxhaven via Immingham showed that most of the external surges increased in height on their way from Aberdeen to Immingham and decreased again on the way to Cuxhaven so that they are smaller in Cuxhaven than in Aberdeen (Gönnert [9]). When the development in height of each external surge is connected with the respective meteorological situation it shows that:

- external surges will increase, if the cyclone tracks southwards and influences the water body of the North Sea (and sometimes produces a storm surge).
- external surges decrease (or stays more or less same level), if the cyclone tracks northwards and has no relevant influence on the water body of the North Sea (figure 5).

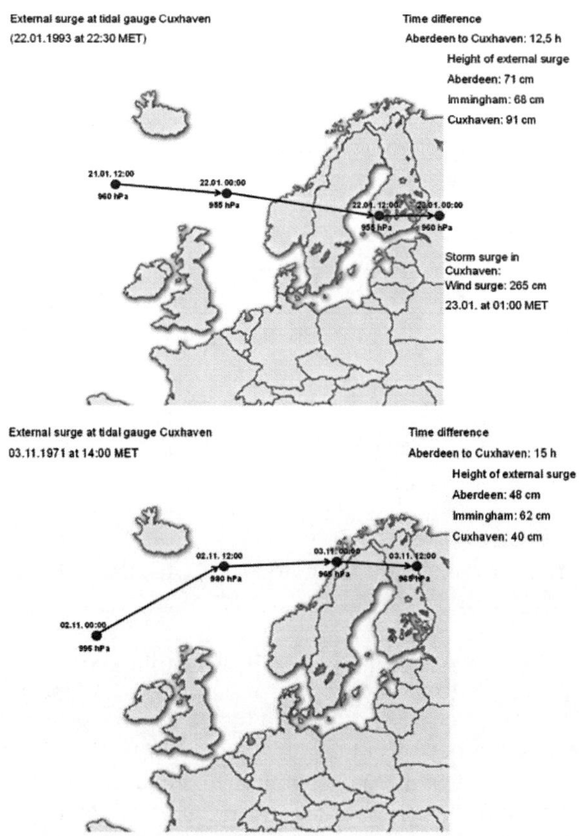

Figure 5: Different tracks of cyclones and their influence on external surges.

www.witpress.com, ISSN 1743-3541 (on-line)

When the cyclone features a southern track the water level in the North Sea can for instance be affected by effects of air pressure and air pressure variations over the sea as well as additional wind. To avoid that the peak of the external surge combined with the extreme storm surge contains alterations of the water level due to these influences, only the external surges are taken into account that occurred without a coinciding storm surge. The average decrease of these external surges between Aberdeen and Cuxhaven is 30 percent.
The highest observed external surge reached 1.08 m at tidal gauge Aberdeen. If reduced by 30 percent on its way to Cuxhaven, it arrives there with a peak of about 80 cm. This equates the peak of the external surge that was observed during the storm surge of 1962.

3.3 Calculation of an extreme storm surge event

The extreme storm surge is calculated by adding the highest observed values of each storm surge component in due consideration of their non-linear interaction investigated in this study and adding the resulting value to the mean high tide.

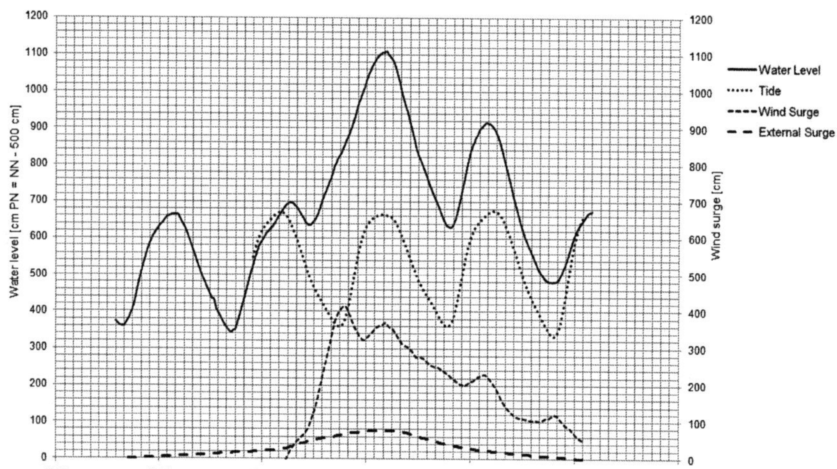

Figure 6: The extreme storm surge and its components at tidal gauge Cuxhaven.

The wind surge curve that is a constituent of the constructed storm surge curve is characteristic for a severe storm surge event at the respective tidal gauge. The storm surge curve is composed by the wind surge curve, the external surge curve and an astronomical tide curve that reflects the mean tide condition plus the detected effect of the astronomical dissimilarity. Figure 6 shows the constructed extreme storm surge at tidal gauge Cuxhaven.

To gain the storm surge curve at tidal gauge St. Pauli/Hamburg the propagation of the extreme storm surge determined at tidal gauge Cuxhaven upstream to Hamburg will be calculated by numerical modelling.

4 Summary

This paper presents an empirical approach to determine extreme storm surges at different tidal gauges. Statistical and numerical methods are brought in to validate the empirical methods. The approach takes into account the storm surges components wind surge, tide and external surge. It focuses on (i) the analysis of the highest event of each component and (ii) the analysis of the interaction between tide and surge and the interaction between surge and external surge to (iii) calculate an extreme storm surge event based on the result of the analyses. The consideration of the non-linear effects between the storm surge components leads to a lower water level than in case of linear superposition.

Since the resulting value is added to the mean high tide, climate change can be taken into account by adding the potential sea level rise to the mean tide.

The presented approach is developed at tidal gauge Cuxhaven because of its continuous time series that reaches back to 1900. It is then transferred to tidal gauge Hörnum. The implementation at this second tidal gauge showed that, provided that an adequate series of data is available, the approach can be transferred to other study areas.

References

[1] Tang, Y.M., Grimshaw, R., Sanderson, B. &Holland, G., A Numerical Study of Storm Surges and Tides, with Application to the North Queensland Coast. *Journal of Physical Oceanography*, **26**, pp. 2700-2711, 1996.

[2] Horsburgh, K.J. & Wilson, C., Tide-surge interaction and its role in the distribution of surge residuals in the North Sea. *Journal of Geophysical Research*, **112 (C8)**. Art. No. C08003. 10.1029/2006JC004033, 2007.

[3] Jones, J.E. & Davies, A.M., Influence of non-linear effects upon surge elevations along the west coast of Britain. *Ocean Dynamics*, **57**, pp. 401-416, 2007.

[4] Lundbak, A., The North Sea Storm Surge of February 1, 1953. Its Origin and Development. *GeografiskTidsskrift*, **54(1)**, pp. 8-23, 1955.

[5] Timmerman, H., On the Importance of Atmospheric Pressure Gradients for the Generation of External Surges in the North Sea. *Deutsche HydrographischeZeitschrift*, **28(2)**, pp.62-71, 1975

[6] Keers, J.F., An empirical investigation of interaction between storm surge and astronomical tide on the east coast of Great Britain. *Ocean Dynamics*,**21(3)**, pp. 118-125, 1968

[7] Siefert, W., Sturmflutvorhersagen für den Tidebereich der Elbe aus dem Verlauf der Windstaukurven in Cuxhaven. *Mitteilungen des Franzius-Instituts für Grund- und Wasserbau der TU Hannover*, **30**, pp. 1-142, 1968

[8] Schmitz, H.P. & Habicht, D., Barotropic numerical experiments on external surge generation at the edge of the north-western European shelf. *GerlandsBeiträgezurGeophysik*, **97/5**, pp. 422-437, 1988

[9] Gönnert, G., Sturmfluten und Windstau in der Deutschen Bucht. Charakter, Veränderungen und Maximalwerte im 20. Jahrhundert. *Die Küste*, **67**, pp. 185-365, 2003

[10] Banner, F.T., Collins, M.B. &Massie, K.S., The North-West European shelf seas: the sea bed and the sea in motion V. 2, Elsevier, Amsterdam, 1979

[11] Pugh, D.T., *Changing Sea Levels: Effects of Tides, Weather and Climate*. Cambridge Univ. Press, Cambridge, 2004

[12] Dick, S., Die Sturmflut am 3. Dezember 1999 – Fallstudien mit dem Windstaumodell des BSH. Presentation at the workshop on storm surges organized by the Federal Waterways Engineering and Research Institute, Hamburg, 16.11.2000, Head of the Directorate Forecasting, Federal Maritime and Hydrographic Agency, Hamburg, Germany

Pacific Island Land Ocean Typhoon (PILOT) experiment

S. J. Boc & K. K. Hathaway
U.S. Army Corps of Engineers, Engineer Research and Development Center, Coastal and Hydraulics Laboratory, Vicksburg, USA

Abstract

The PILOT program sponsored by the U.S. Army Corps of Engineers is designed to acquire data necessary to depict and better understand the physics of the wave–reef interaction. Field sites have been instrumented on the islands of Guam, Oahu and St. Croix. Guam and St. Croix were selected primarily because of their potential of being affected by tropical cyclones annually. In addition, they offer different reef characteristics with favourable physical and logistical settings to study waves, currents, and water levels over reefs. Data were collected from a nearby deep-water directional wave buoy, coastal water-level and meteorological stations, shallow water wave gauges on the reef flats, and a laser-based bathymetric survey. Two storm events were recorded during a St Croix deployment (10 August to 11 November 2008); the first, a swell event, and the second, a wind-sea event, were investigated for different wave transformations over the reef. The first event occurred 30 September–1 October 2008, with a large swell recorded at the offshore Waverider buoy and low wind speeds (~7 mph E-ENE) measured at the St Croix airport. Incident wave heights peaked at 2.7 m with peak periods of approximately 15 s, and an incident direction of 47 degrees (compass direction from true north). This event produced the largest waves observed over the reef in this deployment. Mean water level (setup/surge) was about 0.2 m above normal. The second event occurred with the passing of Hurricane Omar on 16 October 2008. Incident waves (wind seas) peaked at 4.4 m with peak periods of nearly 8 s and directions from about 36 degrees. Although the incident waves were substantially higher, 4.4 m vs. 2.7 m, the peak wave heights over the reef were lower, 0.55 m vs. 0.72 m. Wave energy on the reef was primarily in the infragravity band (20 s–120 s). Surge and setup were similar and only about 0.2 m.

Keywords: reefs, waves, surge, wave setup, hurricane, wave-reef interaction.

WIT Transactions on Ecology and the Environment, Vol 149, © 2011 WIT Press
www.witpress.com, ISSN 1743-3541 (on-line)
doi:10.2495/CP110151

1 Introduction

The Pacific Island Land Ocean Typhoon Experiment (PILOT) Experiment follows from the recommendations stated in the Islands Task Force Report. More specific objectives address topics identified by the FEMA/USACE ad-hoc committee. The PILOT program, sponsored by the US Army Corps of Engineers (USACE), is designed to acquire data necessary to depict and better understand the physics of wave–reef interaction during cyclonic-storm events. Field data are collected to support development of predictive models for coastal inundation and flooding along shorelines of islands lined with fringing reefs. This research will be applied to the development of an effective emergency management system that addresses elevated sea-levels and storm waves associated with land-falling tropical cyclones in island environments. Four field study sites were selected in the Atlantic and Pacific Oceans for this purpose. The two Pacific sites were established at Ipan, Guam and the North Shore of Oahu, HI. The Atlantic sites were on north and south facing shores of St. Croix, USVI. Data presented here are from the north shore of St Croix.

2 Storm events – 2008

Two storm events were recorded during the St Croix (Figures 1 and 2) deployment of the Sea-Bird wave and tide recorders, 10 August to 11 November 2008 (Figure 3) in Pelican Bay, which is located near Christiansted on the island's north shore.

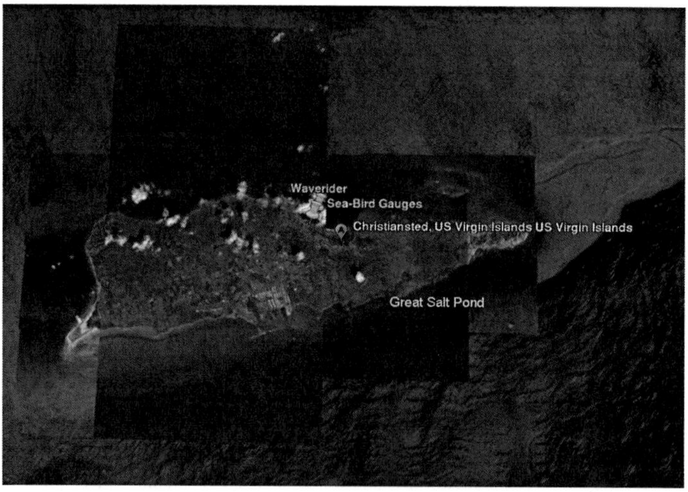

Figure 1: St. Croix north shore site.

Figure 2: Gauge locations at the St. Croix north shore site.

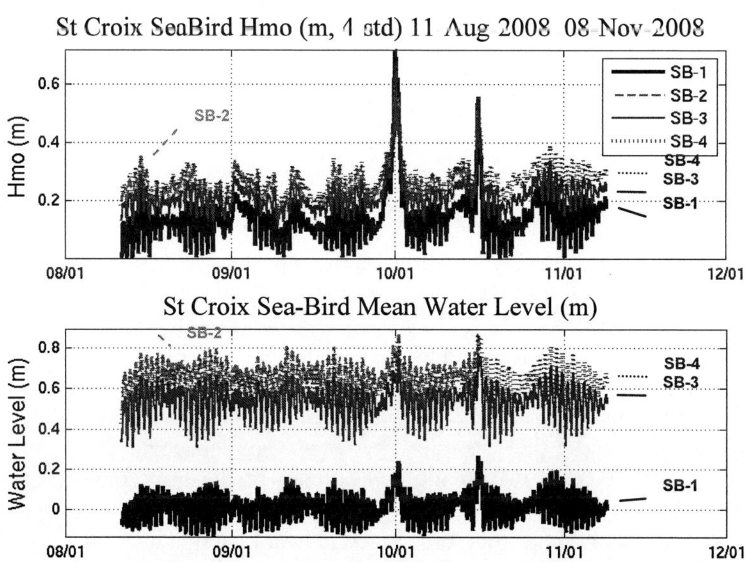

Figure 3: Sea-bird wave and water level.

2.1 Swell event

The first event occurred 30 Sep–1 Oct 2008, with incident large swell and low wind speeds (~7 mph E-ENE) measured at the St Croix airport. Incident wave heights peaked at 2.7 m with peak periods of approximately 15 s, and directions from 47 degrees (compass direction from true north) as measured from the Waverider buoy moored in deep water (255 m) just seaward of the instrumented reef. This event produced the largest waves observed over the reef in this deployment (Figure 3). Mean water level (setup/surge) was about 0.2 m above normal (Figure 4).

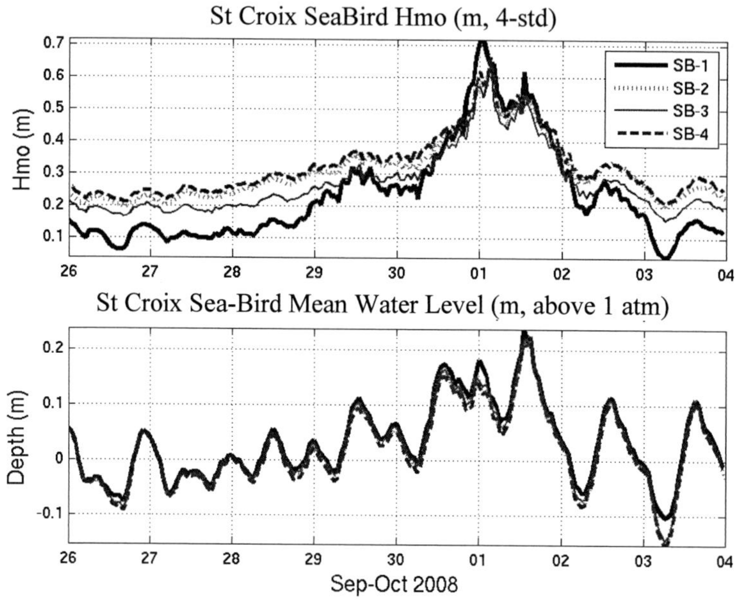

Figure 4: Swell event, 30 Sep to 1 Oct 2008.

A sample of 2-Hz timeseries for the cross-shore gauges near the storm peak shows a dominance of infragravity energy (periods > 20 s, frequencies < 0.05 Hz), increasing towards shore (Figure 5). Incident wave amplitudes have nearly disappeared at the shoreward most gauge.

Wave spectra were computed from 2048 s records using the Welch method with 50% segment overlap of 1024 s segments. A 3-point spectra band average was applied to smooth the spectra. The large segment length and small band averaging were selected to provide some detail in the infragravity band, but resulted in low degrees-of-freedom (dof ~= 14). Users of this spectra data may wish to perform additional smoothing in the incident band to increase the dof.

On 28 September 2008 (0000UTC), prior to swell arrival, spectra for the seaward gauge (#4) were swell frequency dominated; however, the shore gauge (#1) energy was primarily in the infragravity band (Figure 6). It should be noted the wave heights were quite small 0.1 m to 0.25 m at this time. During the swell event, the spectra show infragravity energy dominating with possible distinguishable spectral peaks at 120 s and 93 s period for gauge #4 and #1, respectively. Peak period in the infragravity band were generally higher at the shoreward most gauge #1 (Figures 8 and 9).

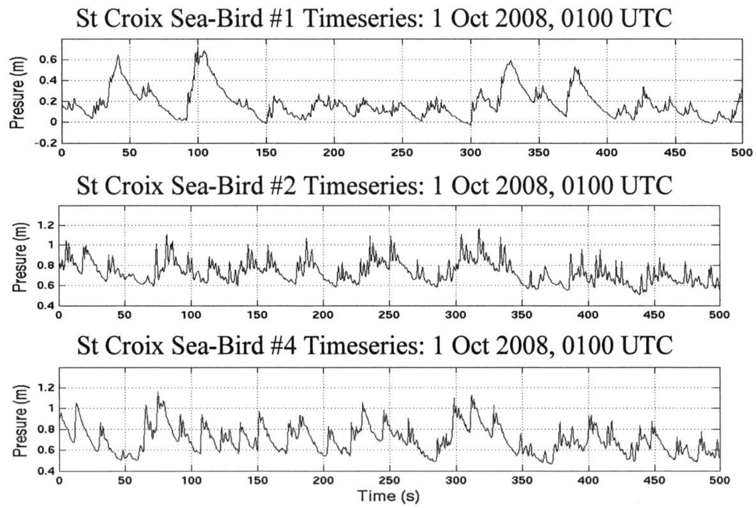

Figure 5: Timeseries near peak of swell event.

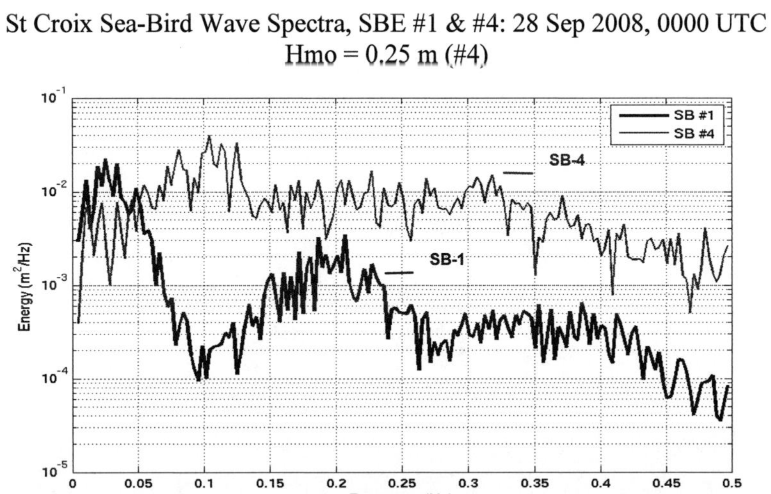

Figure 6: Sea-bird spectra prior to swell event (#1 closest to shore).

2.2 Hurricane Omar

The second wave event occurred with the passing of Hurricane Omar on 16 October 2008. Incident waves (wind seas) peaked at 4.4 m with peak period near 8 seconds and incident directions from about 36 degrees true north as measured at the offshore Waverider buoy. These storm waves were compared to the previous swell event. Although the incident waves were substantially higher during Omar, 4.4 m vs. 2.7 m, the peak wave heights over the reef were lower, 0.55 m vs. 0.72 m (Figure 3). Surge and setup were similar and only about 0.2 m in height (Figure 8). Note that SB-2 stopped functioning at the peak of the storm; only these three lines are plotted in Figure 10 after the storm peak.

St Croix Sea-Bird Wave Spectra, SBE #1 & #4: 1 Oct 2008, 0100 UTC
Hmo = 0.47 m (#4)

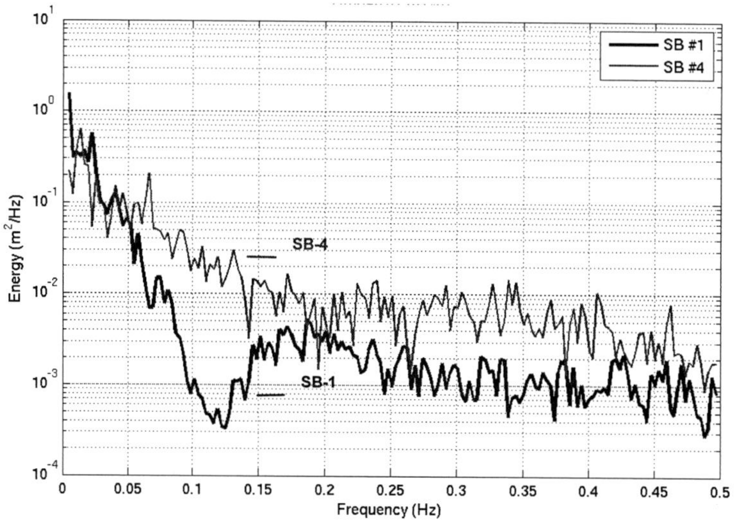

Figure 7: Infragravity dominated spectra during swell event.

St Croix Sea-Bird Infragravity Peak Periods

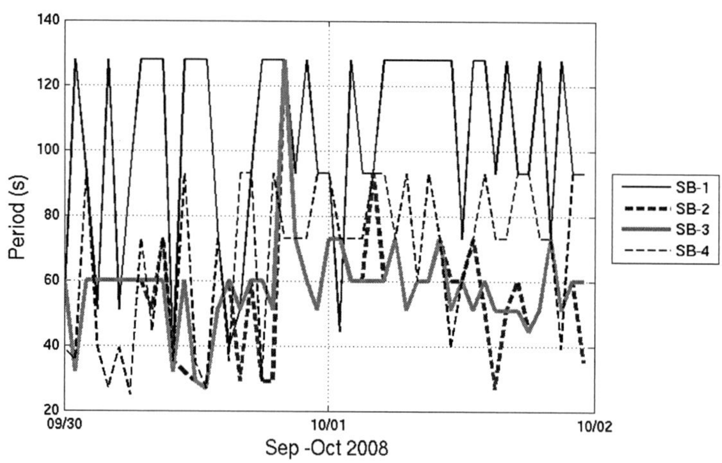

Figure 8: Peak periods in infragravity band in swell event.

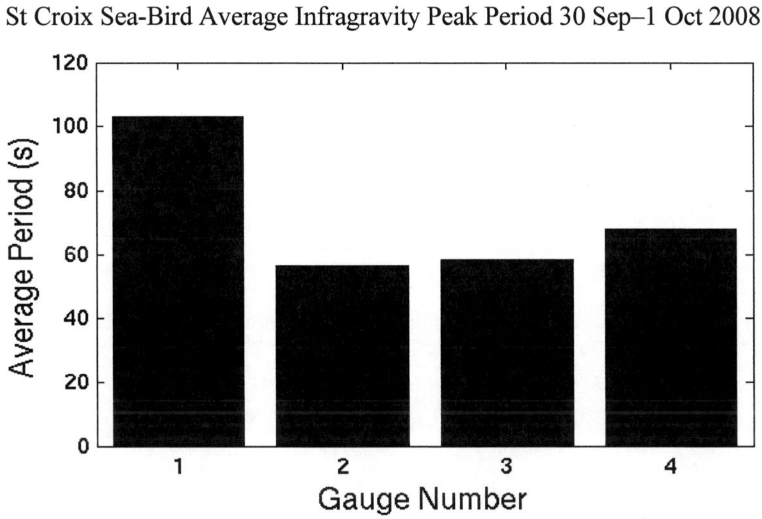

Figure 9: Average infragravity peak period during swell event.

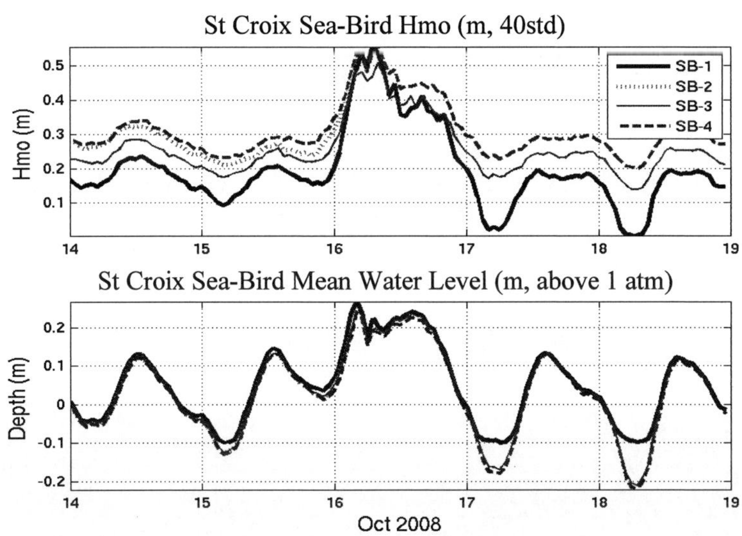

Figure 10: Sea event from Hurricane Omar, 16 October 2008.

A sample of 2 Hz timeseries near the peak of the storm are shown in Figure 11. Infragravity waves were not as obvious in the timeseries as during the swell event since the incident wind-wave energy was not as fully dissipated. Time series from the shoreward most gauge (#1) does have a long-period component of about 100 s.

Spectral analysis used the same method as in the swell event. At the peak of the storm (16 Oct, ~0500 UTC) the shore gauge spectra was infragravity dominated and had a wind-wave peak near 0.15 Hz (~7s) (Figure 12). At that time, the offshore gauge spectra were nearly "white" with little structure in the sea-swell band.

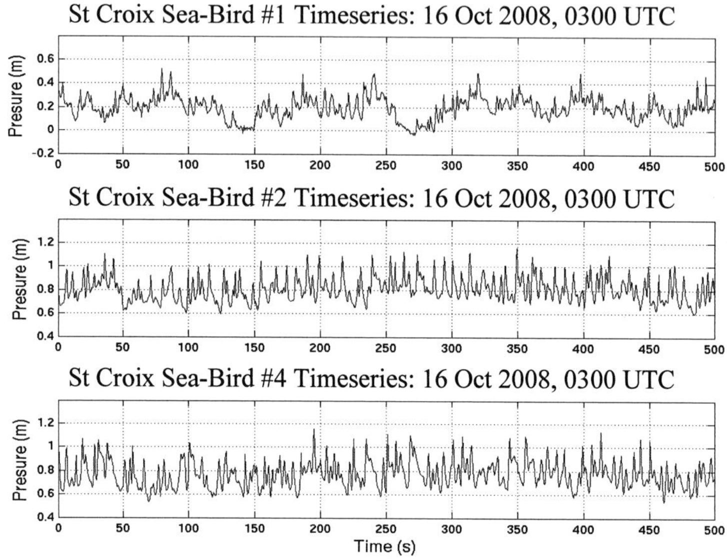

Figure 11: Sample timeseries from wind-sea incident waves.

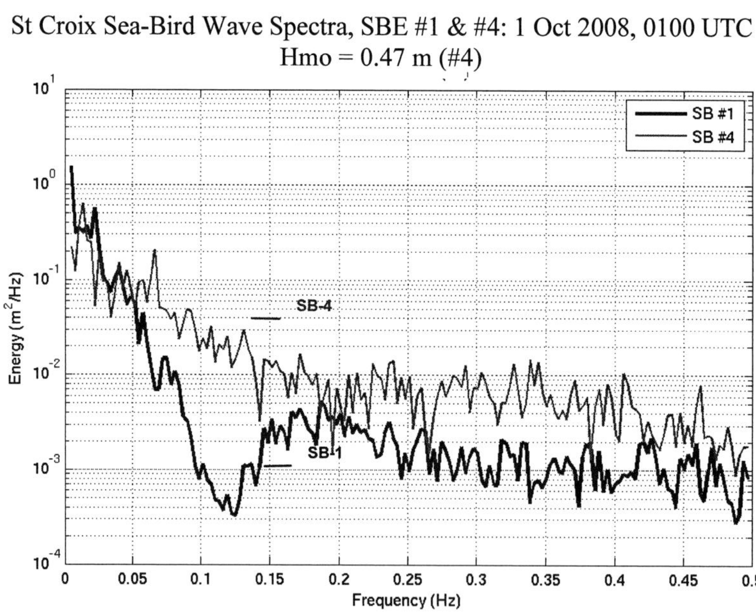

Figure 12: Shoreward most Sea-bird spectra at H. Omar peak wave heights.

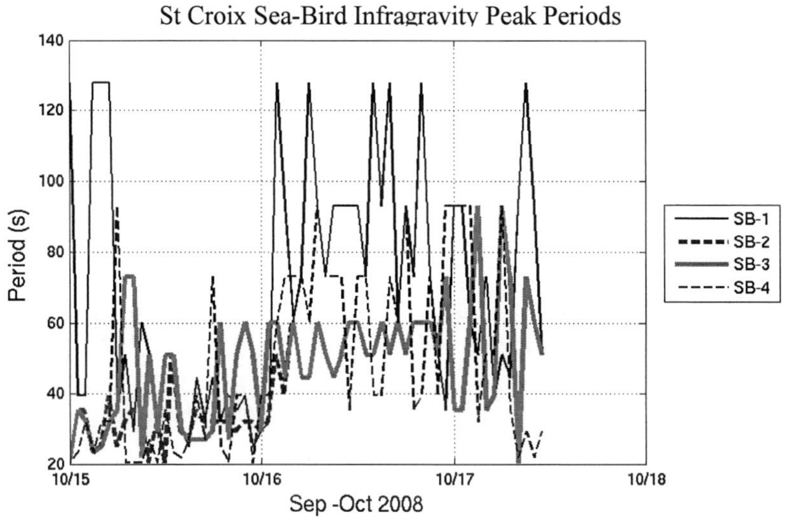

Figure 13: Peak periods in infragravity band during H. Omar.

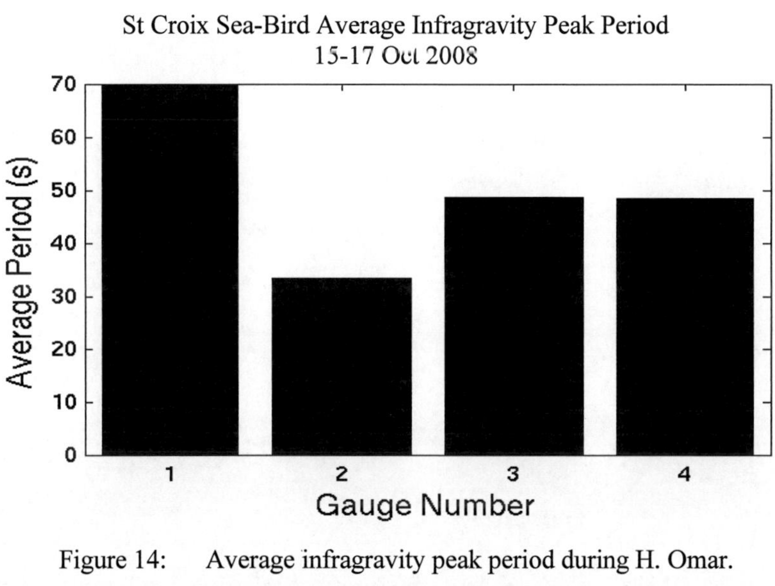

Figure 14: Average infragravity peak period during H. Omar.

Infragravity wave peak periods were generally less in the H. Omar storm than during the swell event (compare Figures 13 and 14 with 8 and 9).

Figures 15 and 16 show a comparison of incident wave spectra (Waverider) with the shoreward most Sea-Bird gauge during the two wave events. It is interesting to see the Sea-Bird spectra show little energy of the peak incident wave frequencies. There may be possible harmonic peaks of the incident peak frequency. These are likely waves that broke at the reef face and reformed at double frequencies.

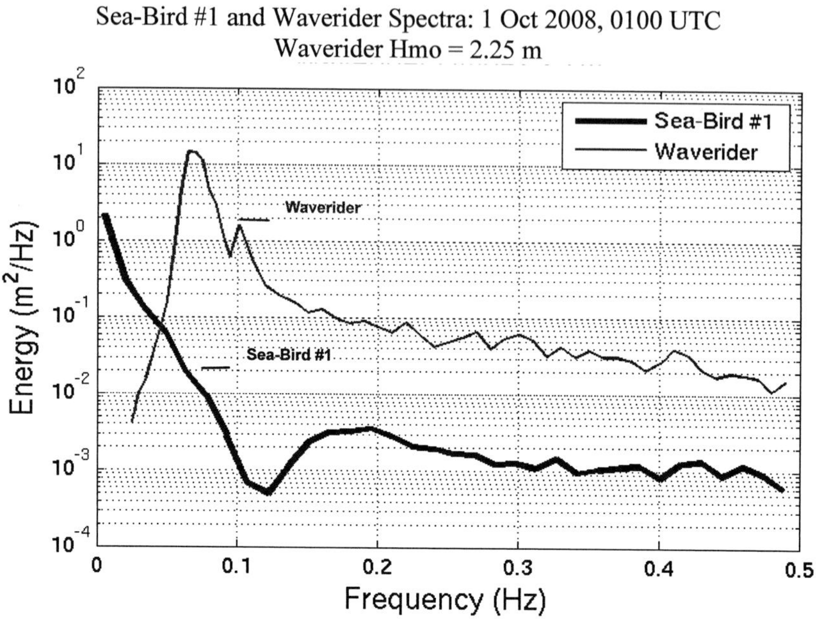

Figure 15: Waverider and SB-1 spectra in swell event.

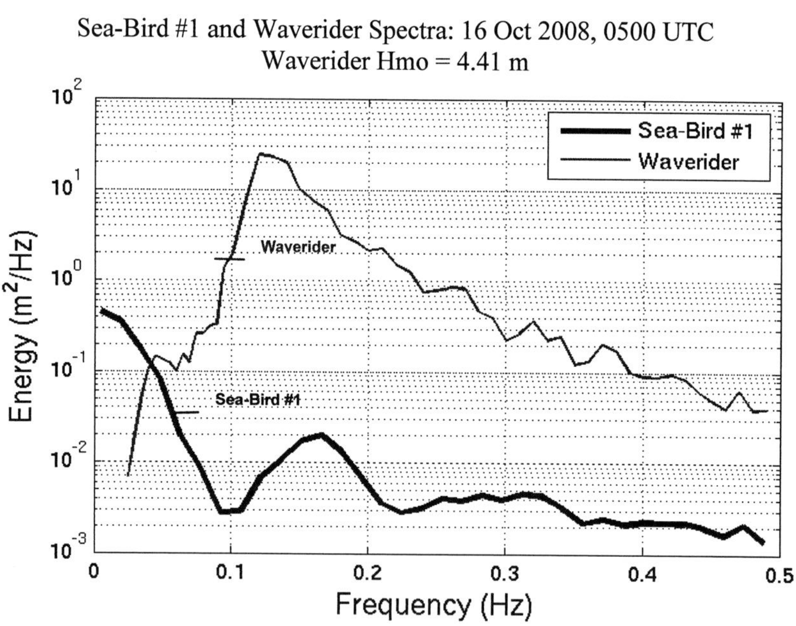

Figure 16: Waverider and SB-1 spectra during H. Omar.

3 Summary

This preliminary examination of the St Croix reef data was consistent with other reef studies (Lowe et al. [1]) where incident band wave energy is highly dissipated over the reef. Several factors are involved in the reduction of incident energy, including non-linear wave-wave interactions transferring energy to higher and lower frequencies, reflection from the steep fore-reef face, wave breaking, and bottom friction as waves propagate over the rough coral surface. All the instruments recorded highly dissipated incident band energy during the two documented storms, and an increase in infragravity energy that dominated the spectra. These long period waves [O (100 s)] could have been generated by incident infragravity energy, possibly by association with wave groups (bound infragravity waves), or by time-varying setup at the reef face caused by breaking wave groups. If the forcing is near the resonant mode of the reef, then water level variations can be enhanced (Pequignet et al. [2]). For this St Croix reef, rough estimates of the quarter-wave resonant mode for an open basin would be O (100 s). Further analysis of the data will look at incident wave grouping for correlations with reef infragravity motions and better estimates of reef resonant frequencies.

This data along with other PILOT site data will be used to improve the parameterizations of wave processes over reefs (Nwogu and Demirbilek [3]). Of particular interest is determining runup during storms to improve modeling of extreme water levels and island inundation (Vetter et al. [4]). Wave runup processes are enhanced with setup at the shoreline, which can have significant temporal and spatial variations over reefs (Jago et al. [5]). Setup observations from the shoreward most gauge, located at the nominal shoreline, provided data for correlating with runup. Extreme runup measurements were recorded from anecdotal information of debris locations and water damage via conversations with the locals, from which elevations were surveyed to the gauge datum. Future work will examine data from other storms and the other PILOT sites, in particular the southern St Croix site with a fringing reef and lagoon.

References

[1] Lowe, R.J., Falter, J.L., Bandet, M.D., Pawlak, G., Atkinson, M.J., Monismith, S.J., and Koseff, J.R. Spectral wave dissipation over a barrier reef. *J. Geophysical Research*, Vol 110, C04001, 16pp, 2005.

[2] Pequignet, A.C., Becker, J.M., Merrifield, M.A., and Aucan, J. Forcing of resonant modes on a fringing reef during tropical storm Man-Yi. *Geophysical Res. Letters*, Vol 36, L03607, 2009.

[3] Nwogu, O., and Demirbilek, Z. Infragravity wave motions and runup over shallow fringing reefs. *J. Waterway, Port, Coastal and Ocean Engineering*, Vol. 136, No. 6, 2010.

[4] Vetter, O., Becker, J.M., Merrifield, M.A., Pequignet, A.-C., Aucan, J., Boc, S.J., and Pollock, C.E. Wave setup over a Pacific Island fringing reef. *J. Geophysical Research*, Vol. 115, C12066, 13pp, 2010.

[5] Jago, O.K., Kench, P.S., and Brander, R.W. Field observations of wave-driven water-level gradients across a coral reef flat. *J. Geophysical Research*, Vol. 112, C06027, 14pp, 2007.

Control of salt water intrusion due to sea level rise in the coastal zone of Bangladesh

M. J. A. N. Bhuiyan & D. Dutta
School of Applied Sciences and Engineering, Monash University, Australia

Abstract

One potential impact of a global warming and rise in sea level would be an increase in the salinity of coastal zones, which would potentially threaten drinking water and aquatic ecosystems. There is a need to be better prepared to respond and adapt to these changes. This report examines the potential impacts of accelerated sea level rise on salinity in the South West region of Bangladesh. The study aims at determining the sensitivity of river salinity to upstream discharge and downstream sea level. To do that, the minimum upstream flow will be determined to protect the salinity intrusion from sea level rise. A one dimensional river salinity transport model has been developed and coupled with an existing surface and river flow model. The model has been proved to be robust with calibration and verification against observed data. The model outputs indicate a significant change of river salinity in the coastal zone. The results also indicate that a considerable advance in seawater intrusion can be expected in the coastal aquifer if current rates of sea level rise continue. The topography of the area is flat and gently sloping towards the Bay of Bengal. The consequences of salinity intrusion in the coastal area especially in the South West region of Bangladesh would be significant on many sectors like land fertility, agriculture, availability of fresh water, existence of the Sundarban forest etc. Pond culture in the coastal area will be affected by intrusion of salt water into ponds, unless embankments are made around them. Shrimp culture in the coastal area is a lucrative business. Increase in salinity is likely to jeopardize the shrimp culture.
Keywords: flow, river, salinity, salt water intrusion, sea level rise.

WIT Transactions on Ecology and the Environment, Vol 149, © 2011 WIT Press
www.witpress.com, ISSN 1743-3541 (on-line)
doi:10.2495/CP110141

1 Introduction

One of the future challenges from climate change is the sea level rise and its effect on salt water intrusion. The coastal zones of Bangladesh are threatened by salt water intrusion due to tidal impact from existing sea level. The intrusion length is increasing as the climate is changing and corresponding sea level rise (SLR). The coastal zones of Bangladesh, especially the South West region is the worst to be affected by sea level rise. The current rate of sea level rise will shift the saltwater further upstream in the coastal river. Bashar and Hossain [1] found that the situation is worsening with low river flow because of upstream withdrawal, increase usages and subsidence. As a result, it is becoming a great challenge to supply required fresh water to the coastal rivers. The adjacent aquifer will be contaminated through leakage from the river water. Existing fresh water fishes will migrate and decrease the availability. Therefore, saltwater intrusion due to sea level rise should be predicted and necessary measures need to be taken to prevent this intrusion.

A vast network of rivers in the South West region of Bangladesh is dependent on the water supply through the Gorai River, fig. 1. Water from this river is used for irrigation, urban water supply and industry. Most importantly, the Gorai river flow opposes away the saline water front in the Passur river. Monsoon flow is adequate for repelling the salinity intrusion but the dry season flow is most critical in controlling salinity in a large part of the South West region of Bangladesh.

Considerable attention has been focused on mathematical models to study the salt water intrusion and developed scenarios to control the intrusion. Various models have been developed to investigate salt water intrusion. However, only few models have been developed to study the control of salt water intrusion. This paper will discuss the salinity intrusion in the South West region of Bangladesh. Also the study will be used to analyse the position and movement of salinity front line and suggest new methodology to protect salt water intrusion. Here, the scenario used is the increased upstream river fresh water discharge.

2 Location and geomorphology

The study area covers part of the Ganges river basin which is located in the South West region of Bangladesh, fig.1. The area is bounded by the Ganges river in the north, Arial Khan-Bishakandia-Bishkhali river in the east, the Bay of Bengal in the south, and by the international boundary between Bangladesh and India in the west. The available data for analysis corresponds to 3 upstream river flow measuring stations, 4 river water level measuring stations and 2 river water quality measuring stations. The area is in a humid climate with an annual rainfall ranging from 1500 to 2100 mm; 90% of which occurring in the wet season (May-Sept) (Nobi and Gupta [2]). The topography of the region shows land slope from north to south with low gradient so the stream flow in the river occurs from upstream river to the Bay of Bengal. Major part of the area is protected by the polders (embankment or levee) from river flooding.

From the coastline, salt water intrudes into the river system and traversed few hundred kilometers towards the inland waters. The salinity level in the river water fluctuates seasonally depending on the amount of fresh-water flow. The river system is highly saline during low flow period in dry season. The Gorai-Madhumati River, major augmenting river system reduces the flow during the dry season. Detail information about hydrogeological and meteorological data of the area of interest, such as salinity, soil type, rainfall are available in reports such as Nobi and Gupta [2], Haque [3], Mirza and Sarker [4] and Mondal *et al.* [5].

The region receives freshwater through two river system: Gorai-Modhumati-Nabaganga and Kapotakshi-Sibsa (Figure 1). The southern part of the region is dominated by tidal river systems: Arpangasia, Sibsa, Passur and Baleswar. The tidal wave propagates into the region through these tidal rivers. These tidal rivers contribute salinity intrusion in the region. Fresh water flows from upstream rivers protect salinity intrusion and lower the salinity concentration. The river system of the region starts to be affected by sea salinity in November and reaches to a maximum in April and May (Mirza and Sarker [4]).

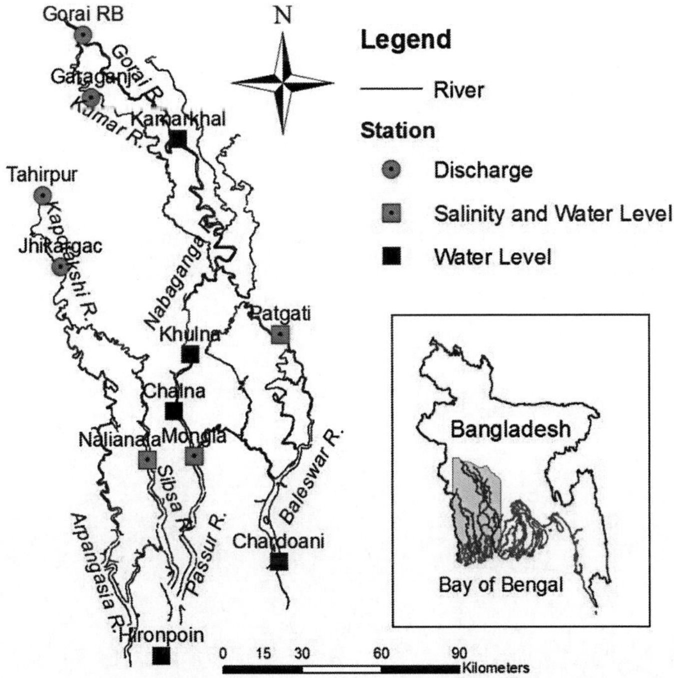

Figure 1: River network and stations in the South West region of Bangladesh.

3 Mathematical model

The mathematical model has been described here in two different sections; flow model and solute transport model. The flow model is a hydrodynamic model which comprises river and overland flow section. The salinity transport model computes salinity along with the river flow component.

3.1 Hydrodynamic model

The hydrodynamic model was originally developed at the Public Work Research Institute (PWRI) of Japan. The model has two components: river flow and overland flow. The model has been widely applied for flood modelling and risk analysis in Asian river system (Bhuiyan *et al.* [6]; Dutta and Bhuiyan [7]). For simulating surface flow processes, the study area is first discretized into square grids. Unsteady equations are derived from continuity and momentum equations for one dimensional as well as two dimensional flows. In river flow component, the finite difference equation for one dimensional channel flow is solved in each section of the channel for water level and discharge. It uses explicit solution scheme for river flow calculation. The form of momentum equation described in Dutta *et al.* [8] is as below;

$$A\frac{dQ}{dt}+Q^2\frac{d\beta}{dx}-2\beta Q\frac{dA}{dt}-\frac{\beta Q^2}{A}\frac{dA}{dx}+gA^2\frac{dH}{dx}+\frac{A}{\rho}T_r=0 \tag{1}$$

$$T_r=\rho gA\frac{Q^2}{[\Sigma\frac{A}{n}R^{2/3}]^2} \tag{2}$$

where A is the river cross-section, Q is river discharge, T_r is river bottom shear, H is water level, R is hydraulic radius.

In overland flow component, fundamental equations of two dimensional unsteady flows are constructed from continuity equation and equation of motion. Equation 3 represents the continuity equation and equations 4 and 5 represent the equation of motion in x and y direction respectively (Dutta *et al.* [8]).

$$\frac{\partial h}{\partial t}+\frac{\partial M}{\partial x}+\frac{\partial N}{\partial y}=0 \tag{3}$$

$$\frac{\partial M}{\partial t}+\frac{\partial uM}{\partial x}+\frac{\partial vM}{\partial y}+gh\frac{\partial H}{\partial x}+\frac{1}{\rho}\tau_x(b)=0 \tag{4}$$

$$\frac{\partial N}{\partial t}+\frac{\partial uN}{\partial x}+\frac{\partial vN}{\partial y}+gh\frac{\partial H}{\partial y}+\frac{1}{\rho}\tau_y(b)=0 \tag{5}$$

where H is the water level from datum, h is depth of water, u is flow velocity in x direction, v is flow velocity in y direction, g is gravitational acceleration, ρ is density of water, M is discharge flux in x direction ($M = uh$), N is discharge flux in y direction ($N = vh$), $\tau_x(b)$ is bottom shear stress in x direction, $\tau_y(b)$ is bottom shear stress in y direction.

The main characteristic of the model is the link between unsteady calculation in river channel and calculation of flood in river basin to reproduce the flood inundation phenomenon in large scale over the whole river system. The relation between stages in river channel and height of levee decides the points and scale of flood over the levee with unsteady calculation in river channel.

3.2 Salinity transport model

A salinity transport model has been developed to investigate transport processes within the river system through estimating the advection dispersion process and coupled with the hydrodynamic model. The transport by dispersion of solute along the longitudinal direction involves a mathematical representation in the form of following one dimensional, partial differential equation (Fischer *et al.*, [9]; Orlob [10]; Henderson-Sellars *et al.* [11]; Young and Wallis [12]), usually known by Fickian Diffusion Equation or Advection Dispersion Equation,

$$\frac{\partial C}{\partial t} + v\frac{\partial C}{\partial x} = D\frac{\partial^2 C}{\partial x^2} \tag{6}$$

where C is the concentration of the solute; v is cross-sectional average longitudinal velocity; and D is longitudinal dispersion coefficient. The dispersion coefficient can be specified as a function of the flow velocity calculated by the following equation,

$$D = av^b \tag{7}$$

where D is the dispersion coefficient, v is flow velocity, a is dispersion factor, b is a dimensionless exponent. MIKE 11 model uses same concept when running advection dispersion simulation (DHI [13]).

The finite difference method is a well-established numerical method which has been used here to approximate the partial derivatives in equation 6. If the concentration at a particular time step is calculated based on the concentration at previous time step and in approximating the advection and dispersion terms of the transport equation, the temporal discretization is said to be explicit, leading to the following form of the finite difference form (Zheng and Bennett [14]):

$$C_i^{t+1} = C_i^t + \frac{D\Delta t}{(\Delta x)^2}(C_{i+1}^t - 2C_i^t + C_{i-1}^t) - \frac{v\Delta t}{\Delta x}[(1-\alpha)C_i^t + \alpha C_{i+1}^t - (1-\alpha)C_{i-1}^t - \alpha C_i^t] \tag{8}$$

where C is the salinity concentration; i is river section; t is time; D is dispersion coefficient; v is velocity, Δx is distance between section, α is spatial weighting factor.

3.3 Model setup

The surface water system of interest is approximately 120 X 269 km^2, and is covered with a grid of 500 m square size in the model set up. The model contains 129120 cells: nx = 240, ny = 538, where n denotes the number of cells, x denotes East-West direction and y denotes North-South direction, fig.1. The model simulation time step to calculate surface flow is 2 seconds but the results have been compared with 3 hourly data. The existing polder heights have been used to protect against flooding in the surface. Here simulation result only from river flow model has been discussed but the combined river and surface flow model have been used for better performance and water budget calculation.

For this study, hydrological and salinity data were collected from a number of agencies in Bangladesh. Temporal scale discharge data were derived from the Institute of Water Modelling (IWM) data set. Station wide salinity during high and low tide for various periods have been supplied by the Bangladesh Water Development Board (BWDB). Salinity time series data has been collected from IWM, Dhaka, Bangladesh.

3.4 Calibration and verification

The combined model has been calibrated and verified against measured data at some selected stations in the river. The hydrodynamic model result has been compared against observed water level and salinity transport model against observed river salinity. The details of result comparison have been tabulated in Table 1. Here the water level unit is meter above PWD datum and salinity unit is parts per thousand. The PWD datum is 0.46 m below mean sea level (Tingsanchali and Karim [15]. Further details of time series data for different stations are available at Bhuiyan and Dutta [16].

Table 1: Summarized model results from calibration and verification.

Model setup/ Parameter	Station	Observed			Model simulated		
		Max	Min	Avg	Max	Min	Avg
Calibration/ Water level	Kamarkhali	1.83	1.18	1.43	1.84	1.34	1.58
	Patgati	1.8	0.38	1.18	1.88	0.36	1.09
	Mongla	2.5	-1.6	0.52	2.39	-0.94	0.56
	Nalianala	2.47	-1.80	0.33	2.82	-1.56	0.60
Verification/ Water level	Kamarkhali	2.87	1.72	2.30	2.86	1.77	2.33
	Patgati	2.12	1.08	1.65	2.42	0.66	1.60
	Mongla	2.64	-1.17	0.92	2.67	-0.63	0.94
	Nalianala	2.98	-1.40	1.06	3.01	-1.03	1.06
Calibration/ Salinity	Mongla	12.4	11.7	12.0	13.8	11.8	12.6
	Nalianala	19.4	15.7	17.7	19.8	15.6	18.1
Verification/ Salinity	Mongla	15.1	9.4	13.4	14.6	11.3	13.2
	Nalianala	19.9	8.6	15.6	20.2	15.2	17.7

The model performance has been checked using some statistical indicators, which have been summarized in Table 2. Here the indicators used are: relative root mean squared error (RRMSE); mean absolute error (ABSERR); the Nash-Sutcliffe modeling efficiency index (EF); the goodness-of-fit (R^2) and the % of deviation from observed stream flow (PBIAS). The equations used when determining these indicators are taken from Stehr *et al.* [17]. The statistical indicators have been derived for water level and salinity at Kamarkhali, Patgati, Mongla and Nalianala station during calibration and verification period. The closer the values of RRMSE and ABSERR to zero, and those of R^2 and EF to unity, the better the model performance is evaluated (Abu El-Nasr *et al.* [18]). For PBIAS, the optimal value is 0; a negative value indicates an overestimation of observed values, whereas a positive value indicates underestimation.

Table 2: Statistical analysis of model performance.

Model setup/ Parameter	Station	PRMSE	ABSERR	EF	R^2	PBIAS
Calibration / Water level	Kamarkhali	0.14	0.15	0.32	0.43	-11.04
	Patgati	0.18	0.16	0.62	0.79	8.11
	Mongla	0.68	0.40	0.78	0.80	-7.91
	Nalianala	1.87	0.90	-0.22	0.15	-77.67
Calibration / Salinity	Mongla	0.23	0.64	-9.71	0.42	-5.13
	Nalianala	0.22	0.74	0.35	0.81	-2.56
Verification / Water level	Kamarkhali	0.05	0.06	0.97	0.96	1.42
	Patgati	0.19	0.22	-0.13	0.53	2.98
	Mongla	0.65	0.50	0.66	0.66	-3.05
	Nalianala	1.12	0.96	0.01	0.16	-0.21
Verification / Salinity	Mongla	0.29	0.89	0.53	0.57	2.29
	Nalianala	0.90	2.76	0.20	0.64	-13.72

The hydrodynamic model results at Kamarkhali are well calibrated and verified where the model computed water levels have been compared against observed values. The water level comparison at Patgati station shows very good correlation and good matching with observed level in phase and amplitude of tide. Tidal water level at Nalianala station shows poor correlation during low tide period but better during high tide. In spite of this discrepancy, the simulated salinity is close to the observed data at this station. Computed salinity at Patgati shows good match but due to limited number of observed data, calculation of statistical measures was not possible.

3.5 Scenarios

The combined model has been simulated for different scenarios to observe the sensitivity of river salinity to upstream discharge as well as downstream sea level. The following scenarios have been used;

Scenario 1: Present condition with no sea level rise

Scenario 2: Constant sea level rise of 59 cm

Scenario 3: Same sea level rise with 200 m^3/s discharge through Gorai river, 50 m^3/s discharge through Kumar-Nabaganga River and 10 m^3/s discharge through Kapotakshi River.

Scenario 4: Same sea level rise with 100 m^3/s discharge through Gorai river, 10 m^3/s discharge through Kumar-Nabaganga River and 5 m^3/s discharge through Kapotakshi River.

4 Result analysis

Salinity values were extracted from the simulated results for generation of salinity profile for different river system. Then the results for different scenarios were compared with that of current condition to determine the salinity changes at particular stations and salinity intrusion along particular rivers.

Salinity long profile along Kumar-Nabaganga-Rupsa-Passur river system is shown in fig. 2. This Figure also indicates the increase in salinity due to the sea level rise along the river system. The salinity long profile along Kapotakshi-Arpangasia river system is shown in fig. 3.

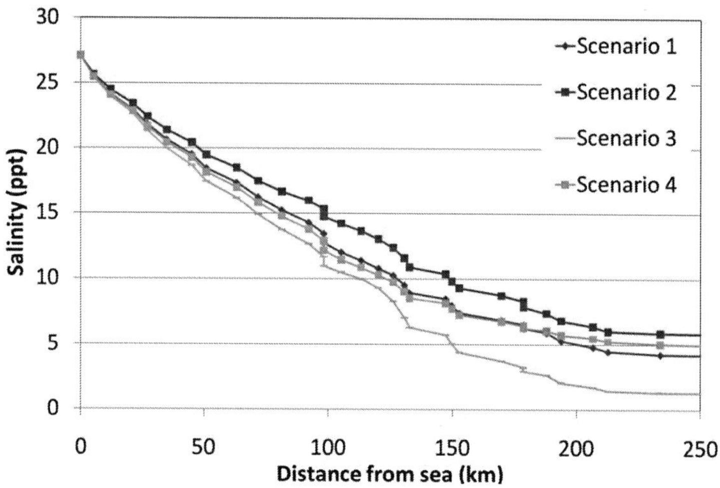

Figure 2: Salinity along Kumar-Nabaganga-Rupsa-Passur river system.

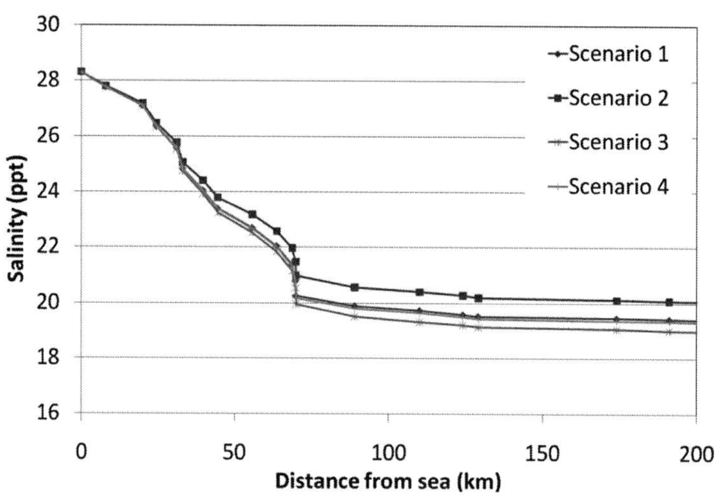

Figure 3: Salinity along Kapotakshi-Arpangasia river system.

For sea level rise of 59 cm, the 5 ppt and 10 ppt salinity front moved further inland by 25 and 7 km respectively from present condition. Also the dry period maximum salinity has increased due to sea level rise. The increased salinity has been protected with increased upstream river flow in other scenarios. The simulated salinities at Mongla and Nalianala stations for different scenarios have been shown in Table 3.

Table 3: Simulated maximum salinity for different scenario.

Station	Scenario 1	Scenario 2	Scenario 3	Scenario 4
Mongla	15.3	16.7	13.8	14.8
Nalianala	20.8	21.6	20.1	20.6

From the simulated model results, it is observed that the salinities in the river system have increased due to sea level rise. Again, using increased freshwater system in the upstream station, the previously increased salinity can be reduced and salinity front can be pushed back to downstream.

5 Conclusion

Combined use of surface flow and salinity transport in coastal river systems is one of the most significant strategies for optimal development of water resources maintaining the salinity intrusion level within the desirable limit. This requires an assessment of flow and salinity intrusion in the river network. The river flow, surface flow and salinity transport models are coupled and solved interactively to account for the dynamic exchange of flow and solute transport between the rivers and the flood plain.

The results showed intrusion of salinity in South West region due to sea level rise. Sea level rise of 59 cm will increase salinity at Mongla and Nalianala station by 1.4 and 0.8 ppt. However, if minimum flow of 100 m^3/s and 10 m^3/s can be maintained at Gorai Railway Bridge and Garaganj station, it will restrict the salinity at Mongla up to a maximum of 14.8 ppt. Similarly, maintaining minimum flow of 5 m^3/s at Jhikargacha will restrict the salinity at Nalianala up to 20.6 ppt.

The study focused the salinity impact on river due to sea level rise only. Other consequences of climate change may accelerate or delay the consequences of sea level rise. For example, if droughts become more severe in the future then there will be a reduction in upstream flow. This will further worsen the situation and increase the salinity.

References

[1] Bashar, K. E. & Hossain, M. A., Impact of Sea Level Rise on Salinity of Coastal Area of Bangladesh. *9th International River Symposium*. Brisbane, Australia, 2006.

[2] Nobi, N. & Gupta, A. D., Simulation of regional flow and salinity intrusion in an integrated stream-aquifer system in coastal region: southwest region of Bangladesh. *Ground Water*, **35(5)**, pp. 786-796, 1997.

[3] Haque, S. A., Salinity problems and crop production in coastal regions of Bangladesh. *Pak. J. Bot.*, **38(5)**, pp. 1359-1365, 2006.

[4] Mirza, M. M. Q. & Sarker, M. H., *Effects on water salinity in Bangladesh, in The Ganges water diversion: environmental effects and implications, Water science and technology library,* Kluwer Academic Publishers, Netherlands, pp. 81-102, 2005.

[5] Mondal, M. K., Bhuiyan, S. I. & Franco, D. T., Soil salinity reduction and prediction of salt dynamics in the coastal ricelands of Bangladesh. *Agricultural water management*, 47(1), pp. 9-23, 2001.

[6] Bhuiyan, M. J. A. N., Dutta, D., Gupta, A. D. & Babel, M. S., Flood Simulation and Its Socio-Economic Impact Analysis in Meghna Delta, Bangladesh Under Climate Change Conditions. *International Symposium on Floods in Coastal Cities under Climate Change Conditions*, Bangkok, Thailand, June, 2005, pp.1-10, 2005.

[7] Dutta, D. & Bhuiyan, M. J. A. N., Flood Characteristics under Changing Climatic Conditions in Meghna Delta, Bangladesh. *16th IASTED International Conference APPLIED SIMULATION AND MODELLING*, Palma de Mallorca, Spain, August 29 – August 31, 2007, pp. 162-167, 2007.

[8] Dutta, D., Alam, M. J., Umeda, K., Hayashi, M. & Hironaka, S., Urban Flood Modeling in Lower Mekong Basin: A Physically Based Distributed Modeling Approach. *International Conference on Sustainable Water Resources Management in the Changing Environment of the Monsoon Region*, Colombo, Sri Lanka, 2004.

[9] Fischer, H. B., List, E. J., Koh, R. C. Y., Imberger, J. & Brooks, N. H., *MIXING in Inland and Coastal Waters*. Academic Press Inc. (London) Ltd, 1979.

[10] Orlob, G. T. (ed.), Mathematical Modelling of Water Quality: *Streams, Lakes and Reservoirs*. J. Wiley, Chichester, 1983.

[11] Henderson-Sellars, B., Young, P. C., & Ribeiro da Costa, J., Water quality models: rivers and reservoirs, *in Proc. 1988 Int. Symp. On Water Quality Modelling of Agricultural Non-Point Sources; United States Dept of Agriculture*, Agricultural Research Service; 1990; Publication No. ARS-81, pp. 381-420, 1990.

[12] Young, P.C., & Wallis, S.G., Solute transport and dispersion in channels, forthcoming in: K.J. Beven and M.J. Kirby (eds), Channel Networks, *J. Wiley*, Chichester, 1992.

[13] DHI, *MIKE 11 – A Modelling System for Rivers and Channels*. Short introduction tutorial. DHI Water and Environment, pp. 5-27, 2003.

[14] Zheng, C. & Bennette, G. D., *Applied Contaminant Transport Modeling, Theory and Practice*. Van Nostrand Reinhold, a division of International Thomson Publishing Inc. New York, 1995.

[15] Tingsanchali, T. & Karim, M. F., Flood Hazard and Risk Analysis in the Southwest Region of Bangladesh. *Hydrological Processes* 19(10): pp. 2055-2069, 2005.

[16] Bhuiyan, M. J. A. N. & Dutta, D., Adaptation Strategies For Sea Level Rise Impact On Coastal Cities: A Case Study, South Western Coastal Region Of Bangladesh. *International Symposium on Coastal zones and Climate Change: Assessing the Impacts and Developing Adaptation Strategies, Monash University*, Australia, April 12 – 13, 2010, pp. 303-313, 2010.

[17] Stehr, A., Debels, P., Romero, F. & Alcayaga, H., Hydrological modelling with SWAT under conditions of limited data availability: evaluation of results from a Chilean case study / Modélisation hydrologique avec SWAT en conditions de données peu disponibles: évaluation des résultats d'une application au Chili. *Hydrological Sciences Journal*, **53(3)**, pp. 588-601, 2008.

[18] Abu El-Nasr, A., Arnold, J. G., Feyen, J. & Berlamont, J., Modelling the hydrology of a catchment using a distributed and a semi-distributed model. *Hydrol. Processes,* **19(3)**, pp. 573–587, 2005.

Section 5
Coastal processes and navigation

The satellite positioning evolution in coastal processes

A. Angrisano, S. Gaglione, C. Gioia & M. Vultaggio
Department of Applied Sciences, Parthenope University of Naples, Italy

Abstract

In order to guarantee future generations, a fair and consistent level of environmental quality is necessary to ensure sustainable development of coastal resources.

Adopting a static mode approach towards sea-level rise conflicts with sustaining a dynamic coastal system that responds to perturbations via sediment movement and long-term evolution. In the monitoring of this long-term evolution ancient technologies has played a very important role. Among these, satellite positioning is becoming more and more relevant.

This paper presents a brief survey of the evolution of GNSS (Global Navigation Satellite Systems) application in coastal process. The Global Position System (NAVSTAR GPS) is the best known of these satellite navigation systems. Global Navigation Satellite System (GNSS) is the standard generic term for satellite navigation systems that provide autonomous geo-spatial positioning with global coverage. A GNSS allows small electronic receivers to determine their location (longitude, latitude, and height) to within a few metres using the signals as transmitted by the GNSS satellites. In the same process the receivers also calculate the precise time of the signal reception and as such GNSS receivers can be used as highly accurate clocks.

Currently, the United States NAVSTAR Global Positioning System (GPS) is the only fully operational GNSS. The Russian GLONASS is a GNSS in the process of being restored to full operation. The European Union's Galileo positioning system is a next generation GNSS in the initial deployment phase. China has indicated it may expand its regional Beidou (also called COMPASS) navigation system into a global system. India's IRNSS, is also a next generation GNSS. However, its status and future is unclear as India seems to have entered into close cooperation with the Russians on the GLONASS system. Satellite Positioning, for this reason, is very popular and its accuracy can support a large

WIT Transactions on Ecology and the Environment, Vol 149, © 2011 WIT Press
www.witpress.com, ISSN 1743-3541 (on-line)
doi:10.2495/CP110151

number of navigation and timing applications. However, it is generally recognized that these systems lack the accuracy, integrity, and availability to satisfy the more critical applications. This has led to the development of techniques to augment the basic GPS service.

The goal of this paper is to introduce an overview of the aforesaid technologies and to present the improvement outcome from their integration (multi constellation) and their augmentation (SBAS – Satellite Based Augmentation Systems) in terms of positioning accuracy; very important in the definition of long-term sediment movement and system integrity.
Keywords: GNSS, augmentation.

1 Introduction

The increasing attention paid to protecting the environment and the consequent need of mapping and surveying areas at risk, such as the coasts subject to erosion and sedimentation, have fostered the use of GNSS (Global Navigation Satellite Systems) systems in this area. Satellite navigation systems have not only revolutionized navigation, but also geodetic survey. By means of satellite range measurements, it is possible to obtain positioning accuracies previously impossible to obtain, especially for long baselines. With such systems it is possible both the control and the relief of the areas at risk, such as ridges subject to landslides. The advantage of GNSS systems is the simplicity of using them and speed of survey; in addition, along with the suitable augmentation techniques, they offer accuracy comparable to those reached by traditional means of survey.

2 GNSS systems

2.1 GPS overview

The Global Positioning System (GPS) is a space-based radionavigation system which is managed for the Government of the United States by the U.S. Air Force (USAF), the system operator; it is a dual use, military and civil system whose primary mission is to provide Position, Navigation and Time (PNT) services. GPS positioning is based on range measurement from known satellite positions to unknown user positions. The current GPS consists of three major segments: a space segment, a control segment and a user segment.

The space segment is a constellation of artificial satellites, 24 in the baseline configuration but more satellites can be in orbit [2], distributed in 6 orbital planes inclined at 55° relative to the equatorial plane.

The control segment provides command, control, and maintenance services to the space segment and supports the PNT and the user segment is made up of all the system users, equipped with a GPS receiver-processor able to receive and decode the GPS signal to obtain PNT information.

GPS satellites broadcast signals on two carrier frequencies: L1 (1575.42 MHz) and L2 (1227.60 MHz). The carriers are modulated by two kinds of Pseudo-Random-Noise (PRN) codes, the Coarse-Acquisition (C/A) code with a 1.023 MHz chip-rate and a period of one millisecond, and the Precise (P) code with a 10.23 MHz chip-rate a period of seven days. The C/A code is available on the L1 carrier only and the Precise (P) code on both L1 and L2.

PRN codes are referred to as "ranging code", because they are used to estimate the propagation interval of the signal from the satellite to the user (and hence the pseudo-range). The PRN codes are binary signals having spectral characteristic similar to random sequences, but actually are deterministic [4]. GPS system uses the Code Division Multiple Access (CDMA), i.e. all satellites use the same frequencies and each satellite transmits a different PRN code, so that at receiver level it is possible to distinguish the tracked satellite.

The C/A code is for civil use, it can be used by every user equipped with GPS receiver/processor and its use is associated with Standard Positioning Service (SPS). The P code instead is for military purpose, its access is restricted to authorized users only and its use is associated with Precise Positioning Service (PPS).

To meet the increasing demands of improved performance and to compete with the other developing GNSS systems, a GPS modernization program started in 2000 with the cancellation of the Selective Availability SA [11]. The program involves the deployment of new ground stations and the lunch of satellites of new generation able to broadcast additional signals. In detail, the new additional signals are a military signal (M-code) on both carriers, further codes on L1 carrier (L1C) and on L2 (L2C) for civil use and a third civil frequency L5.

The L2C code is designed to meet commercial needs, allowing us to provide higher accuracy through dual-frequency ionospheric correction (associated to the existing C/A code on L1) [10]. L5 is designed to meet demanding requirements for transportation and safety-of-life. The satellites transmitting these types of signals are GPS IIF, whose 1st launch was made 27 May 2010; 24 satellites of them are expected to be operational in 2018. L1C is designed with international partners for GNSS interoperability and will be transmitted by GPS Block IIIA satellites; 24 satellites of them are expected to be operational in 2021. Therefore, future civilian GPS users will be able to receive three signals on L1, L2 and L5.

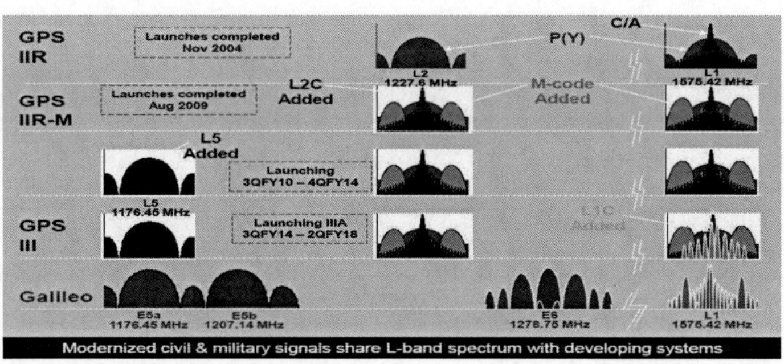

Figure 1: GPS signal modernization.

L2C, together with the present L1 C/A-code signal and the future modernized civil signal L1C, will provide an alternative to augmented single frequency GPS for precision users. Many private and government precision applications could potentially benefit from multiple frequency GPS, e.g. centimeter accuracy is important to many land and marine surveying applications including planning, zoning, land management, cadastral surveying, harbor and port mapping, aids to navigation, coastal resources management, mapping and surveys of sensitive habitats.

2.2 GLONASS overview

GLONASS system has an operational principle similar to GPS system: the user equipment is able to measure pseudo-ranges and pseudo-range rates with respect to space vehicles, and uses this information to compute the position, velocity and for synchronization with a reference time scale.

Also GLONASS system consists of three segment as GPS, space, control and user segment, with similar tasks.

The space segment is a constellation of artificial satellites that, in the complete configuration are 24, placed in three orbital planes whose ascending nodes are 120° apart. Theoretically 8 satellites are equally displaced on each plane, with a displacement of argument of latitude among the planes of 15°. The orbits are planned to be circular with an inclination of 64.8° and a radius of 19100 km, corresponding to an orbital period of 11^h15^m with ground tracks repeating every 17 orbital periods. The satellite constellation, if completely deployed, would provide a continuous and worldwide coverage for users on the Earth surface or in the near-Earth space [3]. The GLONASS space segment at November 11th 2010 includes 20 operating satellites, 4 in maintenance and 2 spares, and currently satellites of the "Glonass-M" generation are in orbit.

The Glonass-M satellites transmit signals on two frequency bands: L1 ~ 1.6 GHz and L2 ~ 1.25 GHz. On both the bands, the carriers are modulated by two ranging codes, one associated to a standard positioning service accessible to all the users and another associated to precise positioning service restricted to authorized users only. The ranging code associated to the standard service has a 511 KHz chip-rate and a period of one millisecond, the code associated to the precision service has 5.11 MHz chip-rate.

GLONASS system uses the Frequency Division Multiple Access (FDMA) technique, i.e. each satellite transmits on carriers with different frequencies; hence at receiver level, GLONASS satellites can be distinguished by the frequency.

2.3 Galileo overview

The European Union (EU) conceived the establishment of Galileo in 1990s and it was finally approved in a meeting of the EU Transport Council in 2004. The first test-bed satellite system, GIOVE-A was launched in December 2005 and the second one, GIOVE-B was launched in April 2008. A series of in-space signal validations have been carried out and the ground control segment has been under development since that time. GIOVE-A and GIOVE-B are part of the first step of the In Orbit Validation (IOV) phase, including the use of a reduced constellation of four satellites to perform a validation of the system; four satellites are the minimum number necessary to perform positioning and timing.

The four Galileo In Orbit Validation (IOV) satellites will be launched in the following years and Galileo is expected to achieve its Full Operational Capability (FOC) in 2013. Once validation is complete, the full constellation will be deployed. The constellation will contain 30 satellites (27 operational satellites and 3 active spares) in three orbital planes, transmitting on three frequencies. The orbital planes are inclined 56 degrees relative to the equatorial plane. Due to the distribution of the satellites, Galileo will offer a slightly better coverage at high latitude which is very helpful for north European countries. In case of a single satellite failure, an active spare will be used to replace it, and therefore the user will not experience a significant degradation in overall performance [10].

Once complete, Galileo will offer four services:

- Open Service (OS), corresponding to the signals freely available which will be transmitted on two frequencies.
- Commercial Service (CS), corresponding to the encrypted signal which will be transmitted on three frequencies.
- Public Regulated Service (PRS) and Safety of Life (SoL), which will be used by the authorities for safety of life applications.

The Galileo Navigation Signals are transmitted in the four frequency bands. These four frequency bands are: the E5a band, the E5b band, the E6 band and the L1 band.

3 GNSS augmentation

Satellite Navigation systems are characterized by a set of parameters which define the system performance:

- *Integrity*, i.e. the ability to protect the user from inaccurate information in a timely manner.
- *Accuracy,* i.e. the difference between measured and true positions of a vehicle at any given time.
- *Continuity,* the ability to complete an operation without triggering an alarm.
- *Availability,* the ability to be used by the user whenever it is needed [13].

The GPS SPS does not provide suitable performance in terms of the defined parameters in all environment conditions or in every possible application. In severely signal degraded environments, e.g. mountainous or urban areas, GPS signals are blocked by buildings or natural obstacles and the positioning results inaccurate or even impossible owing to the lack of satellites [14]. Moreover GPS SPS is inadequate for critical safety applications like aircraft take-off or landing or for high-precision applications as land survey.

To solve GPS gap, different strategies can be adopted, e.g. the combined use of more satellite systems (as GPS/GLONASS or GPS/GLONASS/Galileo), the integration with other navigation systems (as GPS/inertial sensors) or the use of suitable techniques (DGPS, RTK).

3.1 DGPS

In applications where greater accuracy is required, such hydrographic surveying, differential techniques (DGPS/Differential GPS) can be used to improve the GPS solution. In these areas, however, the surrounding topography or man-made structures may mask GPS signals line-of-sight. These obstructions may result in reduced satellite availability and reliability, degraded geometry, and so reduced horizontal DGPS accuracy. Relative positioning with GPS is achieved by performing simultaneous observation of satellite signal at two sites. The errors of the two receivers are spatially correlated, i.e. a particular satellite will experience common errors on ranging signal if the receivers are near.

The position of one receiver is precisely known and the common errors sources can be canceled or greatly reduced performing the difference between the observations; different approaches can be adopted and the more common are the double differences (DD) among the satellites and receivers, which removes all satellite dependent, receiver dependent and most of the atmospheric errors (if the distance between the two receivers is not too large).

A reference receiver, or base station used as a reference station (RS), computes corrections for the measurement of each satellite in view. Because individual pseudo-ranges must be corrected prior to the formation of a navigation solution, DGPS implementations require software in the reference receiver that can track all SVs in view and form individual pseudo-range corrections for each SV [5]. These corrections are passed to the remote, or rover, receiver which must be capable of applying these individual pseudo-range corrections to each SV used in the navigation solution.

Applying a simple position correction from the reference receiver to the remote one has limited effect on ranges used for positioning. Both receivers would have to use the same set of SVs measurement in their navigation solutions and to have identical GDOP terms (not possible at different locations) to be identically affected by bias errors [6].

The application of DGPS improves integrity compared to stand-alone GPS positioning. There are two types of DGPS approaches: Local Area DGPS (LADGPS) and Wide Area DGPS (WADGPS). LADGPS system uses only one ground reference station and transmits either scalar corrections or raw measurements to the near users. If the users are close to the reference station, common mode errors can be canceled and better accuracy can be achieved (of course the location of the reference receiver must be precisely surveyed) [7]. LADGPS broadcasts either corrections or raw measurements from the ground reference receiver to its users via any suitable data link. Because LADGPS is trying to cancel the common mode errors for the reference station and the user, accuracy strongly depends on their separation. As the user moves away from the LADGPS reference station, spatial decorrelation in the ephemeris errors, tropospheric and ionospheric errors cause accuracy degradation. The WADGPS approach uses multiple ground reference stations to monitor GPS satellites and generate vector corrections which include satellite clock errors, satellite ephemeris errors and ionospheric delays. Because of these vector corrections,

WADGPS users will not suffer the spatial decorrelation problem characteristic of LADGPS. Approximately two meter accuracy (95%) can be achieved over continental areas [8].

3.2 RTK overview

Centimetre-level accuracy positioning in real-time based on GNSS measurements was developed in the mid 1990s and is nowadays referred to as real-time kinematic (RTK) positioning. RTK is a special form of Differential GPS. The first use of GPS for centimeter position relied on static receivers which collected only carrier data, hours of data, to be post - processed. This was called *static* surveying.

Thanks to improvement in technology, the time computation for this kind of process decreases to a few minutes, having the so called, *rapid-static technique.* This one required only one static initialization of the receiver. From then, as long as the receiver maintained phase-lock, a centimeter accuracy is possible to achieve only in post processing. This was defined *kinematic* surveying. With faster and smaller computers, the desktop processing moved into the GPS receiver itself, providing real time results on field.

There are two kinds of RTK surveys: RTK with static initialization and RTK on-the-fly. Their difference leads only for the absence of the first part of the process procedure for the RTK on-the-fly technique.

However, for high precision applications, the use of carrier phase measurement must be used. This different type of data has an impact on the overall system complexity in terms of:

a) Equipment; the use of at least two double frequency receivers is necessary (for rover and base station). Also a data communication link is necessary (UHF or VHF) in order to broadcast this information between the receivers;
b) Computational complexity: this kind of measurement is, in fact, *ambiguous*, requiring sophisticated data processing software.

The limitation of single base RTK is the distance between reference and rover receivers. Such biases as: orbit error, ionospheric and tropospheric signal refraction depend from this distance. Good performance can be obtained if inter receiver length is less or equal to 10km. On the other hand, Wide Area Differential GPS (WADGPS) and the Wide Area Augmentation System (WAAS) use a network of master and monitor stations spread over a wide geographic area; measurement biases are modelled for any position of the area covered and proper transmitted. For this reason the positioning accuracy will be almost independent of the baseline length.

A good agreement between the above explained systems is the Network RTK; it is a centimeter-accuracy, real-time, carrier phase-based positioning technique capable of operating over inter-receiver distances up to many tens of kilometers with equivalent performance to current single base RTK systems (operating over much shorter baselines). The reference stations must be deployed in a dense enough pattern to model distance-dependent errors to such an accuracy that residual double-differenced carrier phase observable errors can be ignored in the context of rapid ambiguity resolution. Network RTK is therefore the logical outcome of the continuous search for a GPS positioning technique that challenges the current constraints of cm-accuracy, high productivity, carrier phase-based positioning [15].

3.3 EGNOS overview

Real-time satellite positioning at the sub-meter or even sub-centimeter level of accuracy requires the use of the augmentation system. DGPS has a fundamental limitation in that the broadcast corrections are good only for users in a limited area surrounding the base station. Outside this area the errors tend to be decorrelated, rendering the corrections less accurate. An obvious technical solution to this problem would be to use a network of base stations, each with its own communication link to serve its geographic area. However, this would require a huge number of base stations and their associated communication links.

Four space-based augmentation systems (SBASs) have been developed. These are the Wide-Area Augmentation System (WAAS), European Geostationary Navigation Overlay System (EGNOS), Multifunctional Transport Satellite (MTSAT) based Augmentation System (MSAS), and GPS & GEO Augmented Navigation (GAGAN) by India. This SBAS are primarily defined as regional systems, and co-operation/co-ordination among the different systems will enable us to provide a seamless world-wide navigation system. To this aim, the systems have set up some common interoperability requirements and will provide adequate system compatibility. The development teams of SBAS systems meet regularly to develop and maintain common understanding of interoperability requirements and capabilities, and on the identification of the necessary interfaces among SBAS that each conceivable interoperability scenarios may imply [12].

EGNOS is the first step on the European contribution to the Global Navigation Satellite System. EGNOS is an augmentation system to the GPS and GLONASS Satellite Navigation systems, which provides and guarantees navigation signals for precision and kinematic applications.
EGNOS is composed of four segments:

- Ground segment
- Space segment
- User segment
- Support facilities.

The *EGNOS Ground Segment* consists of GNSS Ranging and Integrity monitoring Stations (RIMS) which are connected to a set of redundant control and processing facilities Mission Control Centre (MCC) that determines the integrity, differential corrections for each monitored satellite, ionospheric delays and generates GEO satellite ephemeris. This information is sent in a message to the Navigation Land Earth Station (NLES), to be uplinked along with the GEO Ranging Signal to GEO satellites; GEO satellites broadcast this data on the GPS L1 frequency with a modulation and coding scheme similar to the GPS one. All ground Segment components are interconnected by the EGNOS Wide Area Communications Network (EWAN).

The *EGNOS Space Segment* is composed of Geostationary transponders with global Earth coverage. The EGNOS AOC system is based on the use of the EGNOS GEO satellites AOR-E at 15.5°W, ARTEMIS at 21.5°E and IOR-W at 25°E, and the ESA ARTEMIS navigation transponders.
The *EGNOS User Segment* consists of an EGNOS Standard receiver, to verify the Signal-In-Space (SIS) performance, and a set of prototype User equipment used to validate and eventually certify EGNOS for the different applications being considered.

EGNOS *support facilities* include the Development Verification Platform (DVP), the Application Specific Qualification Facility (ASQF) and the Performance Assessment and System Checkout Facility (PACF). Those are facilities needed to support System Operations and future Qualifications.

4 Conclusions and results

The goal of this work is to verify and to compare the satellite positioning Augmentation Systems discussed. This study is focused on a measures accuracy analysis where the reference solution for comparison, computed by a post processing geodetic method (DGPS), is:

Table 1: WGS84 reference coordinate.

Latitude	Longitude	Height
40°49'24.37413''N	*14°12'58.11149''E*	*122.659 mt.*

24 hours of data were collected, at 1 Hz frequency, in the Parthenope University ECDN (ESV/The Data Collection Network) Station on November 30th 2010 by a Septentrio PolaRx 2 receiver. These data are processed by PEGASUS v 4.5.0 software, developed by EUROCONTROL, currently considered as a benchmark from the European research community involved in the same network.

Accuracy and precision indicate different attributes on how close a solution is to the reference. Accuracy refers to the ability of a system to be close to the reference value, while precision refers to the ability of the system to repeat its solution. In the Gaussian distributions domain, good accuracy would correspond to a zero mean, while high precision would correspond to a low standard deviation. Thus it is possible to have a precise, but inaccurate system (and vice versa), when the distribution is centered on a bias with a very small deviation [16].

In the following figures the solution obtained by a GPS stand alone processing method is shown.

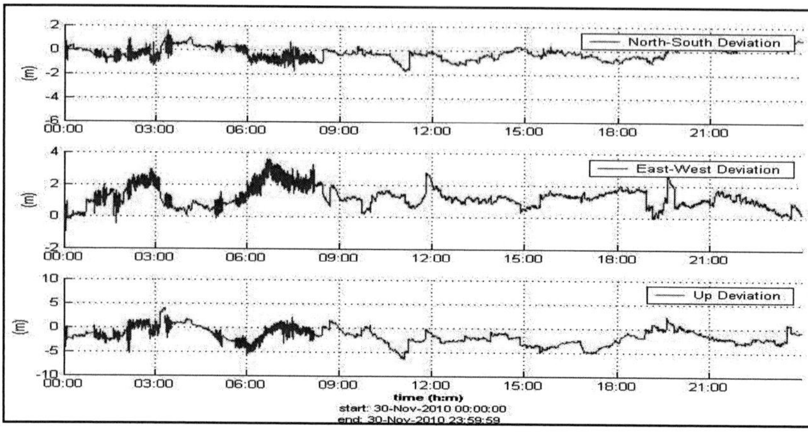

Figure 2: Deviation from the reference coordinate with GPS stand-alone.

The distribution of the error is:

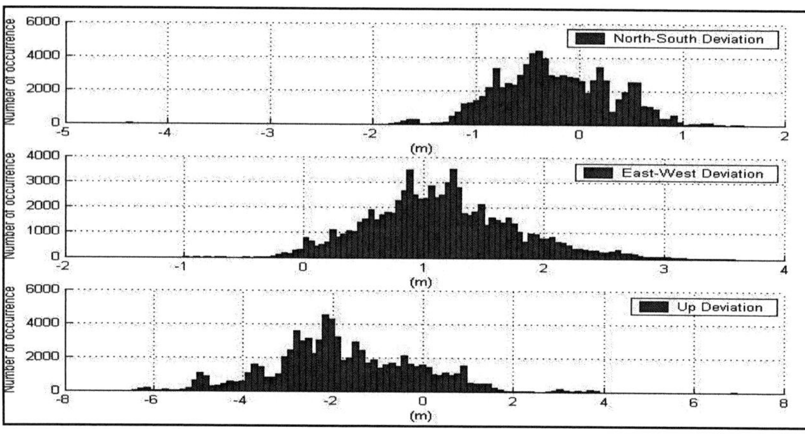

Figure 3: Noth Est Up deviation with GPS stand-alone.

In the following figures the solution obtained by GPS measures augmented by EGNOS correction is shown.

Figure 5 shows how the augmented solution is more accurate than the GPS stand alone one. From a qualitative analysis it is clear that in this case the error distribution is closer to the zero mean and from a quantitative analysis (see table 2) the precision is also improved.

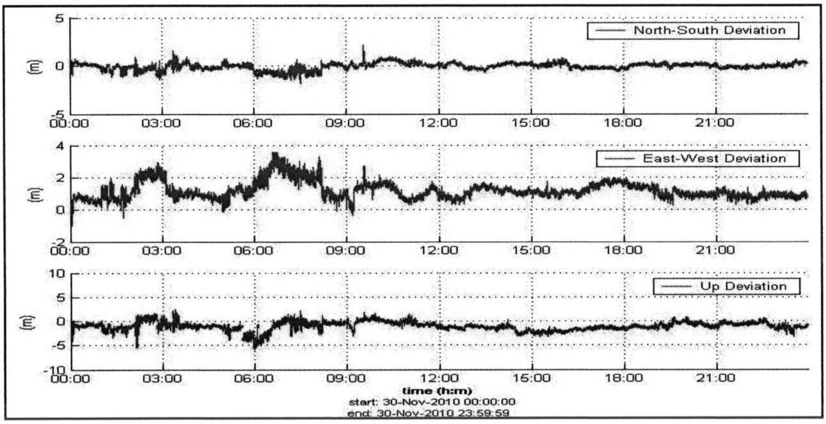

Figure 4: Deviation from the reference coordinate with SBAS.

The distribution of the error is:

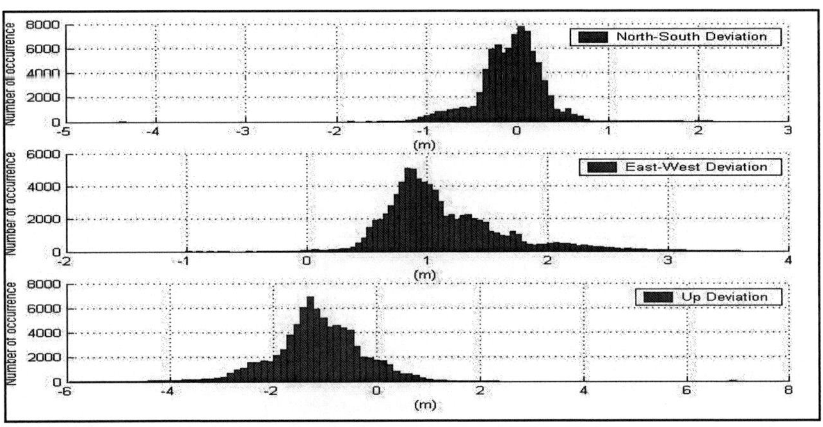

Figure 5: Noth Est Up deviation with SBAS.

The accuracy and precision of each result are summarized in Table 2.

Table 2: Error distributions standard deviation and variance.

Method	*Std_N/S [m]*	*Std_E/W [m]*	*Std_U/D [m]*	*Var_N/S [m]*	*Var_E/W [m]*	*Var_U/ D [m]*
GPS	1.0158	1.2894	2.2134	1.0319	1.6625	4.8994
SBAS	0.7226	0.7031	1.0949	0.5222	0.4944	1.1988

The most important improvement obtained by the application of the SBAS correction leads in the Height solution that is of great interest within the geodynamic study.

The good GPS stand-alone solution is due to the optimal measurement condition. Infect, as aforesaid, it was processed data from a static receiver in a good location (open sky). A next step of this research it will be the improvement assessment in worst condition (urban canyon, non static mode).

References

[1] IS-GPS-200 (2004) Navstar GPS Space Segment/Navigation User Interfaces, Revision D, ARINC Research Corporation, El Segundo, CA.

[2] Global Positioning System Standard Positioning Service Performance Standard (2008), US Department of Defense, 4th Edition, September 2008.

[3] ICD-GLONASS (2008) Global Navigation Satellite System GLONASS Interface Control Document, version 5.1, Moscow.

[4] Kaplan, E.D. (2006), Understanding GPS: Principles and Applications (second edition), Artech House Mobile Communications Series.

[5] Tunissen, P.J.G. Kleusberg A. (1998), GPS for Geodesy. Springer, Germany.

[6] Skaloud J. (1999) Optimizing Georeferencing of Airborne Survey System by INS/DGPS, PhD Thesis, University of Calgary, Alberta.

[7] Zhang Y. (2005) High performance differential global positioning system for long baseline application, PhD Thesis, Ohio University.

[8] Differential Global Positioning System (DGPS) for Flight Testing.

[9] Galileo Open Service Signal In Space Interface Control Document (OS SIS ICD). (2006)

[10] Cao W. (2009) Multi-frequency GPS and Galileo Kinematic Positioning with Partial Ambiguity Fixing, University of Calgary, Alberta.

[11] Zhang W. (2009) Triple Frequency Cascading Ambiguity Resolution for Modernized GPS and GALILEO, University of Calgary, Alberta.

[12] Mohinder S. Grewal, Global Positioning System, Inertial Navigation, and Integration. Second edition Wiley (2007).

[13] M.G. Petovello, G. Lachapelle, P.C. Fenton, (31st Annual Precise Time and Time Interval Meeting), an Overview of GPS Augmentation Systems.

[14] Angrisano, A., Gaglione, S., Pacifico, A. and Vultaggio M. (2009): Multi-Constellation System as Augmentation to GPS Performance in Difficult Environment or Critical Applications, Proceedings ENC-GNSS 09, Naples, Italy, 3-6 May 2009.

[15] Van Diggelen F., GPS and GPS+GLONASS RTK, ION-GPS, September 1997.

[16] Bancroft J. (2007) A Study of Accuracy Metrics (ENGO 585).

Longshore current velocities prediction: using a neural networks approach

T. M. Alaboud & M. S. El-Bisy
Civil Engineering Dept., College of Engineering and Islamic Architecture, Umm Al_Qura University, Makkah, K.S.A.

Abstract

Accurate current conditions prediction and supplement is an important task in the successful development and management of coastal zone infrastructure. Longshore current velocities forecasting is currently made by adopting conventional numerical and analytical models. In fact, longshore current velocities prediction in the conventional numerical and analytical models require a large amount information apart from historical wave observations and topography maps and are complex and tedious to apply specifically when point-forecasts at specific locations are needed. Therefore, this paper presents an application of a neural network for forecasting and supplementing the daily and monthly longshore current velocities. The neural network was trained using back propagation and cascade correlation algorithms. The data of five stations along the Damietta promontory on the Egyptian Nile delta coast were used to test the performance of the neural network model. The results indicated that the neural network can efficiently forecast longshore current velocities. Neural network forecasting was also found to be more accurate than traditional statistical time-series analysis.
Keywords: longshore current velocities, forecasting, neural networks, ARIMA model.

1 Introduction

Longshore currents play the main role in sediment transport along the coast. The longshore current combined with the agitating action of the breaking waves, are the primary factors in causing sand movement. Studies indicate that the greatest

doi:10.2495/CP110161

percentage of sand transport along a coast occurs shoreward of the breaker zone [1].

The strength of the littoral current in terms of the wave and beach characteristics for a relatively straight shoreline has been studied by several investigators, [2–6], and the subject is so important that many analytical solutions of the governing equations are published almost yearly. As analytical solutions are made more general, however, their evaluation becomes mathematically and computationally complex.

The alternative approach to analytical models of the longshore current is to solve the governing equations numerically. The numerical models are more computationally efficient than the complex analytical solutions. While simulations space wise information yielded by these models is advantageous, they require excess information apart from historical wave observations and topography maps and are complex and tedious to apply specifically when point-forecasts at specific locations are needed [1].

In a situation when a considerable amount of data about the studied processes is available, the use of traditional techniques may hardly be justified. In order to achieve optimal result it is important to use all available data. One of the options is to use regression – and auto-correlation – based statistical method like ARIMA models. However, with the recent advancements in artificial intelligence, data mining and soft computing, there is a choice of better techniques that may help to solve the problem. One of them is neural networks (NNs) which may be considered as a class of models which form a numerical modeling point view, are a general framework for representing non-linear mappings between multi-dimensional spaces in which the form of the mapping is governed by a number of adjustable parameters. It's by the modification of the adjustable parameters that the NNs model "learns" or identifies the mapping [7].

The present study aims to producing a reliable neural methodology applicable to the tasks of site-specific longshore current velocities estimations. The data base of Damietta promontory in Egypt was used to test the performance of the different models.

2 Time series forecasting models

2.1 ARIMA model

Introduced by Box and Jenkins [8], the ARIMA model has been one of the most popular approaches to forecasting. In an ARIMA model, the future value of a variable is supposed to be a linear combination of past values and past errors, expressed as follows:

$$\begin{aligned} y_t = \theta_0 + \phi_1 y_{t-1} + \phi_2 y_{t-2} + ... + \phi_p y_{t-p} \\ + \varepsilon_t - \theta_1 \varepsilon_{t-1} - \theta_2 \varepsilon_{t-2} - .... - \theta_q \varepsilon_{t-q} \end{aligned} \tag{1}$$

where y_t is the actual value and ε_t is the random error at time t, ϕ_i and θ_j are the coefficients, p and q are integers that are often referred to as autoregressive and moving average polynomials, respectively. Basically, this method has three phases: model identification, parameter estimation and diagnostic checking.

In the identification phase, data transformation is often needed to make the time series stationary. Stationary is a necessary condition in building an ARIMA model that is useful for forecasting. A stationary time series has the property that its statistical characteristics such as the mean and the autocorrelation structure are constant over time. When the observed time series presents trend and heteroscedasticity, differencing and power transformation are often applied to the data to remove the trend and stabilize the variance before an ARIMA model can be fitted. Once a tentative model is specified, estimation of the model parameters is straightforward. The parameters are estimated such that an overall measure of errors is minimized. This can be done with a nonlinear optimization procedure.

The last phase of model building is the diagnostic checking of model adequacy. This is basically to check if the model assumptions about the errors, ε_t, are satisfied. Several diagnostic statistics and plots of the residuals can be used to examine the goodness of fit of the tentatively entertained model to the historical data. If the model is not adequate, a new tentative model should be identified, which is again followed by the steps of parameter estimation and model verification. Diagnostic information may help suggest alternative model(s).

This three phase's model building process is typically repeated several times until a satisfactory model is finally selected. The final selected model can then be used for prediction purpose.

2.2 Neural networks model

NNs are flexible computing frameworks for modeling a broad range of nonlinear problems. One significant advantage of the NN models over other classes of nonlinear model is that NNs are universal approximators which can approximate a large class of functions with a high degree of accuracy. Their power comes from the parallel processing of the information from the data. No prior assumption of the model form is required in the model building process. Instead, the network model is largely determined by the characteristics of the data.

Single hidden layer feedforward (FF) network is the most widely used model form for time series modeling and forecasting [9]. The model is characterized by a network of three layers of simple processing units connected by acyclic links. The relationship between the output (y_t) and the inputs ($y_{t-1}, y_{t-2,\ldots\ldots,}\ y_{t-m}$) has the following mathematical representation:

$$y_t = \alpha_0 + \sum_{j=1}^{n} \alpha_j f\left(\sum_{i=1}^{m} \beta_{ij} y_{t-i} + \beta_{0j} \right) + \varepsilon_t \qquad (2)$$

where m is the number of input nodes, n is the number of hidden nodes, $\alpha_j (j{=}0,1,\ldots,n)$ and B_{ij} ($i{=}0,1,\ldots,m$; $j{=}\ 0,1,\ldots,\ n$) are the model parameters often called the connection weights. α_0 and B_{oj} are weights of arcs leading from the bias terms. The logistic function is often used as the hidden layer transfer function, that is,

$$g(x) = \frac{1}{1 + \exp(-x)} \tag{3}$$

Hence, the ANN model of eqn. (2) in fact performs a nonlinear functional mapping from the past observations ($y_{t-1}, y_{t-2,\ldots\ldots,} y_{t-m}$) to the future value y, i.e.,

$$y_t = f\left(y_{t-1}, y_{t-2}, \ldots, y_{t\text{-p}}, w\right) + \varepsilon_t \tag{4}$$

where w is a vector of all parameters and f is a function determined by the network structure and connection weights. Thus, the neural network is equivalent to a nonlinear autoregressive model. Note that eqn. (2) implies one output node in the output layer which is typically used for one-step-ahead forecasting.

Many alternative training processes are available, out of which the present study adopted two schemes, back propagation (BP) and cascade correlation (CC).

2.2.1 Back propagation algorithm

The goal of any training algorithm is to minimize the global (mean sum squared) error E, defined below:

$$E = \frac{1}{2}\sum \left(O_n - P_n\right)^2 \tag{5}$$

where O_n is the observations, and P_n is the predictions for any n output node. The summation has to be carried out over all output nodes for every training pattern. A pair of input and output values constitutes a training pattern.

The BP algorithm calculates the error using eqn. (5), and then it adjusts the weights, first in the output layer, and then distributes it backward from the output to hidden and input nodes. This is done using the steepest gradient descent principle where the change in weight is directed towards negative of the error gradient, i.e.

$$\Delta w_n = \alpha \Delta w_{n-1} - \eta \frac{\partial E}{\partial w} \tag{6}$$

where w is the weight between any two nodes; Δw_n, Δw_{n-1} are the changes in this weight at n and n-1 iteration, α the momentum factor, and η is the learning rate.

2.2.2 Cascade correlation algorithm

The cascade correlation (CC) is another network that modifies its own architecture as training progresses. CC network starts with a minimal topology, consisting only of the required input and output units. This net is trained until no further improvement is obtained.

Next, one hidden node is added to the net in a two-step process. During the first step, a candidate unit is connected to each of the input units, but is not connected to the output units. The weights on the connections from the input units to the candidate unit are adjusted to maximize the correlation between the candidate's output and the residual error at the output units. The residual error is the difference between the target and the computed output, multiplied by the derivative of the output unit's activation function, i.e., the quantity that would be propagated back from the output units in the back propagation algorithm. When this training is completed, the weights are frozen and the candidate unit becomes a hidden unit in the net.

The second step in which the new unit is added to the net now begins. The new hidden unit is then connected to the output units, and the weights on the connections being adjustable. Now all connections to the output units are trained. Training of this node is based on maximization of overall correlation S given by:

$$S = \sum_{o}\left[\sum_{p}\left(V_p - \bar{V}\right)\left(E_{p,o} - \bar{E}_o\right)\right] \tag{7}$$

where V_p is the output of the new hidden node for pattern p, $\bar{V}$ is the average of V_p over all patterns, $E_{p,o}$ is the network output error for output node o on pattern p training pattern, and $\bar{E}_o$ is the average of network error over all patterns.

The second step in which the new unit is added to the net now begins. The new hidden unit is then connected to the output units, and the weights on the connections being adjustable. Now all connections to the output units are trained. The addition of hidden nodes continuous until the desired learning is over.

Details of the concept of neural networks and the description of various training algorithms can be seen in References [10–12].

2.3 Performance evaluation

To examine how close of the predict to the real value of longshore current velocities, four indices, mean absolute percentage error *MAPE*, root mean square error *RMSE*, scatter index *SI*, and correlation coefficient *R*, were employed to evaluate the performance of neural networks and ARIMA model:

$$MAPE = \left[\frac{1}{N}\sum_{i=1}^{N}\left|\frac{P_i - O_i}{O_i}\right|\right] \times 100 \tag{8}$$

$$RMSE = \sqrt{\frac{1}{N}\sum_{i=1}^{N}(P_i - O_i)^2} \tag{9}$$

$$SI = \frac{RMSE}{\bar{O}} \times 100 \tag{10}$$

$$R = \frac{\sum_{i=1}^{N}\left(P_i - \bar{P}\right)\left(O_i - \bar{O}\right)}{\sqrt{\sum_{i=1}^{N}\left(P_i - \bar{P}\right)^2 \sum_{i=1}^{N}\left(O_i - \bar{O}\right)^2}} \tag{11}$$

where O_i is the observed value, P_i is the predicted value, N is the total number of data points in validation, $\bar{O}$ is the mean value of observations, and $\bar{P}$ is the mean value of predictions.

3 Data acquisition

There are twelve stations along the two promontories of the Egyptian Nile delta coast for measuring the longshore currents. Seven stations lie along both sides of Rosetta promontory and five stations along Damietta promontory. The longshore current measurements are taken two times per day. The longshore currents data was recorded in the surf zone, which is bounded by the shore and the breaker lines at depths ranging between 1.2 m to 1.5m. The measurements were taken by Coastal Research Institute, CRI over a period of 16 years ranging from 1982 to 1997. The observations from five stations along Damietta promontory were used in this study.

3.1 Pre-processing of data

NNs have been shown to be able to process data from a wide variety sources. They are however, only able to process the data in a certain format. Furthermore the way the data is presented to the network affects the learning of the network. Therefore, a certain amount of data processing is required before presenting the training patterns to the network. In this study a linear scaling was used.

One of the reasons for pre-processing the output data is that a sigmoidal transfer function is usually used within the network. Upper and lower limits of output from a sigmoid transfer function are generally 1 and 0 respectively. Scaling of the inputs to the range $[-1,+1]$ greatly improves the learning speed, as these values fall in the region of sigmoid transfer function where the output is most sensitive to variations of the input values. It is therefore recommended to normalize the input and output data before presenting them to the network. Scaling data can be linear or non-linear, depending on the distribution of the data. Most common functions are linear and logarithmic functions. A linear normalization function within the values of zero to one is:

$$S = (V - V_{\min})/(V_{\max} - V_{\min}) \tag{12}$$

where S is the normalised value of variable V, V_{min} and V_{max} are variable minimum and maximum values respectively.

4 Results and discussion

4.1 Monthly forecasts

4.1.1 NNs model

The determination of the optimal network architecture for a given task remains an open research question. Network size is determined by input number m, the number of hidden layers and the number of neurons in the hidden layers. If different network sizes have similar values for the error function, the smallest network size is optimum.

Many techniques are proposed to determine optimum neural networks such as the ad hoc approach, the dynamic approach, and the distribution approach. In this study trails were used to characterize the deterministic and random properties of the data to determine optimum neural networks.

The longshore current data of Damietta promontory pertaining to the first thirteen years (from 1982 to 1994) was used for training while the observations of the last three years (from 1995 to 1997) were reversed for testing the network. It was found that the minimum input number of dynamic variables needed to train the networks equal three preceding values. At the length of input sequence (m) equal 3, there is a maximum statistical independence between distance observations of data series, so, $m = 3$ maximize the network ability to learn from the data and should be the optimum number of nodes in the input layer. To verify the optimum input number, *RMSE* corresponding to various input number of nodes is calculated. It is found that neural networks corresponding to $m = 3$ give minimum *RMSE*.

To finally determine the optimum size of neural networks based on the determined input numbers, the networks with one hidden layer were used for training and test by changing neuron size in the hidden layer. It is found that the network with one layer and five neurons can make RMSE converge to lower values. In the case of the alternative CC scheme, hidden layers and neurons get automatically fixed during training as explained in the sec. 2.2.2.The neural network models were trained using backpropagation and cascade correlation and all computations were performed within the MATLAB version 6.1.0.450 release 12.1 (The Math Works Inc. May18, 2001). Figure 1 represents the topology of the optimum network trained using BP algorithm. In this figure $y_t, y_{t-1}, ..., y_{t-3}$denote longshore current velocity at time t, t-1, ..., t-3, respectively.

For the developed BP network mentioned earlier, the results tested instances for monthly longshore current velocities were shown in Figure 2. The predictions can be seen as fairly close to the corresponding actual measurements. All ups and downs in the observed time series may well be modeled well in the predicted series. It can be observed that a maximum absolute percentage error of 9.18%, a minimum absolute percentage error of 0.17% and the MAPE of 4.87% were obtained between the predicted and the observed series. The correlation coefficient of 0.957 was obtained for longshore current velocities prediction.

The longshore current velocities forecasting as above based on FF-BP was repeated by using the CC scheme of training. . This was done to see whether the training improves by adopting a different algorithm. It was found that a maximum absolute percentage error of 8.34%, a minimum absolute percentage error of 0.18% and the MAPE of 4.43% were obtained between the predicted and the observed series. The correlation coefficient of 0.965 was obtained for longshore current prediction. Results of the testing of this network indicated almost same performance of the PB training method.

www.witpress.com, ISSN 1743-3541 (on-line)

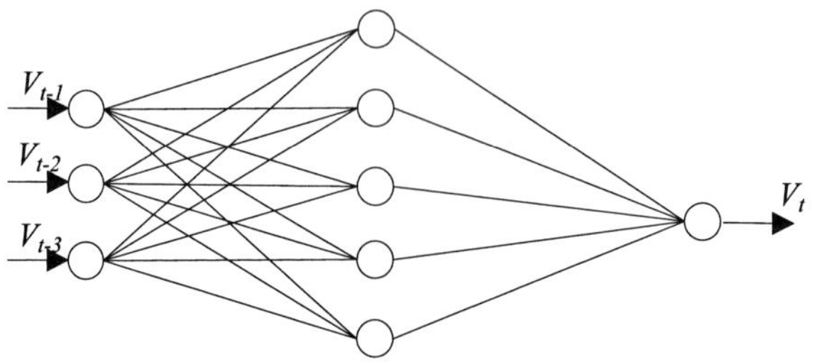

Figure 1: Structure of the artificial neural network.

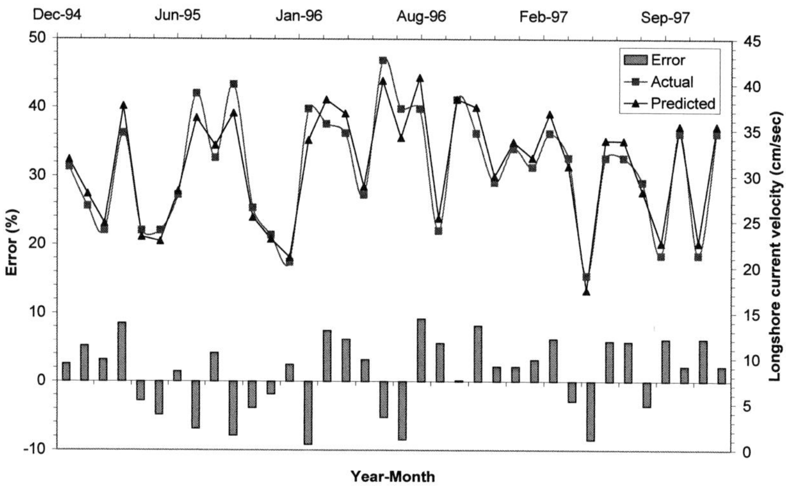

Figure 2: Time series plot of monthly longshore current velocities observed and predicted by FF-PB.

4.1.2 ARIMA model

The ARIMA model was implemented via SAS/ETS software version 6, a component of the SAS system (SAS Institute, Inc., 1992). This model was used to forecast longshore current velocities. Similar to the case of neural networks the model was calibrated using the first 13 years data and tested on the last 3 years observations. Considerable experimentation was made to arrive at the most satisfactory calibration. Figure 3 represent the time series plot of monthly longshore current velocities observed and predicted. It can be observed that a maximum absolute percentage error of 14.69%, a minimum absolute percentage error of 1.23% and the MAPE of 6.801% were obtained between the predicted and the observed series. The correlation coefficient of 0.931 was obtained for longshore current velocities prediction.

Table 1 compares the forecasting results of different models. Those results indicate that the neural networks model (PB and CC) outperforms the ARIMA model in terms of four indices, revealing that the ARIMA model cannot capture all of the patterns in the longshore current velocities. The neural networks model is, however, can significantly reduce the overall forecasting errors. Furthermore, the results indicate the proposed CC model is superior to PB model. The development of the longshore current velocities scheme in a model free manner seems to have resulted in more flexibility in the data mining approach of neural network than one associated with fixed type of stochastic time series schemes (ARIMA). In terms of RMSE, the percentage improvements of the CC scheme over ARIMA and FF-BB for monthly forecasting were 51.88% and 10%, respectively.

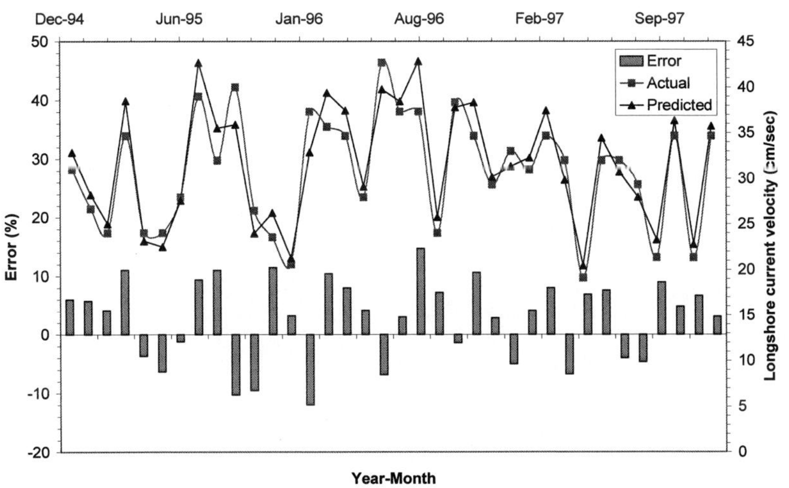

Figure 3: Time series plot of monthly longshore current velocities observed and predicted by ARIMA.

Table 1: Evaluating performance of models.

Forecast	Scheme	MAPE (%)	RMSE	SI (%)	R
Month	FF-BP	4.873	5.461	17.739	0.957
	FF-CC	4.43	4.964	16.126	0.965
	ARIMA	6.801	7.535	24.477	0.931
Daily	FF-BP	8.68	9.404	32.816	0.951
	FF-CC	7.548	8.178	28.536	0.962
	ARIMA	11.677	12.943	45.163	0.927

4.2 Daily forecasts

Here also, like in the case of monthly longshore current velocities forecasting FF network trained using BP and CC algorithms were developed using the first thirteen years data and tested using three years data (as well as ARIMA schemes for compression purpose). Figure 4 shows the performance of neural networks with respect to testing a set of longshore current data belonging to the last 3 years of measurement, while Figure 5 shows the same for the ARIMA model. Here also, like the previously described monthly predictions, the neural network

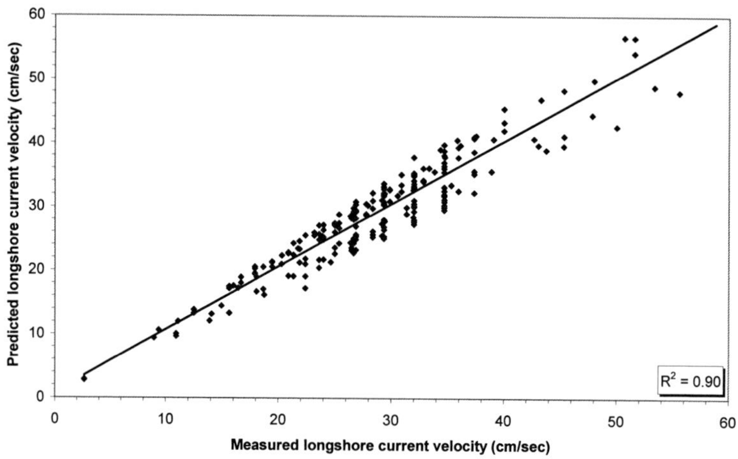

Figure 4: Scatter plot comparing observed and predicted daily longshore current velocity by FF-BP.

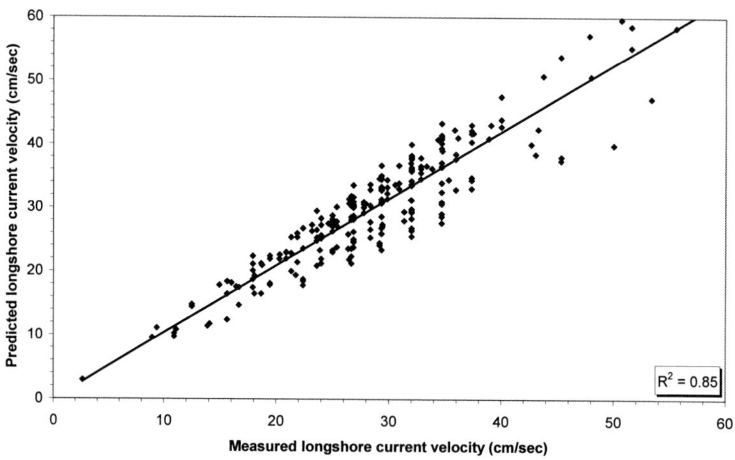

Figure 5: Scatter plot comparing observed and predicted daily longshore current velocity by ARIMA.

model can be seen to perform more satisfactorily than the ARIMA model. The CC scheme found that 58.27% and 15% decrease in RMSE over ARIMA and FF-BB, respectively.

Table 1 shows that forecasting accuracy decreased as the interval of forecasting reduced from one month to one day. As the noise in the time series increases, neural network performance generally gets worse. Similar to the monthly forecasting here also the neural network approach was more satisfactory than the traditional stochastic ARIMA model. Although both neural networks and ARIMA models perform worse as data becomes noisier, neural networks seem to be more robust as the differences between the two models are getting larger.

5 Conclusions

Longshore currents play the main role in sediment transport along the Egyptian Nile delta coast. The longshore current prediction in the conventional numerical models require a large amount information apart from historical wave observations and topography maps and are complex and tedious to apply specifically when point-forecasts at specific locations are needed. In this study, the neural mythology was implemented to forecast longshore current velocities. For this, the data were used, which were collected by Coastal Research Institute (CRI) for a period of sixteen years from 1982 to 1997 at Damietta promontory of Nile delta coast.

In this study two different NN training algorithms (PB and CC) and ARIMA were used in order to show the significance of suitable selection of the appropriate NN training algorithm.

A compression of results between feed forward neural network model (BP and CC) and the ARIMA model would show that the neural network perform much better than ARIMA. The four indices (MAPE, RMSE, SI, and R) were all the lowest for the neural network model. The results indicated that the ARIMA model does not capture the entire pattern of longshore current velocities. Among the PB sand CC networks employed for forecasting the longshore current velocities, the CC scheme was found to be superior to the PB scheme. The development of the longshore current velocities scheme in a model free manner seems to have resulted in more flexibility in the data mining approach of neural network than one associated with fixed type of stochastic time series schemes (ARIMA).

Forecasting accuracy decreased as the interval of forecasting reduced from one month to one day. This is expounded that the over fitting was made by large training patterns.

References

[1] El-Bisy, M.S., Longshore current prediction with neural networks. *Ain Shams Univ, Eng. Bulletim,* **40 (4)**, pp. 489-503, 2005.

[2] Thornton, E. B. & Guza, R.T., Surf zone longshore currents and random waves: field data and models. *Journal of Physical Oceanography*,**16**, pp.1165-1178,1986.

[3] Gourlay, M.R., Nonuniform alongshore currents. *Proc. 15th Coastal Engineering Conf.*, American Society of Civil Engineers, pp.701-720, 1976.

[4] Gourlay, M.R., *Nonuniform Alongshore Currents and Sediment Transport; A One-Dimensional Approach*. Research Report No. CE31, Department of Civil Engineering, University of Queensland, St. Lucia, Australia, 1982.

[5] Basco, D.R., *Surf Zone Currents. Miscellaneous*. Report No. 82-7 (I),US Army Engineer Waterways Experiment Station, Coastal Engineering Research Center, Vicksburg, MS, 1982.

[6] Basco, D.R. &Coleman, R.A., *Surf Zone Currents*. Miscellaneous Report No. 82-7 (II), US Army Engineer Waterways Experiment Station, Coastal Engineering Research Center, Vicksburg, MS, 1982.

[7] Deo, M.C. &Naidu, C.S., Real time wave forecasting using neural networks. *Ocean Engineering*, **26**, pp. 191–303, 1999.

[8] Box, G.E.P. & Jenkins, G.M., *Time Series Analysis Forecasting and Control*. Holden Day, San Francisco, CA, 1976.

[9] Zhang, G.P. & Qi, M., Neural network forecasting for seasonal and trend time series. *European J. of Operational Research*, **160**,pp. 501-514, 2005.

[10] Hoehfeld, M. & Fahlman, S. E., Learning with limited numerical precision using the cascade correlation learning algorithm. *IEEE Transactions on Neural Networks*, **3(4)**, pp. 602-611,1992.

[11] Adams, A. &Waugh, S., Function evaluation and the cascade correlation architecture. *ICNN*, Perth, Western Australia, pp. 1126- 1129, 1995.

[12] Wu, JK, *Neural Networks and Simulation Methods*. Marcel Dekker, New York, 1994.

Section 6
Sediment transport and erosion

Temporal and spatial variability of bottom sedimentation for survey periodicity

P. C. Chu & N. S. Wheatley
Department of Oceanography, Naval Postgraduate School, Monterey, California, USA

Abstract

Bedforms determine survey periodicity. Sediment grain size, tides, currents, and wind-generated waves are influential in bedform formation. To investigate if sediment properties change over time, localized grab samples for a three-year period in San Francisco Bay were analyzed. The analysis showed little variability in sediment characteristics at a given location. A weighted suitability model based on the United Kingdom Hydrographic Office (UKHO) model has been constructed. Three layers were developed including sediment grain size, interpolated from 174 grab samples, tidal and current data from over 50 current stations and ripple height inferred from wind generated wave height. A weighting for each layer was determined. Regions indicating the presence of bedforms were assigned a low survey periodicity; as bedforms reduced, survey periodicity was increased.
Keywords: survey periodicity, sediment, bedforms, San Francisco Bay, suitability model.

1 Introduction

To retain maritime security, an up-to-date database of route surveys for mine or maritime improvised explosive device (IED) countermeasures are essential [1, 2]. Bedforms are an integral part of the survey periodicity problem. Sediment grain size, tides, currents, and wind-generated waves are influential in bedform formation.

San Francisco Bay (Fig. 1) is a large, shallow, dynamic estuary located in California on the west coast of the U.S. It is a major international shipping port, with large container facilities, which makes it a significant, economically

WIT Transactions on Ecology and the Environment, Vol 149, © 2011 WIT Press
www.witpress.com, ISSN 1743-3541 (on-line)
doi:10.2495/CP110171

important port. It is an extremely busy waterway used by both commercial and recreational vessels. Approximately 40% of water drainage from the central coast rivers enters the Pacific Ocean through the Golden Gate channel. This represents a high freshwater discharge rate approximate 800 m^3/s, and has the potential to carry a significant amount of sediment into the area. The San Francisco Bay area is subject to a complex semi-diurnal tidal regime, this leads to temporally and spatially variable currents that can exceed 2.5 m/s. This leads to a diverse and complex pattern of bedform formations, which were first mapped using side-scan sonar in the late 1970's, and are now mapped using high resolution multi-beam surveys [3].

In this study, the Golden Gate region is investigated in detail (Fig. 1). A comparison study of localized sediment grab data in the same positions for a three year period is assessed and analyzed. This data is then compared to the Naval Oceanographic Office (NAVOCEANO) High Frequency Environmental Acoustics (HFEVA) sediment database, and an assessment of the validity of this database is made. Multi-beam data, obtained by the USGS is examined and the impact of these findings on the mine warfare route survey periodicity assessed.

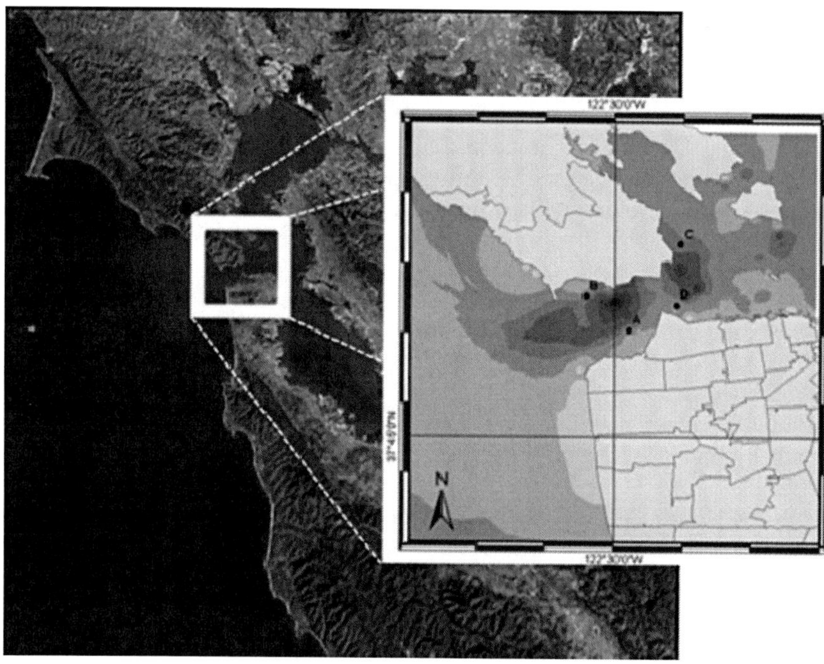

Figure 1: San Francisco Bay and sample locations for study.

2 Sediment

In the winter of 2007, 2008, 2009, sediment samples were collected in the vicinity of the Golden Gate region of San Francisco Bay, onboard the R/V Point

Sur in four locations (A, B, C, D) (Fig. 1). The samples were all collected using a double trap Van Veen sediment grab (Fig. 2), deployed off the stern of the ship using a crane. The Van Veen grab is a light weight stainless steel sampler designed to take samples of soft bottom sediment. Water is able to flow through the grab as it is lowered. When it hits the seabed, the doors of the grab close due to tension on the cable, they remain closed while the grab is raised and recovered on deck. Upon recovery of the grab a representative sample of the sediment was then collected in a quart mason jar. The jar was then sealed, labeled and stored, for laboratory processing [4].

Figure 2: Van Veen grab on board R/V point Sur.

The sediment sample analysis was conducted in the oceanographic laboratory at the Naval Postgraduate School. Laboratory analysis can be broken down into phases. The first phase involved emptying the contents of each jar into a standard plastic Rubbermaid basin; the sample was rinsed with fresh water while being agitated. The sample was then left to settle – the time this took depended on the consistency of the sample, with silty samples taking much longer. The samples were generally left overnight; this allowed all the sediment to return to the bottom, leaving clear water on top. Following the settling period, any particulates or biologic material floating on the water was removed. The fresh water was then decanted out, being careful not to pour out any sediment. If necessary, this process was repeated.

The rinsed sediment was then transferred into a pre-weighed 8×8 inch, Pyrex casserole dish. Sediment was transferred by pouring, scraping using a spoon, and rinsing by squeezing a fine stream of water into the bowl. Once transferred, the sample was placed in the laboratory oven overnight to dry. The oven was set at approximately 90° C. Once the sample was completely dry, it was weighed and prepared for the sieving process. The dried sample was broken up, in some cases this could be achieved by using a spoon. However it was necessary to use a hammer to break up some of the more difficult samples. These tended to be the finer samples that had become like baked clay. The broken up sample was then place in a pre-weighed plastic bag. The bagged sample was weighed and the result recorded.

The next phase, the sieving phase was achieved by using a Ro-Tap automated sieve. A 100 ml glass beaker was weighed, a quantity of the sample was added to the beaker and it was re-weighed, both weights were recorded. This was the part of the sample to be analyzed. The Ro-Tap sieve used in this experiment utilized 14 sieves ranging from 2.00 mm to 0.070 mm in mesh diameter. The sample was poured into the top sieve (2.00 mm), and then sieved through the column of sieves for 15 minutes. The sample collected in each sieve was carefully collected, by pouring it onto a sheet of card and removing any residue from the sieve with a wire brush, it was then transferred into a pre-weighed plastic bag. The bag and sample was then weighed and the results recorded. A loss of less than 1% of the sediment weight had to be achieved if the result was to be deemed accurate.

The sediment grain size varies drastically from smaller than 0.0002 mm to larger than 256 mm. The Udden-Wentworth scale,

$$\varphi = -\log_2(X) \tag{1}$$

is used with $\varphi = -8$ corresponding to $X = 256$ mm and $\varphi = 12$ corresponding to X = 1/4096 mm. The sediment can be classified by calculating the percentage of each sample within each range, then calculating the mean grain size for that sample and converting this to φ units. Usually, the sediment size distribution, $f_i = f(X_i)$, must first be calculated. From this, the mean grain size, standard deviation, and skewness

$$\overline{\varphi} = \sum_i f_i \varphi_i, \quad \sigma = \sqrt{\sum_i f_i (\varphi_i - \overline{\varphi})^2}, \quad \alpha = \sum_i \frac{(\varphi_i - \overline{\varphi})^3}{\sigma^3}, \tag{2}$$

can be calculated. Table 1 shows the $\overline{\varphi}$-values of the sediment samples collected in 4 locations (A, B, C, and D) marked in Fig. 1 during three cruises in 2007, 2008, and 2009. Fig. 3 shows the breakdown of percentage sample mass for the location-A each year.

Table 1: Sediment classification based on $\overline{\varphi}$ values for positions A–D.

	2007	2008	2009
A	2.48 Fine Sand	2.52 Fine Sand	2.51 Fine Sand
B	2.31 Fine Sand	2.29 Fine Sand	2.11 Fine Sand
C	1.47 Medium Sand	1.75 Medium Sand	1.90 Medium Sand
D	2.25 Fine Sand	2.17 Fine Sand	2.20 Fine Sand

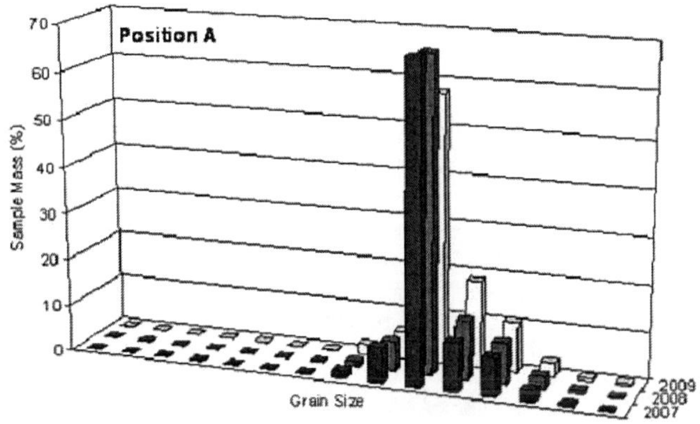

Figure 3: Variation of sediment size distribution in three years: 2007 (dark black), 2008 (light black), and 2009 (white) at the location-A from largest grain size (left) to smallest grain size (right).

3 Bedforms

When sediment begins to move bedforms start to form. A flat bottom can become deformed, with a series of undulations. As water flow increases, drag will be increased, and this increases in the shear stress available at the bed to create grain movement [5]. In laboratory investigations, the sequence of bedforms with increasing flow intensity is: Flat bed, Ripples, Dunes, High Stage Plane Bed, followed by Antidunes. If the average current velocity, water depth and sediment size are known factors, then the expected bedforms can be predicted.

Table 2 shows the estimated ripple heights from waves and characteristics for positions A to D. The wave conditions were obtained from marine gridded climatology data provided by Fleet Numerical Meteorology and Oceanography (METOC) Detachment in Ashville. Values were calculated by re-analysis of data from 1857 to 1997. Location-A results show all the ripples classed as orbital, the ripple height varies from 0.3 cm to 0.4 cm. This indicates a limited amount of variability at position A over the time period. Location-B results also show the ripples are classified as orbital in all cases. The ripple heights vary from 2.5 cm to 3.1 cm. Although this position has more variability it remains at less than 1 cm, so cannot be deemed significant. Location-C results, again, classify the ripples as orbital, while the ripple height varies from 1.9 cm to 3.1 cm. Although the variability is slightly larger than the other two positions the range of ripple heights remains relatively small and inconsequential. Location-D shows the largest variability, all ripples remain orbital, but heights range from 1.4 cm to 3.4 cm. The range of phi values is from 2.17 to 2.25, which is not a large range, however the depth at which the grab samples were obtained is more variable for this position, which could explain the variability in results. The difference of 2 cm ripple height over a three year period is not large enough to be a significant problem.

The ripple heights for each position show a degree of variability, although not on a large scale. The variation at each location is in the order of a centimeter, the estimated ripple heights from waves are all relatively small and would be inconsequential for mine burial at these positions. However, this does not take into account the currents in this region. Although the ripple height is assessed as too small to bury a mine it still remains an important issue in the mine warfare survey periodicity problem. Smaller ripples in the order of centimeters can cause a significant problem in mine detection due to scattering of acoustic rays [4].

Table 2: Ripple characteristics at locations A–D.

		2007	2008	2009	Mean	SD
A	φ	2.48	2.52	2.51	2.50	0.02
	Depth (m)	60	63	63	62	1.73
	Orbital Diameter (mm)	0.019	0.015	0.015	0.016	0.0023
	Orbital Velocity (m/s)	0.017	0.013	0.013	0.014	0.0023
	Ripple Height (cm)	0.4	0.3	0.3	0.33	0.0577
	Ripple Classification	Orbital	Orbital	Orbital		
B	φ	2.31	2.29	2.11	2.23	0.11
	Depth (m)	37	35	38	36.6	1.52
	Orbital Diameter (mm)	0.128	0.151	0.118	0.132	0.0169
	Orbital Velocity (m/s)	0.115	0.135	0.106	0.119	0.0148
	Ripple Height (cm)	2.7	3.1	2.5	2.77	0.3055
	Ripple Classification	Orbital	Orbital	Orbital		
C	φ	1.47	1.75	1.90	1.71	0.21
	Depth (m)	38	41	35	38	3.00
	Orbital Diameter (mm)	0.118	0.093	0.151	0.120	0.0291
	Orbital Velocity (m/s)	0.106	0.083	0.135	0.108	0.0261
	Ripple Height (cm)	2.5	1.9	3.1	2.5	0.6
	Ripple Classification	Orbital	Orbital	Orbital		
D	φ	2.25	2.17	2.20	2.21	0.04
	Depth (m)	40	45	34	39.67	5.508
	Orbital Diameter (mm)	0.101	0.067	0.164	0.116	0.0492
	Orbital Velocity (m/s)	0.090	0.060	0.147	0.099	0.0442
	Ripple Height (cm)	2.1	1.4	3.4	2.3	1.01
	Ripple Classification	Orbital	Orbital	Orbital		

4 Comparison to existing database

The Maritime environment is an extremely important factor in determining the route survey periodicity. The United Kingdom Hydrographic Office (UKHO) model included seabed sediment types, sediment deposition, bottom texture, gas presence and vessel traffic. The data was obtained from many different sources, bottom texture and bottom contacts data was taken from the UKHO Route Survey Database (RSDB) and processed in Microsoft Excel, allowing it to be imported easily into ArcGIS. The British Geological Survey (BGS) supplied the seabed sediment type data in a digital map. In order to determine total suspended matter, satellite data from the NASA MODIS satellites were used. The density of fishing vessels was obtained from the Centre for Environment, Fisheries and Aquaculture Sciences (CEFAS). Gas presence was taken from the UKHO Geological Database (GEODB) and vessel traffic was supplied by NAVOCEANO. Table 3 shows our 2009 data to the NAVOCEANO HFEVA database.

Table 3: 2009 sediments samples compared to NAVOCEANO database data.

	Phi	Wentworth Sediment Classification	HFEVA Database
A	2.51	Fine Sand	Fine Sand
B	2.11	Fine Sand	Fine Sand
C	1.90	Medium Sand	Medium Sand
D	2.20	Fine Sand	Fine Sand

5 Accuracy and errors

There are issues involving the accuracy and errors associated with this investigation. Although, during the collection and laboratory processing, as much care as possible was taken to limit or eliminate errors. During the collection phase, the bridge of the R/V Point Sur was given the positions of previously collected samples, the ship aimed to stay in station at these locations as accurately as possible during the deployment and retrieval of the grab. However, from comparing the positions over the three years, it can be noted that although the positions are extremely similar, they are not exactly the same. This is reflected in the depths used in calculating ripple height and is the main reason for the variation in the ripple height.

In order to gain a better representation of sediment type, it would be preferable to take a selection of samples at each position, so that the average result could be used, rather than relying on one sample. This would allow any erroneous sediment samples to be excluded, or have a minimal effect on the results used for comparison. The laboratory procedure for sediment analysis was carried out in such a way to minimize error. In order to be deemed a valid result less than 1% sample loss could occur during the sieving process. There were problems that occurred that could introduce error. Finer samples proved problematic after the baking phase. The aim was to break up these samples as much as was possible, however, this proved difficult at times and could have caused a skew in results indicating a sample was coarser than it actually was. Every care was taken to avoid this.

During the sieving phase, care had to be taken to ensure that all of the sediment samples were removed from each sieve – at times this could be difficult and was achieved by using a wire brush or a sharp pencil to poke any remaining sediment grains from the sieve. The sieves available for the Ro-Tap sieve ranged from 2.00 mm to 0.070 mm. This limited the sediment classification range, from fine gravel to very fine sand, in the case of these sediment samples this range appeared adequate.

6 Determination of survey periodicity

To determine the survey periodicity for San Francisco Bay, a weighted suitability GIS model, utilizing a similar methodology to the UKHO model [6], was developed. The model is established based on the fact that waves, tides, currents and sediment size affect bedform formation and sediment processes; this in turn will affect the survey periodicity requirement. The concept of the weighted suitability model used here is summarized in Fig. 4. It utilizes three main input layers; predicted bedform type (green), predicted bottom current (blue) and predicted wave generated ripple height (red). Each of these layers can be thought of as a sub-model, similar to those used in the UKHO model.

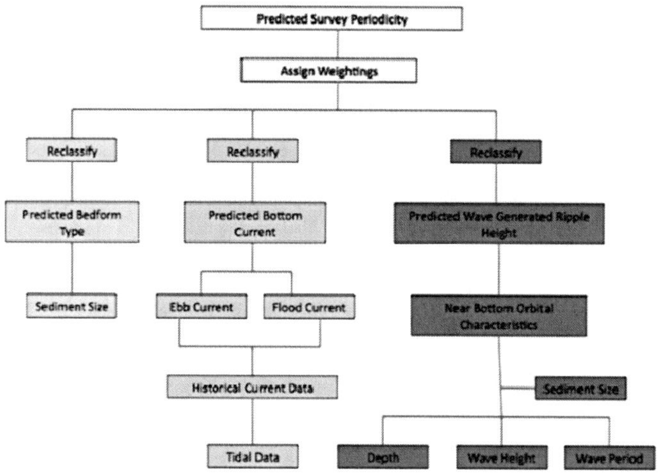

Figure 4: Flow chart showing the three layers used to predict survey periodicity.

In order to predict the bedform type, 174 grab samples were obtained from the USGS. The grab samples were taken during surveys dated between 2004 and 2008. In addition the grab samples detailed in Section 2 were also included. The data included latitude, longitude, depth, and sediment grain size. The dataset was compiled in excel and entered into the GIS software program ArcMap (Fig. 5). Tidal data, from NAVOCEANO predictions is examined and the variability in this region (not shown). Historical current data, provided by NAVOCEANO is analyzed. Using linear wave theory and climatological data, the estimated wave generated ripple heights are calculated. Sediment data

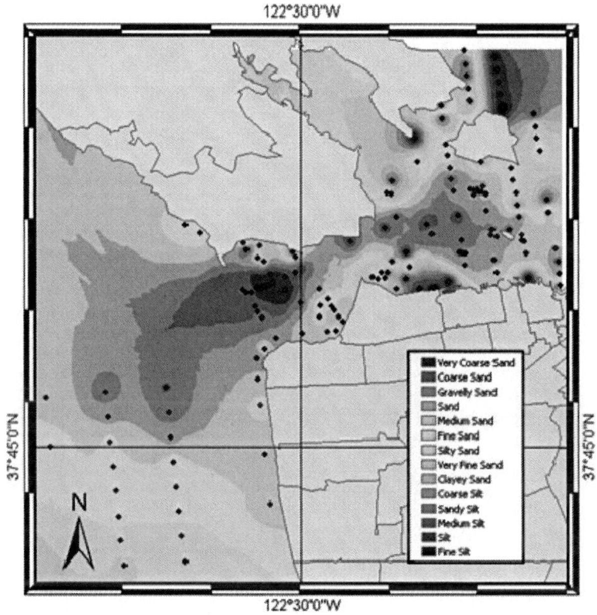

Figure 5: Sediment type constructed from grab samples in San Francisco Bay.

obtained from grab samples provided by USGS is utilized. This data is weighted and combined, and a model for survey periodicity is obtained. Using linear wave theory, the depth, mean wave height and mean wave period, the near bottom orbital velocity and near bottom orbital diameter were calculated. Applying the Wiberg and Harris model, and including the sediment grain size data the predicted ripple height was calculated. These data were then combined in order to obtain wave generated ripple height layer. Fig. 6 shows January ripple heights. Smaller ripple heights are shown in pink, with the greatest ripple heights shown in orange and brown, as according to the color scale shown. It can be seen that the greatest ripple heights occurred seaward of the San Francisco Bay region in January, this coincided with the larger wave heights [4].

Due to the importance of flood and ebb dominated currents in bedform formation, further interpolation of this data was conducted. The currents were interpolated into a raster dataset and separated into ebb dominated and flood dominated regions. Bottom ebb currents were assigned a negative value, and flood currents were assigned a positive value and the residual differences between the two calculated. Ebb dominated regions are indicated in red and flood dominated regions are shown in green. In order to determine the survey periodicity, the weighted option layers were classified. With 45% weighting for the sediment layer, from the background theory it was extremely apparent that sediment grain size was particularly important in sediment transport mechanisms and in bedform formation mechanisms. Currents were weighted at 35%, this demonstrates the importance of currents, in this case due to a particularly strong

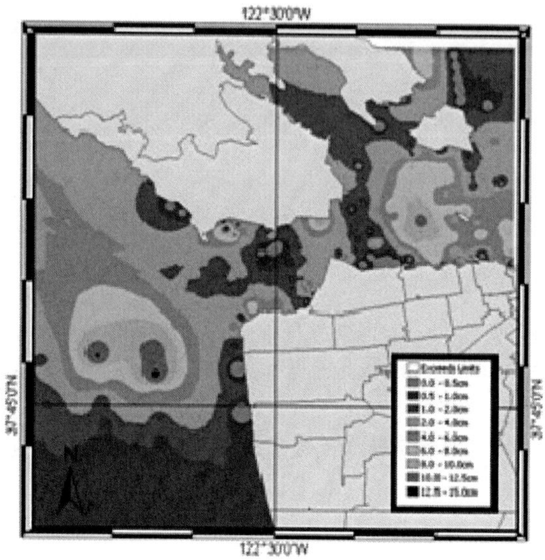

Figure 6: Wave generated mean ripple heights (cm) in January in San Francisco Bay.

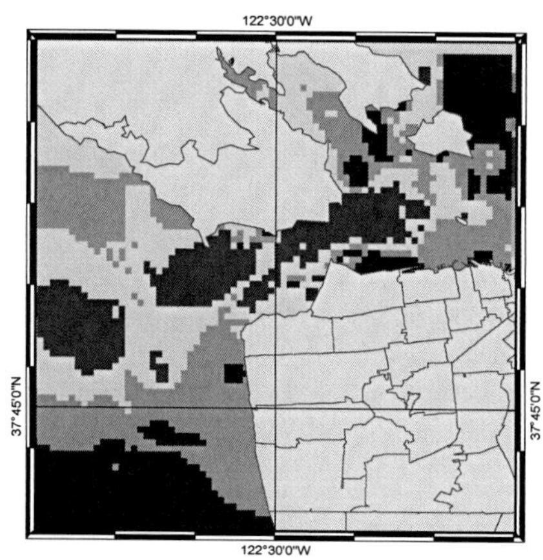

Figure 7: Recommended survey periodicity for San Francisco Bay. The yellow regions should be re-surveyed every 5–7 years, the light blue regions every 7–10 years and the dark blue every 10–15 years.

tidal regime, the importance of currents was also apparent from the background theory of sediment transport [7,8]. Waves had a weighting of 20%, the lower weighting was due to the smaller magnitude of ripple heights due to wave motion.

Fig. 7 shows the model results for the survey periodicity. The red colored regions (priority-1) occupying throughout the Golden Gate Channel and the Alkatraz Shoal, have highest seabed changeability, those that should be surveyed most often (every 3–5 years). This is due to its significant economic and commercial importance. The yellow regions should be re-surveyed every 5–7 years, the light blue regions every 7–10 years and the dark blue every 10–15 years.

7 Conclusions

The route survey periodicity model developed for San Francisco Bay. The sediment size, tides and currents and ripples generated from wind waves are used as input for the survey periodicity model since they are critical in bedform formation and sediment size.

This model is comprised of three input layers: bedform type, bottom current, and wave generated ripple height. Each of these layers can be thought of as a sub-model. Each layer was weighted, the weighting scheme was used, and each layer was re-classified with a scale of 0 to 9, with 0 representing a high degree of change, and 9 representing little change. As these layers had not been used before, the weighting schemes used were based primarily on background theoretical concepts. The bedform type layer weighting of 45%, in all background theory literature sediment size was shown to be the most important factor in bedform type and hence size. The predicted bottom current layer had a weighting of 35%, indicating that currents, in this case due to the tidal regime had a greater importance than waves, which were given a weighting of 20%. A lower weighting was given to waves due to the fact that the ripple heights capable of being generated were much smaller than those generated by currents. Near the mouth of the San Francisco Bay, the seabed changeability is high, a survey interval of 3–5 years is suggested.

References

[1] Chu, P. C., 2009: Mine impact burial prediction from one to three dimensions. *Applied Mechanics Reviews*, American Society of Mechanical Engineers, **62**, 010802.

[2] Chu, P.C., and C.W. Fan, 2007: Mine impact burial model (IMPACT35) verification and improvement using sediment bearing factor method. *IEEE Journal of Oceanic Engineering*, **32** (1), 34-48.

[3] Barnard, P. L *et al.* (2007). Coastal Processes Study at Ocean Beach, San Francisco, California; Summary of Data Collection 2004-2006. US Geological Survey Open File Report 2007-1217. http://pubs.usgs.gov/of/2007/1217/.

[4] Wheatley, N. S. (2009). Determining Route Survey Periodicity for Mine Warfare: Investigation of Bedforms, Waves, Tides, and Currents. *Master Thesis in Physical Oceanography*, pp. 105, Naval Postgraduate School, Monterey, California, USA.

[5] Dyre, K. R. (1986). Coastal and Estuarine Sediment Dynamics, John Wiley & Sons, Chichester.

[6] Armishaw, J. E. (2005). Route Resurvey Model (QRM): Use of GIS Modelling to Review the UK Route Survey Maintenance Schedule. UKHO/MEIC TR/05/01, June 2005. Taunton, Somerset.

[7] Chu, P. C., 2008: Probability distribution function of the upper equatorial Pacific current speeds. *Geophysical Research Letters*, American Geophysical Union, **35**, doi:10.1029/2008GL033669.

[8] Chu, P. C., 2009: Statistical Characteristics of the Global Surface Current Speeds Obtained from Satellite Altimeter and Scatterometer Data. *IEEE Journal of Selected Topics in Earth Observations and Remote Sensing*, **2** (1), 27-32

The hydrodynamic response of beach nourishment in the Cancun Beach, Quintana Roo, México

J. C. Sanchez[1], J. C. Espinal[2] & D. Moran[3]
[1]*Subgerente de Estudios Hidrográficos, GEIC, CFE, México*
[2]*Jefe de Departamento de Modelos Matemáticos, GEIC, CFE, México*
[3]*Departamento de Oceanografía, GEIC, CFE, México*

Abstract

The beach nourishment project implemented in 2009 to rehabilitate the beachface and profiles of Cancun beach had been the most important project of its type in Latin America: the length of rehabilitation (10.5 kilometers), the volume of sand placed ($5.2x10^6$ m^3), besides to have a dry beach in approximately one kilometer on an area where for a long time there was not a dry beach. The environmental conditions were fully fulfilled on the deposition area and also on the sand borrow areas.

The nourishment project was based on the analysis beach profiles data set, interpretation of data collected by means of oceanographic equipment, native and borrow sediment, further to the hydrodynamic and morphodynamic characterization.

This paper presents the analysis of the hydrodynamic response of Cancun Beach nourishment in Quintana Roo, México, supported on historical profiles data set analysis, seabed changes after the nourishment project and coastline evolution, in order to predict the possible changes in the morphodynamic conditions, after nourishment.

Keywords: beach nourishment, Cancun, hydrodynamic, beach profile.

1 Introduction

For different reasons most of the beaches in Mexico are experiencing erosion processes, depending on the particular conditions of each site, this problem will

WIT Transactions on Ecology and the Environment, Vol 149, © 2011 WIT Press
www.witpress.com, ISSN 1743-3541 (on-line)
doi:10.2495/CP110181

be presented for the special case of Cancun, the well known resort area of Mexico.

Cancun beach is important in Mexico environmentally, socially and economically. It is located on the Caribbean Sea and its beach nourishment project has to take into consideration the effect of yearly occurrence of tropical cyclones besides the anthropogenic activities such as the construction of touristic infrastructure on the dunes.

Cancun is a littoral barrier beach with a 12 km beach front. The origin of its native sediment is biogenic with 0.32 mm mean grain size. Historically, several hurricanes have stroke this beach accelerating the erosion process immensely, such as Hurricane Gilbert (1988) and Wilma (2005).

Previous studies suggest that in 1988 Cancun beach was an equilibrium and stable system with an estimated volume of 11 million cubic meters and a dry beach width between 50-60 m. After Hurricane Gilbert (1988), 5.9 million cubic meters were eroded and the beach showed a dry beach width between 35-40 m. Before Hurricane Wilma (2005) there was a 5 million cubic meters volume of sand along the 12 km beachface, with a dry beach width between 25-30 m, and after Wilma remained only 400 000 cubic meters of sand approximately, with a dry beach width between 5-7 m. so 4.6 million cubic meters of sand were completely swept away from the project area. In 2006 the Federal Government placed an emergency sand volume of 2.7 cubic meters in order to rehabilitate partially the main attraction of Cancun: its beautiful and white beach sand. The dry beach width varied between 35 to 40 m this year 2006. Between 2006 and 2009 the combined effects of hurricane waves and the anthropogenic activities continued with the erosion process. The beach front showed a dry beach width between 15-20 m. which was not attractive enough for tourism. Then the hotel owners asked the President of Mexico to help recover Cancun main attraction through the investment in the final project and the complete rehabilitation of the beach, with the placement of 5.2 million cubic meters along a 10.5 km beach front.

In 2009, the most important beach nourishment project in Mexico was carried out to recover the original beach conditions (1988), forming a 10.5 km headland beach and decreasing the sediment loss. The basis of the project is to force the hydrodynamic conditions to have profiles with gentle slope and try to maintain the sediment placed into the system.

The final project started in August 2008 and it was finished in March 2009, before the bidding process. The dredging operations started in November 2009 and were finished in February 2010. All the environmental conditions were completely fulfilled on the placement area and on the sand borrow areas.

2 Study area

Cancun beach is located in the state of Quintana Roo, México. It is a littoral barrier beach with 12 kilometer beachface, semi confined to the North by Punta Cancun and to the South by Punta Nizuc. Limited by the coordinates

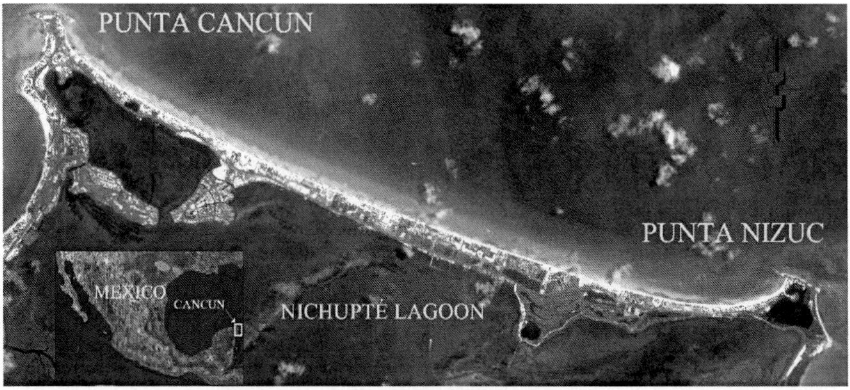

Figure 1: Cancun Beach location.

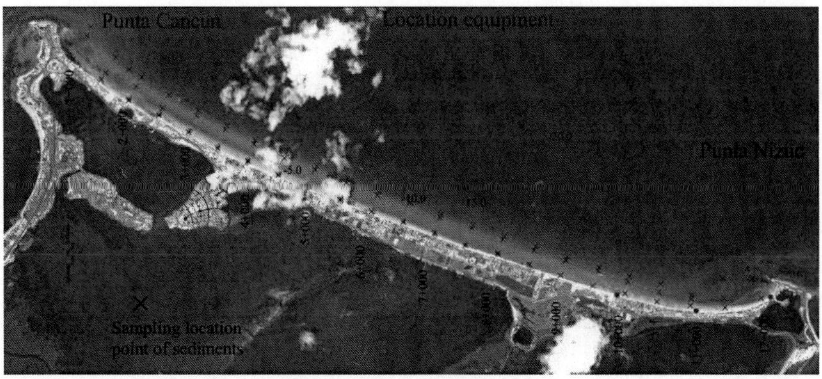

Figure 2: Measurements.

21°,07′,08″N, 86°,44′,25″W, to the North, and 21°,02′,07N, 86°,46′,38″W, to the South, to the West by the Nichupté Lagoon System, as shown in Figure 1.

Site characterization was performed on previous studies and by means of collected data from measurements by the most important parameters such as topobathymetric surveys, native sediment samples, wave current measurements, and sediment samples taken on the borrow areas.

Figure 2 indicates the location equipment for wave measurements, the profile distribution along the beach and the points of sediment samples.

For a detailed characterization study the beach was divided in cross shore sections every 100 meters, starting with the 0+600 km section on Punta Cancun and ending on Punta Nizuc with the 12+000 km section. Topobathymetric surveys covered a 13 kilometers beachfront and reached the 20 meter depth contour.

Its native beach sediment is mainly of biogenic origin, with mean grain size between 0.22 and 0.55 mm [1]. The sediment sampling was made every 500 meters, with a starting point on 1+000 km section and final point in

12+000 km section. In order to cover the whole profile samples were taken on bathymetric curve levels +1.5, 0.0, -1.0, -5.0 y -7.5 meters. All of the sediment samples were analysed in order to obtain the associated granulometric curve for each one, resulting that the lowest mean grain size values were in the northern reach with sand diameters between 0.22 and 0.3 mm. In the Central zone mean grain size diameter varies between 0.4 and 0.65 mm, whereas on the southern reach the mean grain size varies from 0.4 to 0.78 mm. So our project area shows a poorly sorted sample.

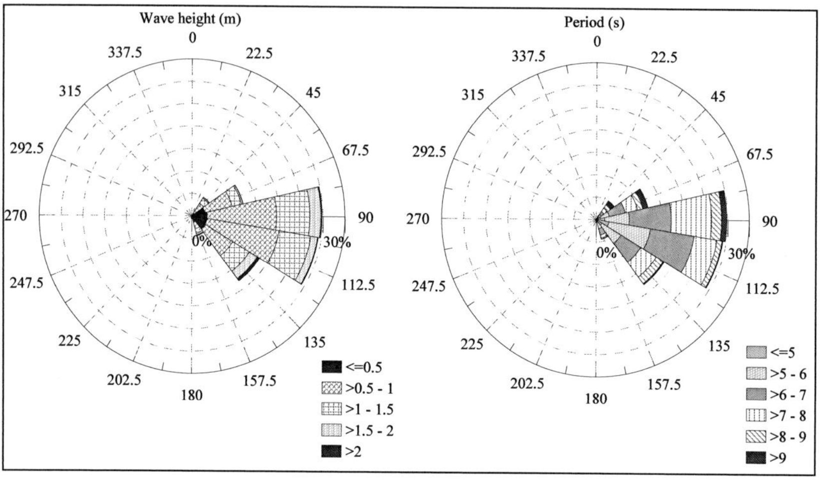

Figure 3: Wave height rose.

The wave climate was determined by means of a wave measurement instrument deployed at 23 m depth in front of the beachface central point, the measurement period started in September 2008 and finished in October 2010. Figure 3 shows the wave rose, observing that during the measurement period the dominant direction comes from the range 78 to 123° (E, SE). The wave heights vary from 0.4 to 2.5 m, meanwhile the mean range wave height is 0.5 to 1.0 meters, and finally the associated periods vary between 4 to 11 seconds.

3 Cancun Beach evolution

According to the morphodynamic state proposed by Wright and Short [2], before Hurricane Gilbert, Cancun beach could be defined as an intermediate beach tending towards the dissipative state. However, over time and due to the constant occurrence of tropical cyclones that every year strike the region, the beaches of Cancun had been evolving toward intermediate and reflective states, depending on the beach face zone. It has been observed that these changes of state are linked with erosion processes, resulting from the transport of the sediment towards the lower profile, forming longshore bars after the hurricane season.

Figure 4: Punta Cancun, after Hurricane Wilma.

Wilma (2005) was the most dangerous meteorological event that hit the region. The hurricane was classified as category five in the Saffir–Simpson scale, and caused a considerably increment of the erosion processes, resulting in a loss of dry beach in some areas and total loss of beach width from the central to the northern zone. Punta Cancun area was the most damaged one, as shown in Figure 4.

In such a situation, an emergency beach nourishment project was performed by CFE [3]. The main object was to restore the beach profile along the 11.6 kilometers between Cancun Point and Nizuc Point. In this stage the nourishment volume to place was restricted by the available federal budget (U.S. $20 Million), resulting in only 2.7 million cubic meters of sand, the final expected design beach width was estimated between 22 and 28 meters.

This emergency measure was not enough to rehabilitate the original beach conditions (1988, before Hurricane Gilbert).

Between 2006 and 2009 the combined effects of hurricane waves and the anthropogenic activities continued with the erosion process. The beach front showed a mean dry beach width between 15–20 m. which was not attractive enough for tourism.

During this period (2006 to 2009) several groins and detached breakwaters were built whose location and geometric characteristics increased the erosion, instead of maintaining the original beach width, as shown in Figure 5.

The beach conditions in 2009 set the stage for considering the option of a new integral project to restore the primeval status without a limited budget.

The main postulates and goals of this project were:

- To build engineering works in harmony with the prevailing environmental conditions, and
- To restore the earliest beach conditions and to keep the touristic attractiveness of Cancun.

Figure 5: Jetties built without project support.

The new integral project is conceived to provide the coastal zone with enough sediment even during tropical cyclone season, because this sand cannot be replaced naturally. Furthermore, with this volume of sediment it is conceived to restore partially the lower profile, which is largely responsible for the stability of the upper profile. And finally, and not less important, to increase the dissipation of wave energy on the study area.

4 Beach nourishment project

Based on the characterization of wave climate, hydrodynamics, sediment transport and morphodynamics studies it is proposed to design a beach nourishment wide enough [2] supported by a groin (closure structure) located on the northern beach, on Punta Cancun.

The closure structure is built to trap the sediment carried by waves and currents action from South to North. This position of the groin was based on the hydrodynamics, sediment transport studies and the incident current conditions. The closure structure on Punta Cancun aims to restore the original continuous headland, which has been degraded by the wave action and the weathering effects.

The nourishment sand volume was calculated according to the historical site characterization. To restore the 10.5 km which showed to be the most damaged, it was needed to place 5.2 million cubic meters of sand. To reach this condition the width of construction had to be ranged from 80 to 100 meters on average. The beach height above mean sea level of design was 3.0 meters.

If another hurricane similar to *Wilma* (category four or higher according to the Saffir–Simpson scale, time of the impact and proximity) strikes the region, the sand loss in the upper profile, and the coastline retreat could be large, but the greatest amount of sediment would remain within the active beach profile. The impact of the hurricane could transport the sediment towards the lower areas of beach profile (offshore) forming longshore bars which in turn play an important role to protect the beachface and keeping the sand within the closure depth.

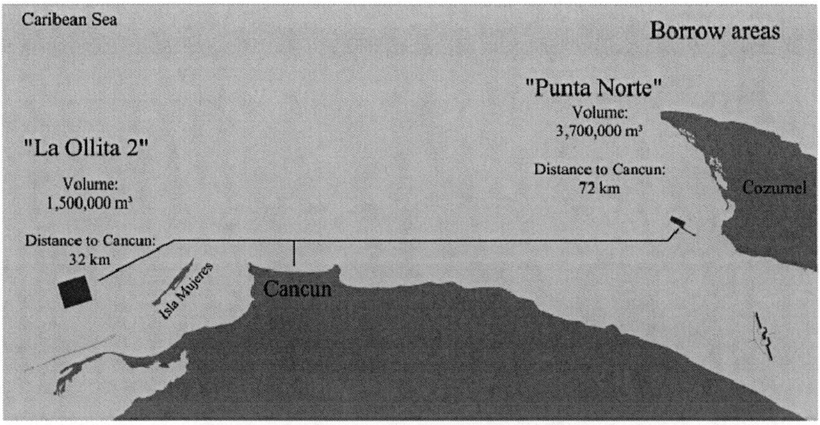

Figure 6: Borrow areas location.

The project included two borrow areas: "La Ollita 2" near Isla Mujeres and "Punta Norte" in front of Cozumel Island (see Figure 6), the main characteristics of the fill sediment in these areas was compatible with the native sand beach.

5 Discussions

A detailed morphological analysis of the study area was performed in order to predict the evolution of the seabed, the beach profiles and finally the coastline position under normal and extreme wave and current conditions.

Planform evolution observation aims to have an idea of the dry beach behavior through field studies. For the suitable calibration of the numerical model, it is necessary to link this information to the beach profile evolution in order to determine the coast line behavior and its morphological evolution due to the changes of profile under normal and extreme wave and current conditions.

5.1 Seabed evolution

From three historical detailed bathymetric surveys performed by CFE in August 2008, February and October 2010, the planform evolution was studied. The bathymetric curves for levels 0.0, -1.0 and -5.0 were selected for each bathymetric, and then a comparison of their position was performed. Figure 7 shows the different depth positions for each bathymetric survey measurement. The seabed analysis for the North Zone "Punta Cancun" is presented in this paper because Punta Cancun was the most damaged area by Hurricane Wilma. The continuous lines represent the coastline position (bathymetric curve level 0.0), the dashes one shows the bathymetric curve level -1.0 m and the composite one indicate bathymetric curve level -5.0 m.

For example in 2008 the coastline position (bathymetric curve level 0.0) was on the basement hotels, the bathymetric curve level -5.0 was located at a distance of 150 meter offshore, on average.

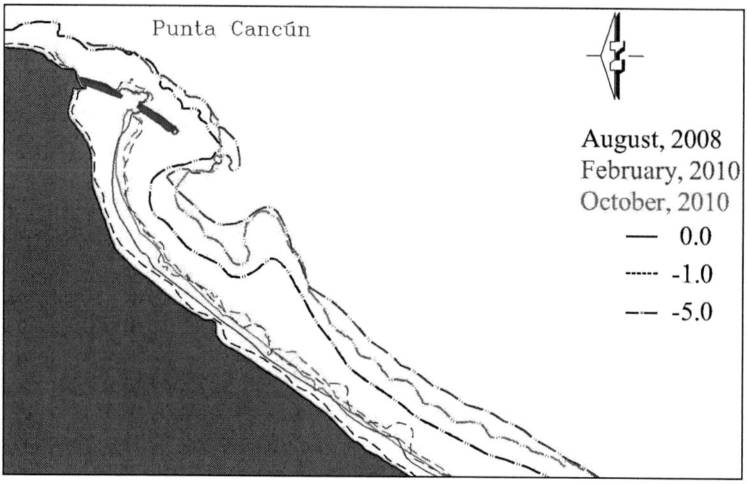

Figure 7: Seabed evolution in Punta Cancun.

As shown in Figure 7, after nourishment the coastline position moves 55 meters offshore on average, consequently also the bathymetric curve level -1.0 moved 55 m offshore. The most notable changes occurred in the position of the bathymetric curve level -5.0 meter, which was located about 230 meters offshore.

The placement of 5.2 million cubic meters of sand also changed the planform slope and therefore the surf zone was moved offshore, forcing to reduce the wave energy approaching the coast, and thus increasing the beach nourishment useful life.

5.2 Planform evolution at Cancun Beach

The planform evolution along the project area was analyzed using historical bathymetric survey measurements. Figure 8 shows the coastline position before and after hurricane Wilma, in circular and filled circular markers, both before the emergency nourishment placement; the squared marker is the coastline position in August 2008, after the emergency nourishment and before the nourishment in 2009. Finally the coastline position after nourishment project is shown in triangular markers.

Before Hurricane Wilma, 2005 the dry beach width varied between cero to 60 meters, with 28 meters on average. This year the authors observed that there were three different beach front zones:

North zone (0+600 km to 4+000 km), the dry beach width was 18 meters. Central zone (4+000 km to 8+000 km) the dry beach width varied between 20 to 30 meters, and on South zone (8+000 km to 12+000 km) the average dry beach width was 22 meters.

After *Hurricane Wilma* hit the region, the beach width loss was 22 meters as a mean value, transforming the three beach zones to two eroded areas: The first

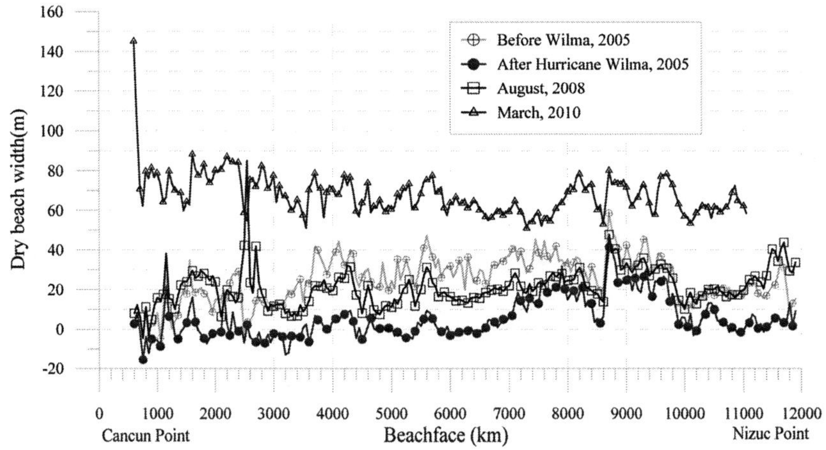

Figure 8: Coastline evolution.

one named "Severe erosion" zone covers from 0+600 km section to 7+000 km section, in this beachfront, 70% of the length without dry beach was recorded. The second area covered from 7+000 km to 12+000 km sections, with a dry beach width of 18 meters as a mean value.

After beach nourishment in 2009, mean dry beach width varied from 50 to 90 meters. Nevertheless, there was a 100 m beach front with dry beach more than one hundred meters and it was located close to the closure structure (Cancun Point).

At present, the North zone shows a 70 meter mean dry beach width. The Center zone has a 60 meter mean dry beach width, and the South zone exhibits a 70 meter mean dry beach width. On South the coastline shows an apparent retreat because of the longitudinal gradient: passing from 80 meter of wide in the nourishment area to 30 meter in the un-rehabilitated area.

5.3 Beach profiles evolution

The analysis of beach profiles has been performed since 2005, starting with measurements before and after the Hurricane Wilma occurrence. The authors present in this paper beach profiles measured after the emergency project: April (immediately after nourishment), August and December 2006. In 2008, the beach was evolving toward accelerated erosive state, reaching the same profile conditions as in 2005.

Figure 9 shows three representative beach profiles of the beach condition: on the northern zone, the profile 0 +800 km, on center zone the profile 5 +500 km, and on south zone 10 +000 km profile.

The analysis of evolution beach profiles from 2005 to 2008 served as a basis for understanding the beach behavior evolution, which is finally taken into consideration for the design conditions in 2009.

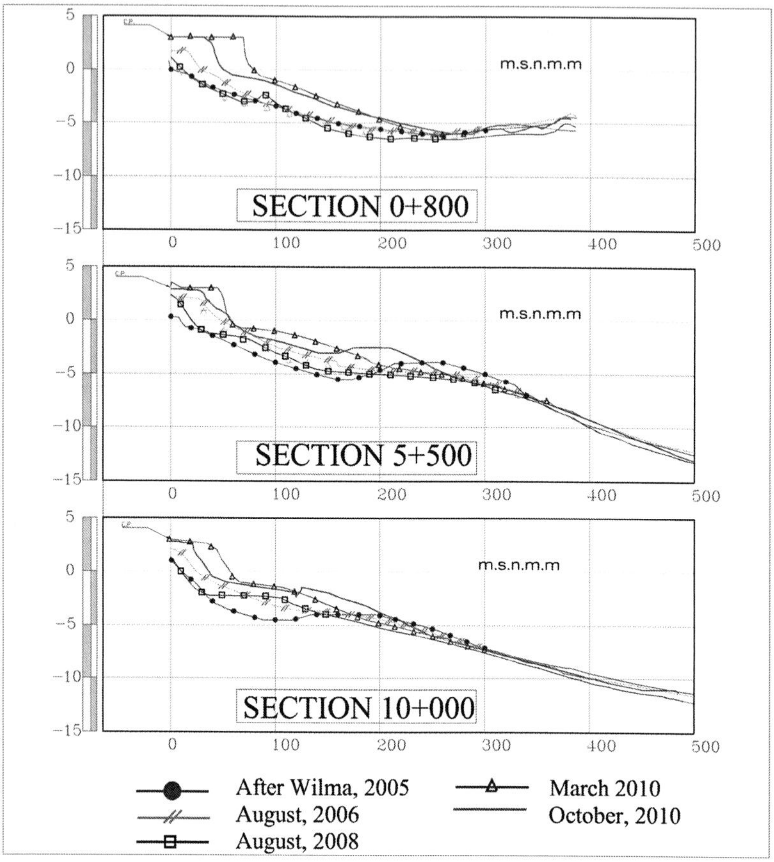

Figure 9: Beach profile at Cancun.

In 2005, after Wilma, the dry beach widths were around 20m, at most, as shown in fig. 8, profile 10 +000 km. However there were areas where the beach was completely lost (0 +800 km), reaching its maximum erosion process in these areas, with very steep slopes. The hotel foundations were a part of the beach profile.

Beach profiles analysis from 2005 to 2008 showed clearly that profiles were still reflective, particularly the profile 0 +800 km remained practically equal during this period of time. The profiles 5+500 and 10+000 km showed that a percentage of the placed material was transported to the subaerial profile. This beach profile behavior leaded the authors to establish a new project for rehabilitation of beaches, which have the ability to maintain a beach width needed for the development of tourist activities and infrastructure protection. This target could be achieved, building a beach with an intermediate morphodynamic state with trend towards a dissipative state.

Profiles from February to October 2010 are all measured after the nourishment project, as shown in Figure 9. It is clear that after 9 months the beach profiles (post-construction) have behaved according to the project expectations, exhibiting dry beach widths between 40 and 70 meters.

So far, the beach profiles evolution has been satisfactory, maintaining a beach width and a mild beach slope within the design predictions.

On the northern zone of the beach, Punta Cancun, it was necessary to support the beach fill with a groin to trap the sediment transport from north to south, and to avoid the sand to reach the coral reefs located on the northern part between Punta Cancun and Mujeres Island.

Finally, on the southern zone, Punta Nizuc, it was not possible to place sand until this point due to severe environmental restrictions imposed by the Mexican Environmental Authorities.

6 Conclusions

The nourishment project was suitable for transforming the beach of Cancun into a headland, confined by Punta Cancun and Punta Nizuc. The closure structure built on Punta Cancun zone creates the hydrodynamic conditions appropriate to trap the placed sediment on this critical area, and created a protected zone, reducing the sand loss, remembering that before nourishment beach it was a zone without dry beach.

The closure groin makes possible to maintain the headland in Cancun beach for all wave climate conditions.

The beach is evolving towards a new intermediate equilibrium state very close to dissipative in some specific cases.

The estimated sand volume for the beach nourishment was enough to restore Cancun beach original conditions.

The fill sand dredged from the two borrow sand pits (La Ollita 2 and Punta Norte) was completely compatible with the native sand.

Taking into consideration the last statements the project and beach nourishment works have been considered as a successful Mexican engineering project.

And last but not least, the project fulfilled completely all the environmental restrictions imposed by the Mexican Environmental Authorities (SEMARNAT, SEMAR, SCT).

References

[1] Morán, Salles, Adame, Sánchez, Espinal, 2006. "Análisis Morfológico de las playas de Cancún". XXII Congreso Latinoamericano de Hidráulica, Ciudad Guayana, Venezuela, octubre 2006.

[2] Wright, L. D., Short, A. D., (1984). "Morphodynamic variability of surf zones and beaches: a synthesis". Marine Geology 56, 93-118.

[3] CFE, 2006. Proyecto Ejecutivo de Restauración y Recuperación de las playas de Cancún, Quintana Roo. Gerencia de Estudios de Ingeniería Civil (GEIC) de la Comisión Federal de Electricidad.

[4] CFE, 2009. Proyecto Ejecutivo para la Rehabilitación de playas de Cancún, Quintana Roo. Gerencia de Estudios de Ingeniería Civil (GEIC) de la Comisión Federal de Electricidad.

Connecting river to sea by a 2-D mathematical model

A. Astaraki & F. Fallah
Civil Engineering Department, Islamic Azad University, Iran

Abstract

A two dimensional mathematical model is presented to simulate flow and suspended sediment transport in rivers. A two dimensional horizontal depth average model computes the flow characteristics. The hydrodynamic model uses a fully implicit finite difference numerical scheme with variable grid size. By solving the equation of sediment transport, the sediment concentration will be calculated. The depth variations could be computed by solving continuity equation of sediment. In this Paper, verification of the model and ability of Model in the field of connecting a river to the sea is shown by figures.

Keywords: mathematical modeling, suspended sediment transport, sea, river.

1 Introduction

In recent decades a lot of numerical and physical models have been developed in the field of river engineering. Physical models are generally very time-consuming, costly and for many practical problems, impossible, while mathematical models are becoming more popular because of low cost, flexibility in design for changing different plans, ability to simulate riverbed deformation under large scale and long term conditions and they also provide a large quantity of information.

However any mathematical model is faced with some problems, for example using concept of equilibrium sediment transport is an important problem in mathematical models. In this model a 2-D mathematical model of flow and suspended sediment transport has been developed for alluvial rivers in which concept of nonequilibrium sediment transport is used. In the concept of equilibrium suspended sediment transport, sediment transport capacity is supposed to be equal with sediment concentration while alluvial rivers are

www.witpress.com, ISSN 1743-3541 (on-line)
doi:10.2495/CP110191

always under nonequilibrium conditions because of the variable inflow, various human activities such as reservoir construction and water diversion, and etc. [1]. Charafi et al. [2] presented a quasi 3-D mathematical model of morphological processes and based on his model mentioned it is not recommended to apply an equilibrium transport formula to compute morphological variations.

Celic and Rodi [3] propounded none of empirical formula does not have universal value because it is impossible to determine boundary between bed load and suspended load, so this model doesn't use the empirical formulas to estimate suspended sediment transport, in fact in this model, the difficulties in most before models like using practical formulas and determining concentrations of bed load has been omitted by using a new algorithm.

2 Equations

2.1 Flow equations

$$\frac{\partial q_x}{\partial x}+\frac{\partial q_y}{\partial y}+\frac{\partial \eta}{\partial t}=0 \tag{1}$$

$$\frac{\partial q_x}{\partial t}=-\frac{\partial q_x q_y/d}{\partial y}-\frac{\partial q_x^2/d}{\partial x}-\frac{d}{\rho}\frac{\partial p}{\partial x}-gd\frac{\partial \eta}{\partial x}+\frac{d}{\rho}\frac{\partial \tau_1}{\partial y}+\frac{1}{\rho}(\tau_{xs}-\tau_{xb})+fq_y \tag{2}$$

$$\frac{\partial q_y}{\partial t}=-\frac{\partial q_x q_y/d}{\partial x}-\frac{\partial q_y^2/d}{\partial y}-\frac{d}{\rho}\frac{\partial p}{\partial y}-gd\frac{\partial \eta}{\partial y}+\frac{d}{\rho}\frac{\partial \tau_1}{\partial x}+\frac{1}{\rho}(\tau_{ys}-\tau_{yb})-fq_x \tag{3}$$

where q_x and q_y are water discharge component in the x and y direction, respectively, $d=h+\eta$ is the total water depth, h is the still water depth, η is the fluctuation of water surface elevation, p is the atmospheric pressure, ρ is the density of water, t is time, g is gravity, $f=2\Omega\sin\phi$, Ω is the angular speed of earth rotation, ϕ is the latitude of study site, τ is the shear stress [4].

2.2 Unsteady nonequilibrium sediment transport equation

$$\frac{\partial hs}{\partial t}+\frac{\partial hus}{\partial x}+\frac{\partial hvs}{\partial y}-\varepsilon_s\frac{\partial^2(hs)}{\partial x^2}-\varepsilon_s\frac{\partial^2(hs)}{\partial y^2}=-\alpha\omega(s-s^*) \tag{4}$$

2.3 Continuity equation of sediment

$$\rho'\frac{\partial z}{\partial t}=\alpha\omega(s-s^*) \tag{5}$$

where s is suspended sediment concentration, s^* is sediment transport capacity, α is adjust coefficient, ε_s is diffusion coefficient for sediment, u, v are depth averaged velocity in direction x and y [1].

2.4 Solving equations

A two dimensional horizontal depth average model computes the flow characteristics. The hydrodynamic model uses a fully implicit finite difference numerical scheme with variable grid size. A program developed in FORTRAN programming language solves the uncouple equation of sediment and current [5].

3 Model verification

In this paper, a comparison has been made between results of this model and a quasi 3-D model of Charafi et al. [2]. Solving network consist of a triangular channel with size of 100*3100meters and a horizontal bed and a dam (100*400 meters). Other main information is given below:

Intel discharge=4000 m^3/s
Water depth in channel=6 m

$$\rho_s = 2650 kg/m^3$$

$$D_s = 200 \mu m$$

Grid size in this model is variable 125 meter in x-direction and 50 meter in y direction which near dam location is smaller. In chart 1 and figure1, 2 results and comparison between this model and the Charafi et al. model have been shown.

Table 1: Computed flow velocity, water depth and concentration in different zones of Charafi et al. work [2] and present work.

Point coordinate (m)		Current Velocity (m/s)		Depth (m)		Sediment Concentration (Kg/m3)	
x	y	Charafi	Present Work	Charafi	Present Work	Charafi	Present Work
1480	500	1.1	1.08	6.13	5.99	0.14	0.10
1520	660	1.35	1.1	6.06	6.004	1.2	0.27
1560	620	1.71	0.73	5.98	6.002	3.2	0.12
1680	540	1.37	0.97	6.06	6.004	0.361	0.182
160	820	.65	0.60	6.24	5.91	0.123	0.106
2320	580	.84	.88	6.02	6.002	0.098	0.074

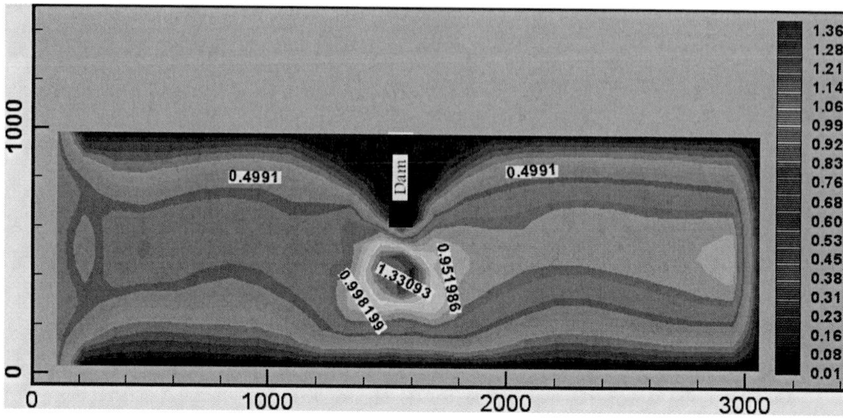

Figure 1: Computed flow velocity (present work).

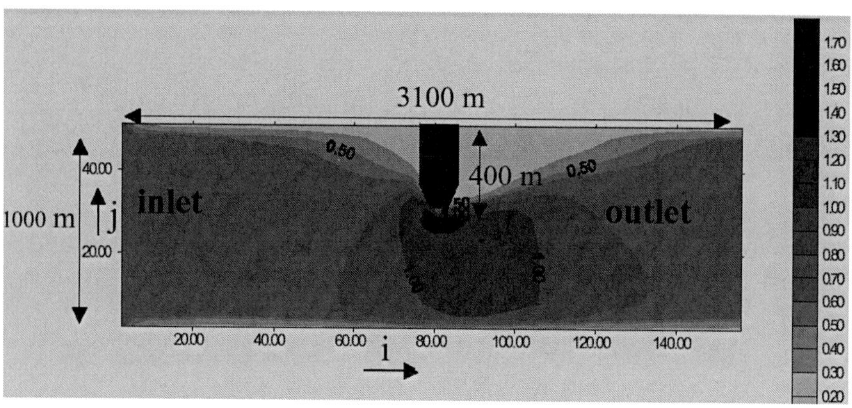

Figure 2: Computed flow velocity (Charafi et al. work) [2].

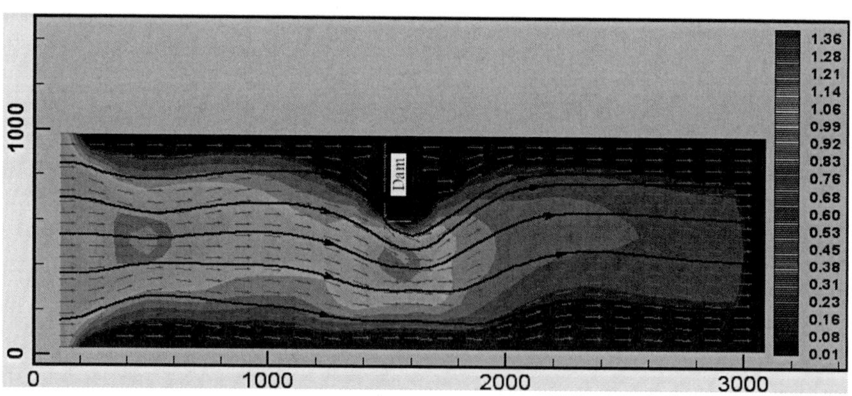

Figure 3: Ability of model in simulation of velocity vectors and flow lines (present work).

4 Simulation of river connection to sea

In this part, ability of model in simulation connecting a river to the sea is shown by figures 5, 6 and 7. Boundary conditions are shown in figure 4 and other characteristics of model is mentioned below. Results of the model in simulation of river connection to the sea are shown in figures 5–8.

Water level=4m
River Discharge= $10\, {m^3}/{ms}$
Constant water level in outlet (Sea water level) =5m
Bed Slope=0.0
Sediment Particle Diameter= 200μm.

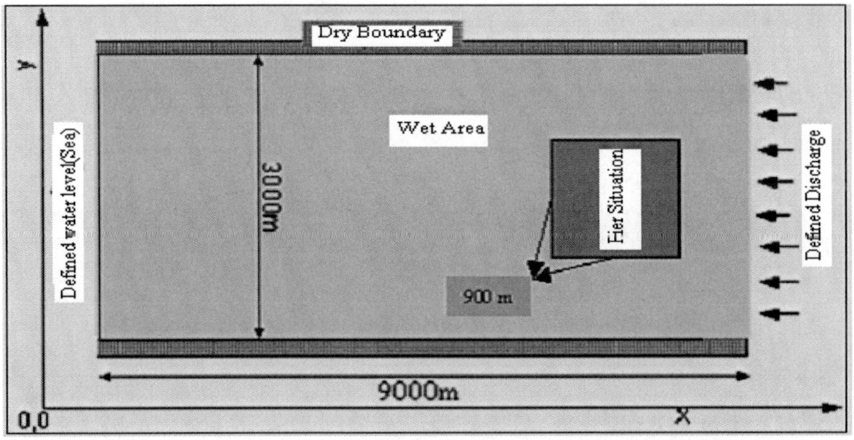

Figure 4: Conditions and details of simulation connecting river to sea.

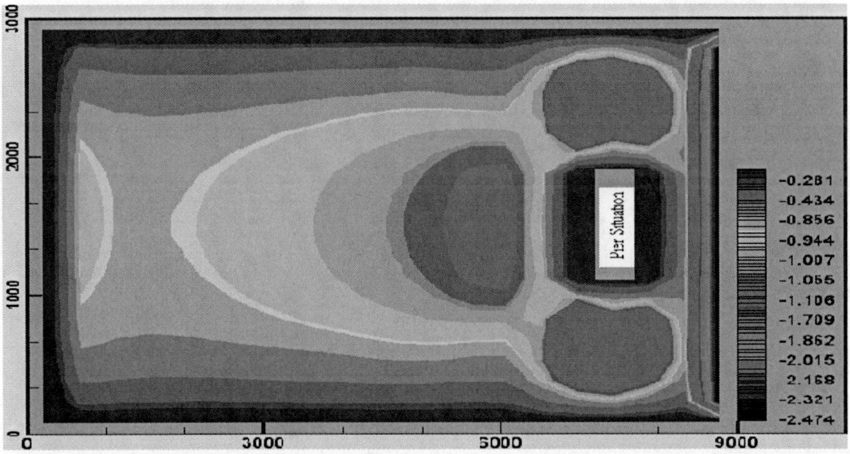

Figure 5: Variation of water velocity in X direction after 75 minutes (m/s).

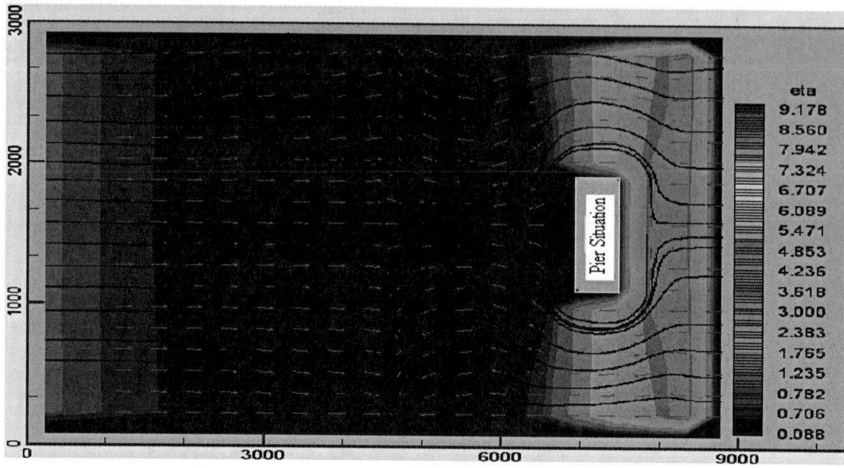

Figure 6: Variation of water level and velocity vectors and flow lines after 9 minutes.

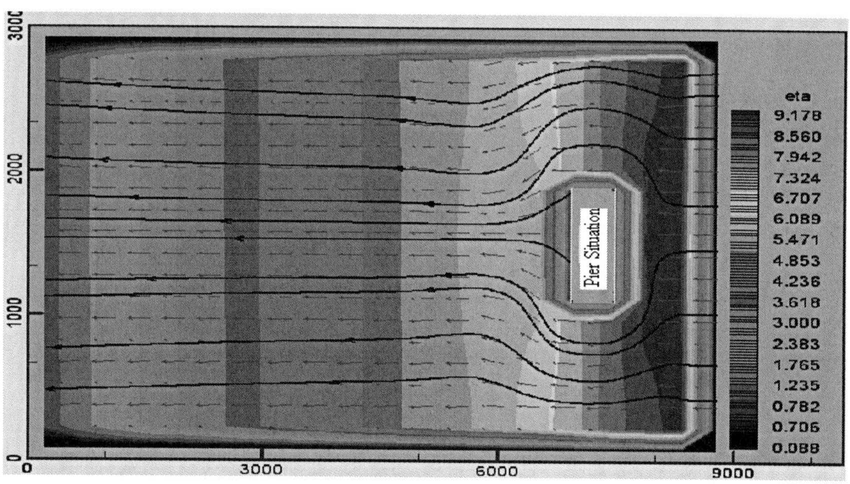

Figure 7: Variation of water level and velocity vectors and flow lines after 75 minutes.

5 Conclusion

In this research, a 2-D mathematical model of depth averaged in shallow water flows have been developed for processes of sediment transport in alluvial rivers. Hydrodynamic model uses a full implicit time-splitting numerical scheme with variable grid size. The sediment transport model uses a simple algorithm that both the difficulty of determining the reference concentration and problem of

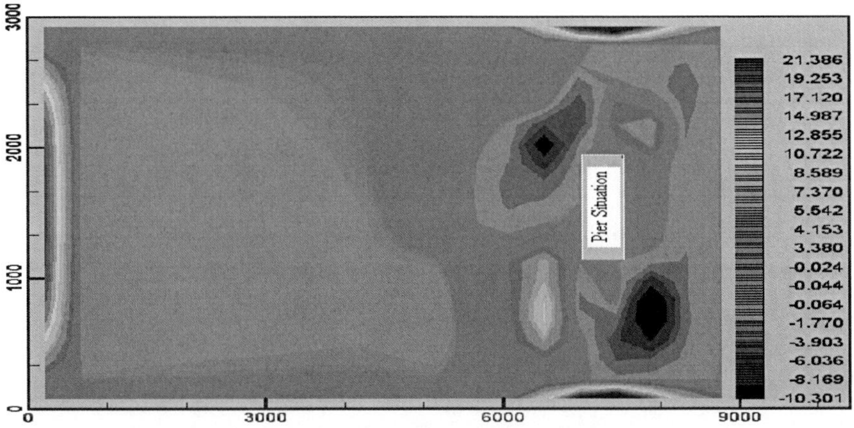

Figure 8: Total variation of bed after 75 minutes (cm).

selecting the sediment transport function are avoided. Results of model are compared with the Charafi et al. model. In this comparison, flow velocities are near to the Charafi et al. model and maximum variation of water depth in two models is 6%. Sediment concentrations in front of dam in two models show large amounts in accordance with other points which is obtained from rectangular elements. Also, ability of the model in simulation of connecting sea to river and total variation of river bed and variation of velocity in directions x and y has been shown by figures.

References

[1] Guo, Q. C., and Jin, Y. C. "Modeling Nonuniform Suspended Sediment Transport in Alluvial Rivers.", Journal of Hydraulic Engineering, ASCE, Vol. 128, No. 9, 2002.

[2] Charafi, My. M. and Sadok, A. and Kamal, A. and Menal, A., "Quasi-Three Dimensional Mathematic Modeling of Morphological Processes Based on Equilibrium Sediment Transport.", International Journal of Modern Physics C, Vol. 11, No. 7., 2000.

[3] Celik, I. and Rodi, W. "Modeling Suspended Sediment Transport in Nonequilibrium Situations.", J. Hydr. Engrg., ASCE, Vol. 114, No. 10, 1988.

[4] Jerome P.-Y. Maa, "An Efficient Horizontal Two-Dimensional Hydrodynamic Model.", Journal of Coastal Engineering, No. 14, 1990.

[5] Astaraki, A., "Two Dimensional Mathematical Modeling of Flow and Sediment Transport in River.", Master of Science Thesis, Civil Department, Shahid Bahonar University of Kerman, Kerman, Iran, 2004.

Section 7
Interaction between coastal defences and processes

Shoreline behind a breakwater for wave energy absorption in Reggio Calabria: comparison between theoretical models and experimental data

G. Barbaro & G. Foti
Department of Mechanics and Materials, Reggio Calabria "Mediterranea" University, Italy

Abstract

This paper deals with models proposed for calculating the shoreline change behind a breakwater. In particular, models developed by Hsu and Silvester, McCormick, Gonzalez and Medina and Hsu et al. are discussed. Shoreline change behind a breakwater is estimated by these models. Results are compared with experimental data measured at the NOEL laboratory of Reggio Calabria, for a caisson breakwater for wave energy absorption made using nine cellular caissons in reinforced concrete. The structure is 16.2 m long and located in a 2 m water depth. Waves in the undisturbed wave field are measured at the same water depth, 20 m away from the structure, by means of two pressure transducers and two ultrasonic probes located on two piles of small diameter (0.05 m). Tide is measured by probes in the undisturbed wave field and by pressure transducers on the breakwater.
Keywords: shoreline, breakwater, modelling, salient, tombolo.

1 Introduction

Caisson breakwaters are widely used for shoreline protection. Their target is to induce accretion of the shore, in order to protect it from erosion processes, and, further, to increase the attraction of the location. With this purpose, they create a safe basin and, at the same time, they modify the cross-shore circulation.

There has been an increasing use of caisson breakwaters for coastal protection during the last two decades.

www.witpress.com, ISSN 1743-3541 (on-line)
doi:10.2495/CP110201

Usually, they are placed in front of, and parallel to, a shoreline. They dissipate sea wave energy and induce a sediment deposit in the safe area. The hydrodynamic and the sediment transport mechanics, close to the breakwater, are not completely known. Based on experimental investigations, some models have been proposed. They focus on the estimation of the shoreline change by considering the occurrence of a salient or a tombolo. There is a tombolo, when the accretion is such that the shoreline reaches the breakwater. Otherwise, there is a salient. Most part of the afore-mentioned models estimate shoreline change by assuming an orthogonal wave train interacting with one barrier (Shinohara and Tsubaki [5]; Rosen and Vajda [6]; Mimura et al. [7]; Ming and Chiew [8]).

Various parameters have been considered for determining shoreline change. For example, the length B of the breakwater, the shoreline-breakwater distance S, the breaking depth d_b, the wave steepness H_0/L_0, the mean sediment diameter D_{50}, the sediment density ρ_s, beach slope S_b.

Some authors (Gourlay [9]; SPM [10]; Dally and Pope [11]; Suh and Dalrymple [12]; Ahrens and Cox [13]) have proposed empirical formulations for estimating the shoreline change behind a breakwater. However, there are different criteria for defining the occurrence of a tombolo or a salient.

2 Previous works about the analysis of a tombolo or a salient behind a breakwater

The following models for the analysis of shoreline change due to a breakwater are reviewed: Hsu and Silvester [1], McCormick [2], Gonzalez and Medina [4], Hsu et al. [3].

2.1 Hsu and Silvester

Hsu and Silvester [1] model gives the shoreline change by considering a polar coordinate system with origin at the tip of the breakwater. The shoreline is symmetric with respect to the centre of the breakwater. It is estimated by the equation

$$r(\theta) = f\left(\theta\right) r(40°) \tag{1}$$

θ being the angular variable. In eq. (1), $40° < \theta < 180°$, $f(\theta)$ is a decreasing function of θ. Then, $r(40°)$ is defined by the equations

$$r(40°) = 0.17B + 1.68S \tag{2}$$

The intersection between the shoreline and a line orthogonal to the breakwater that contains the centre of the breakwater is a cusp.

The authors suggested to correct the shoreline equation, by considering the distance X between the centre of the breakwater and the shoreline:

$$X = 0.6784B(B/S)^{-1.2148} \tag{3}$$

2.2 McCormick elliptic model

Assume a coordinate system with origin at the centre of the breakwater (Fig. 1). Then, assume one breakwater under the action of an orthogonal wave train. McCormick [2] proposed the elliptic equilibrium shoreline, given by the equation

$$\frac{\left(y \mp h\right)^2}{a^2}+\frac{x^2}{b^2}=1 \tag{4}$$

where the x-axis is orthogonal to the shoreline and landward-oriented and the y-axis is parallel to the shoreline, h is the distance between the centre of the breakwater and the centre of the ellipse, a is the semi major axis and b the semi minor axis, S the distance between the breakwater and the original shoreline. The positive and negative sign are referred to, respectively, the right and left side of the shoreline. Foci are at the tips of the barrier.

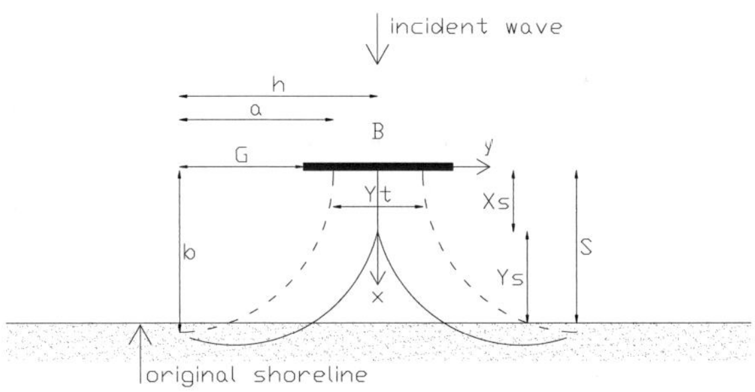

Figure 1: Reference sketch of the elliptic shoreline model (McCormick [2]). The continuous line is a salient, the dashed line is a tombolo.

In eq. (4) h, a and b depend on the wave steepness H_0/L_0 of the incident wave, the beach slope S_b, the breakwater width B, the distance S. McCormick [2] suggested the following equations for estimating the ratio b/S:

$$\frac{b}{S}=1+0.2\varsigma_0 \sin\left(\chi\varsigma_0\right) \tag{5}$$

in which ς_0 is the ratio between the wave steepness and the beach slope:

$$\varsigma_0=\frac{\left(H_0 / L_0\right)}{S_b} \tag{6}$$

and χ is a parameter that depends on the ratio between beach slope and the breakwater width. It is given by

$$\chi = -1.92\left(\frac{S}{B}\right)^2 + 9.92\left(\frac{S}{B}\right) \tag{7}$$

Further, the distance h is given by the equation

$$h = G + 0.5B \tag{8}$$

where G is the distance between the centre of the ellipse and the tip of the breakwater. The semi major axis a is given by

$$a = \sqrt{G^2 + b^2} \tag{9}$$

The ratio G/b depends on ς_0 and on S/B. It is given by

$$\frac{G}{b} = \exp\left[\ln(\mu) + \sigma\ln(\varsigma_0) - \nu\varsigma_0\right] \tag{10}$$

In eq. (10) μ, σ, υ have been defined as

$$\ln(\mu) = 19.4\tanh\left(0.91\frac{S}{B}\right) \tag{11}$$

$$\sigma = 17\tanh\left(0.59\frac{S}{B}\right) \tag{12}$$

$$\nu = 20\tanh\left(0.99\frac{S}{B}\right) \tag{13}$$

McCormick has derived the distance X_s between the salient and the breakwater:

$$X_s = b\sqrt{1-\left(\frac{h}{a}\right)^2} \tag{14}$$

where a is greater than h. Otherwise, there is a tombolo and $X_s = 0$.

2.2.1 Hsu and Jan and Wen modified elliptic shoreline

Hsu et al. [3] modified McCormick model [2] by means of the experimental data of Ming and Chiew [8], Shinohara and Tsubaki [5], Rosen and Vajda [6]. In particular, they proposed the following equations for the estimation of b/S, given ς_0, S/B and D_{50}. That is,

$$\frac{b}{S} = \begin{cases} 1 + 0.2(\varsigma_0)^{1/2}\left(\dfrac{S}{B}\right)^{-5/4} & per\ D_{50} = 0.25\,mm \\ 0.9 + 0.2\varsigma_0 & per\ D_{50} = 0.64\,mm \end{cases} \tag{15}$$

Further, they proposed the following expression of G/b, given μ, σ and υ:

$$\frac{G}{b}=\frac{5}{8}\exp\left[\ln(\mu)+\sigma\ln(\varsigma_0)-\nu\varsigma_0\right] \tag{16}$$

2.3 Gonzalez and Medina

Gonzalez and Medina [4] have investigated shoreline change behind a breakwater by considering, separately, a tombolo and a salient.

Consider the distance between the breakwater and the shoreline, and the breakwater width. If they are smaller than the incident wave length, behind the breakwater there is accretion. The shoreline reaches the breakwater (Figure 2). In this context, the profile depends on the length $2B$ of the breakwater and on the distance Y between the breakwater and the shoreline, and the wave length L, which defines α_{min} in eq. (17). The shoreline length affected by the breakwater $2B_1$, and the attachment width at the breakwater, B_k, can be obtained from Hsu and Evans [15] model and eq. (17).

$$\alpha_{\min}=\arctan\left[\frac{\left(\frac{\beta_r^4}{16}+\frac{\beta_r^2}{2}\frac{Y}{L}\right)^{1/2}}{\frac{Y}{L}}\right] \tag{17}$$

$$\beta_r=2.13 \tag{18}$$

If the breakwater is far from the shoreline and its length is short with respect to the incident wave length, there will be a salient. In this context, the variables used in the context of a tombolo are considered. The salient apex Y_0 (Figure 3) is estimated by the parabolic-shaped formulation of Hsu and Evans [15] and eq. (17).

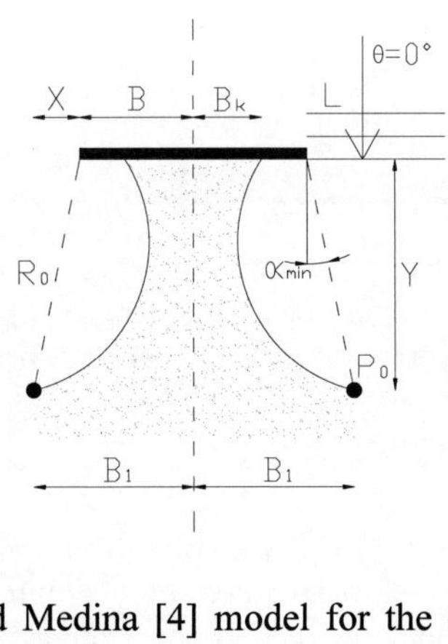

Figure 2: Gonzalez and Medina [4] model for the estimation of a tombolo behind a breakwater.

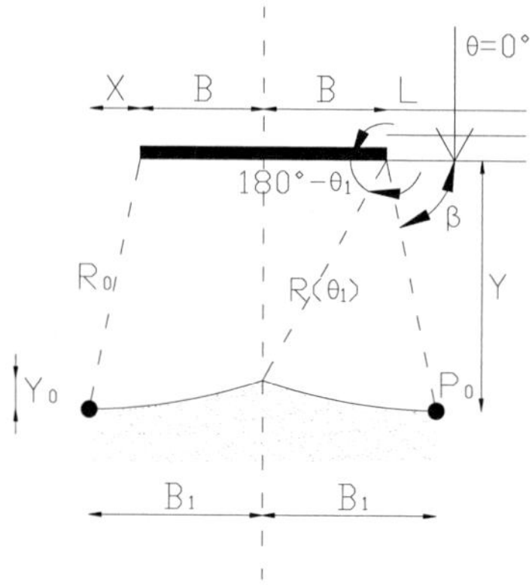

Figure 3: Gonzalez and Medina [4] model for the estimation of a salient behind a breakwater.

Figure 4 shows the graph given by Gonzalez and Medina [4] for calculating shoreline change, that is the lengths $2B_1$ and B_k (for a tombolo), and the length $2B_1$ and apex Y_0 (for a salient).

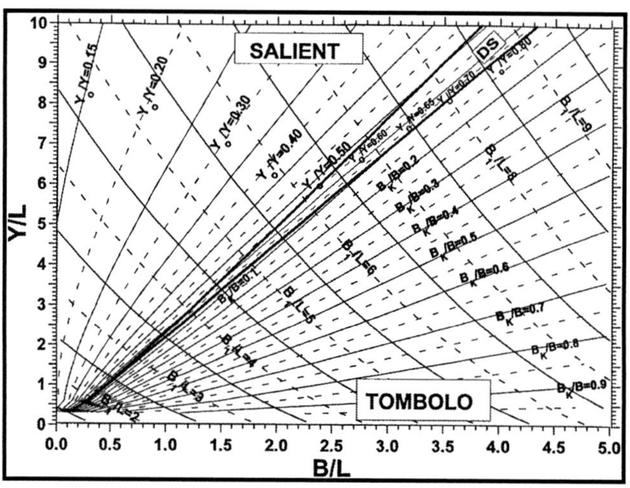

Figure 4: Determination of a tombolo (B_k/B, B_1/L) or a salient (Y_0/Y, B_1/L) given the nondimensional length of the breakwater and the nondimensional distance between the breakwater and the shoreline.

3 Experiment RC 2005

In Reggio Calabria the wind frequently produces waves with dimensions typical of big channels laboratory, which are very stable in time. Waves represent excellent small scale models in the Froude similarity of strong sea storms. We refer to the 1:30 small scale models of strong Mediterranean sea storms or the 1:50 scale models of oceanic storms. This peculiarity depends on a favourable combination of hardly repeatable environmental conditions; first, the high stability of the local wind, the so-called "wind of the Strait", blowing from Messina toward Reggio for many consecutive days; second, the orientation of the coast (see fig. 5), which is naturally protected from the swells that propagate from the Southern mouth of the Strait (which are not small scale Froude models of wind waves, of course); finally, the excursion of tide is very contained, and the water is clean thanks to the passage twice a day of the Strait current. The clearness of the water is another requisite for the experimental activity in the sea. In fact, in the sea, all the assemblages are performed by scuba divers.

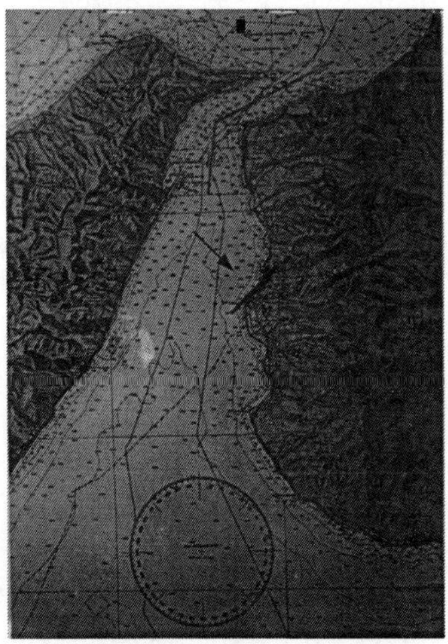

Figure 5: Site of laboratory.

3.1 Description of the experiment RC 2005

On March 2005 the small-scale model of a breakwater embodying an OWC has been built in the NOEL (Natural Ocean Engineering Laboratory) laboratory of Reggio Calabria. The breakwater length is 16.2 m. The water depth is 2 m, with seabed slope of 5%. The distance between the centre of the breakwater and the shoreline is 14.7 m. The mean diameter of the sediment is D_{50} = 6 mm. The plant is the 1:10 scale model (with some modifications) of a hypothesis of breakwater for the North-East Pacific coast. Alternatively, this plant may be thought of as the 1:6 scale model of a breakwater suited for the Mediterranean Sea. The aim of the experiment was to test the plant absorption capability. In the undisturbed wave field, at the same water depth, 20 m far from the structure, 2 piles of small diameter (0.05 m) with two pressure transducers and two ultrasonic wave probes were placed for supporting the gauges for the measurement of the mean energy flux of the incident waves. Tide is measured by probes in the undisturbed wave

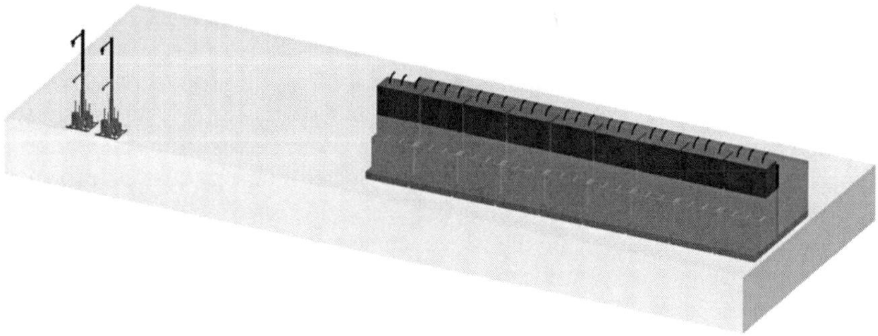

Figure 6: Three-dimensional view of the breakwater used during the RC 2005 experiment and the position of the instruments (Boccotti et al. [16]).

field and pressure transducers on the breakwater. Results of the experiment (Boccotti et al. [16]) showed that it is possible to absorb a large share of the wave energy. Further, it has been measured the shoreline profile behind the breakwater. It has been measured a significant modification of the shoreline.

3.2 Data stored

The duration of each record was of five minutes. The sampling rate was of 10/s for each gauge. In each record we obtained the values of the following variables: H_s, T_p, Ψ^* of the incident waves, with Ψ^* being the narrow-bandedness parameter (Boccotti [17]), the tide level, the average angle of the wave direction, the absorption coefficient, the amplification factor, the significant height of the oscillations of the water column, the significant height of the fluctuating pressure head in the air pocket and the resonance coefficient. Besides these parameters we obtained the frequency spectrum of the time series data for each gauge, the directional spectrum and the regression.

We got 187 records: 96 records of wind waves, 31 records of wind waves superimposed on swells, a sequence of 40 records of swells of average T_p about 5 s and a sequence of 20 records of swells of average T_p about 7 s. The mean significant wave height has been 0.26 m, the peak period has been 4.41 s (Boccotti et al. [16]), the offshore mean dominant wave length has been 30.35 m while the mean dominant wave length at the breakwater depth has been 18.2 m.

4 Comparison between theoretical models and experimental data

4.1 Shoreline measured in Reggio Calabria

The shoreline has been measured three months after the breakwater placing (June 2005), and in September 2010.

Figure 7 shows that the shoreline is not symmetric with respect to the centre of the breakwater. Indeed, the salient apex has moved to the south direction and the erosion has involved, in particular, the North side of the shoreline. This asymmetry is, probably, due to the non-parallel position of the breakwater (with respect to the shoreline) and to the occurrence of oblique wave trains.

Further, Figure 7 shows that there have been significant modifications from 2005 to 2010. In particular: the salient apex is closer to the breakwater and there is still the erosion of the external sides. Table 1 shows the shoreline measured variations. Further, it is shown the shoreline asymmetry. Table 1 shows that, in the 2005–2010 time interval, the asymmetry has increased.

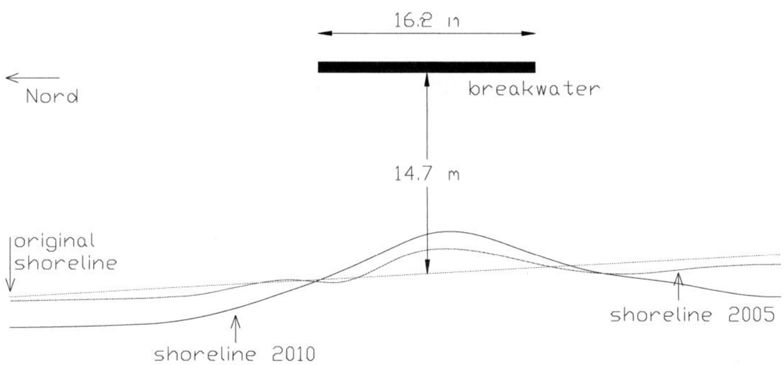

Figure 7: Measured shoreline: the green line is the shoreline before the breakwater placing (March 2005); the red line is the measured shoreline in June 2005 and the black line is the shoreline measured in September 2010.

Table 1: Modifications of the characteristic points of the shoreline. Data have been measured in June 2005 and in September 2010. Both data are compared to the shoreline measured in March 2005.

Position	**Modifications from 2005 to 2010 [%]**
Salient	11
Max erosion North	13
Max erosion South	11
Year	**Asymmetry max erosion North–Sud in the shoreline [%]**
2005	12
2010	14

4.2 Comparison between the theoretical models and the experimental data of RC 2010

The afore-mentioned models have been used for estimating the distance between the salient apex and the breakwater, and the maximum distance between the axis parallel to the breakwater and the shoreline. Eq. (3), eq. (14) and Figure 4 have been used. Table 2 shows the results of the calculation.

Table 2: Salient-breakwater distance and maximum distance between an axis parallel to the breakwater and the shoreline in the North side (N) and in the South side (S). It has been estimated the difference with respect to the 2010 shoreline (a positive quantity means overestimation and negative quantity means underestimation).

Model	**Salient – breakwater distance [m]**	**Difference between theoretical results and experimental data [%]**	**Maximum distance [m]**	**Difference between theoretical results and experimental data [%]**
Hsu and Silvester	9.77	+ 16	18.96	+ 1 (N) + 15 (S)
McCormick	12.09	- 4	14.71	- 21 (N) - 11 (S)
Hsu et al.	10.97	+ 6	13.73	- 27 (N) - 17 (S)
Gonzalez and Medina	11.65	0	17.13	- 8 (N) + 4 (S)
Profilo 2005	12.94	- 11	16.6 (N) 14.8 (S)	- 13 (N) - 11 (S)
Profilo 2010	11.65		18.72 (N) 16.45 (S)	

Table 2 shows that Hsu and Silvester [1] model overestimates both the salient apex and the maximum erosion point in the South side, while it gives a good estimate of the maximum erosion point in the North side.

McCormick model [2] slightly underestimates the salient apex, but it overestimates the maximum erosion points.

Hsu et al. [3] model overestimates the salient and underestimates the maximum erosion points.

Gonzales and Medina [4] model gives a good estimated of the salient apex and the maximum erosion points. There are a small overestimation and a small underestimation, respectively, in the North side and in the South side of the shoreline.

5 Conclusions

Hsu and Silvester [1], McCormick [2], Hsu et al. [3] and Gonzalez and Medina [4] models for the determination of the shoreline change behind a breakwater have been analysed. The theoretical profiles have been compared to the experimental data. Data have been measured at the natural laboratory NOEL (Natural Ocean Engineering Laboratory) of Reggio Calabria. A breakwater for wave energy absorption has been place in 2005. It has been measures the shoreline in June 2005 (three months after the breakwater placing) and in

September 2010. It has been shown that the shoreline is not symmetric with respect to the centre of the breakwater. Indeed, the salient apex is in the South side of the breakwater. This induces erosion in the North side of the shoreline. This asymmetry is due to the non-parallel position of the breakwater (it is intended non-parallel with respect to the shoreline) and to the occurrence of oblique wave trains.

The results have been summarized in Table 2. It has been shows that the best estimation is given by the Gonzales and Medina model. It gives a good estimate of shoreline measured in 2010. The other models are less accurate, but they are still capable of giving a first estimate of the shoreline change.

References

[1] Hsu, J.R.C., Silvester, R., Accretion behind single offshore breakwater. *Journal of Waterway, Port, Coastal and Ocean Engineering*, **116 (3)**. pp. 367–380, 1990.

[2] McCormick, M.E., Equilibrium shoreline response to breakwaters. *Journal of Waterway, Port, Coastal and Ocean Engineering*, **119 (6)**, pp. 657–670, 1993.

[3] Hsu, T.W., Jan, C.D., Wen, C.C., Modified Mc Cormick's model for equilibrium shorelines behind a detached breakwaters. *Ocean Engineering*, **30**, pp. 1887–1897, 2003.

[4] Gonzalez, M., Medina, R., Equilibrium shoreline response behind a single offshore breakwater. *Proceedings Coastal Sediment*, **1**, pp. 933–942, 1999.

[5] Shinohara, K., Tsubaki, T., Model study on the change of shoreline of sandy beach by the offshore breakwater. *Proceedings 10th International Conference of Coastal Engineering*, Tokio, Japan, **1**, pp. 550–563, 1966.

[6] Rosen, D.S., Vajda, M., Sedimentological influences of detached breakwaters. *Proceedings 18th International Conference of Coastal Engineering*, **3**, pp. 1930–1949, 1982.

[7] Mimura, H., Shimizu, T., Horikawa, K., Laboratory study of the influence of detached breakwaters on coastal changes. *Proceedings Coastal Structures*, **2**, pp. 740–752, 1983.

[8] Ming D., Chiew Y.M., Shoreline changes behind detached breakwater. *Journal of Waterway, Port, Coastal and Ocean Engineering*, **126 (2)**, pp. 63–70, 2000.

[9] Gourlay, M.R., Beach processes in the vicinity of offshore breakwaters. *Proceedings of 5th Australian Conference on Coastal and Ocean Engineering*, Perth, Australia, pp. 129–134, 1981.

[10] Shore Protection Manual, 4th Ed., U.S. Army Corps of Engineer, Coastal Engineering Research Center, U.S. Government Printing Office, Washington, D.C., 1984.

[11] Dally, W.R., Pope, J., Detached breakwaters for shore protection. *Technical Report CERC-86-1*, U.S. Army Corps of Engineer, Coastal

Engineering Research Center, Waterways Experiment Station: Vicksburg, Mississippi, USA, 1986.

[12] Suh, K., Dalrymple, R.A., Offshore breakwaters in laboratory and field. *Journal of Waterway, Port, Coastal and Ocean Engineering*, **113 (2)**, pp. 105–121, 1987.

[13] Ahrens, J.P., Cox, J., Design and performance of reef breakwaters. *Journal of Coastal Research*, **7**, pp. 61–75, 1990.

[14] Gonzalez, M., Medina, R., On the application of static equilibrium bay formulations to natural and man-made beaches. Coastal Engineering, **43**, pp. 209–225, 2000.

[15] Hsu, J.R.C., Evans, C., Parabolic bay shapes and applications. *Proceedings of Institution of Civil Engineers*, London, England, **87 (2)**, pp. 556–570, 1989.

[16] Boccotti P., Filianoti P., Fiamma V., Arena F., Caisson breakwaters embodying an OWC with a small opening. Part II: a small scale filed experiment. *Ocean Engineering*, **34**, pp. 820–841, 2007.

[17] Boccotti P., *Wave mechanics for Ocean Engineering*, Elsevier: Amsterdam, 2000.

A model to follow the evolution in time of free beach fill

L. Rebaudengo Landò & S. Stura
Department of Civil Environmental and Architectural Engineering, Genoa University, Italy

Abstract

In this paper the evolution in time of free beach nourishment is dealt with by considering in a linked manner the processes of longshore diffusion of the filled sand and evolution of the cross profile, which presents an initial beach scarp that of steeper than that of equilibrium. After examining existing models relevant to the single processes, a simplified engineering model is proposed that follows the evolution in successive steps. Each step refers to the actual values of wave condition and profile slope that affect both processes. Application of the model to a realistic beach nourishment supplies an indication of the reduction in time of the surface of the sand platform between the initial boundaries, with particular reference to the first months after the filling works. Final considerations are made about the importance each process holds in the evolution.
Keywords: beach fill, free nourishment, longshore diffusion, cross-shore profile evolution.

1 Introduction

This paper takes into consideration the changes that a beach nourishment presents over time after a free beach fill is carried out to increase, for a designed extension, the width of the emerged beach, in order to defend existing coastal structures or to increase the space available as a seaside resort. Free fills, built with sand slightly coarser than the native one to improve the stability of the beach without altering its natural features, are appropriate when wave attacks are predominantly head on. Under this hypothesis the background transport rate has no influences, thus is not taken into consideration in the proposed model. In

WIT Transactions on Ecology and the Environment, Vol 149, © 2011 WIT Press
www.witpress.com, ISSN 1743-3541 (on-line)
doi:10.2495/CP110211

addition the tide oscillation is assumed not significant so that the mean sea level can be taken as fixed.

The most usual building procedure utilizes mechanical equipment which arrange for the spreading and regularisation of the sand. An almost horizontal platform is therefore created with a longitudinal extension (length) usually of the order of some kilometres, a width of the order of several ten of metres, height above the mean sea level (berm height) of 1.5–2 metres, with a slope of about 1/20 and water depth at the toe of 1–2 metres. The initial slope of the beach scarp is steeper than that corresponding to the natural equilibrium profile which in turn depends on the characteristics of the borrowed sand.

The new layout, due to the artificial advance toward the sea of the beach front and due to the new profile, brings about a deep modification of the previous littoral dynamics. There results a rapid reduction in the extension of the platform due both to the strong cross-shore transport toward the sea, with a progressive decrease of the slope of the initial profile (process of evolution of the cross profile), and to a gradient of the longshore transport that, starting from the discontinuity at the unprotected boundaries, spreads the sand towards the lateral beaches (process of longshore diffusion).

Before starting any construction of a beach fill, it is convenient to supply the authority concerned with an indication of the reduction of the platform expected in time. Usually these indications are referred to lengthy times, of the order of years, even if the transformations are more evident in the first period immediately after construction. Indeed, while the process of longshore diffusion evolves slowly, the process of evolution of the cross profile appears quickly, under the effect of the first sea storms. This is also the reason why the models for the description of the two processes are used separately and the diffusion is evaluated with reference to year-averaged wave conditions.

The fate of beach fill over time has been dealt with by several authors. It is worth remembering the EU-COAST3D project which analysed different cases of beach fill in the European context. The results, reported in a special edition of Coastal Engineering (Hamm and Stive [1]), highlighted difficulties in the interpretation of the field observations, as also reported by Kamphuis [2] in a revision on modern Coastal Engineering. Other authors have pointed out the lack of appropriate models to describe the overall evolution of the platform, especially in the immediate term (Elko and Wang [3] and Karasu *et al.* [4]).

In section 2 the models available today for longshore diffusion and for evolution of the cross profile are recalled.

In section 3 the engineering model of simple application proposed here is illustrated which, despite deriving from the models available, is suitable to describe the overall evolution in short times after the end of the construction by taking into account the two processes at the same time with reference to actual sea states. The model can be used either to give a justification of the behaviour found for a beach fill, knowing the characteristics of the recorded sea states, or to formulate a forecast of the changes that a planned beach fill might undergo, starting from a forecast of possible expected sea storms.

In section 4 the proposed model is applied to a littoral stretch and the prevailing importance of the evolution of the cross profile in the first months after the sand supply is highlighted.

2 Reference models

2.1 Models for longshore diffusion

By effect of longshore diffusion of sand a rectangular-shaped platform placed in front of an initially straight shoreline assumes an almost trapezoid ("bell-shaped") shape which extends beyond the initial boundaries and thins down until the littoral comes back to conditions close to the pre-filled ones.

Models for the evolution of the shoreline are usually those, named 1-line, derived from the formula originally introduced by Pelnard-Considère [5] and later developed by several authors, e.g. Dean [6], Hanson and Kraus [7]. The hypotheses are: permanence of the shape of the cross profile, existence of a depth of closure offshore and dependence of the longshore sand transport on the direction of waves relative to the normal to the shoreline. The basic equation is

$$\frac{\partial y}{\partial t}+\frac{1}{h_{*}+B}\frac{\partial Q_{\ell}}{\partial x}=0 \tag{1}$$

in which x and y are the longitudinal and transversal coordinates of a point on the shoreline at time t, h^* is the depth of closure, B the berm height and Q_ℓ the volumetric transport rate of sediment. The angle between the outward normal to shoreline and y axis being small, as usually assumed, the equation of diffusion is obtained

$$\frac{\partial y}{\partial t}=G\frac{\partial^2 y}{\partial x^2} \tag{2}$$

which allows analytical solutions. The longshore diffusivity G is a quantity whose expression depends on that adopted for the volumetric transport rate, being deduced from the derivative of Q_ℓ with x. In an ideal case of an initially rectangular-shaped platform of length ℓ and width Y_0 – an appropriate assumption is a length of the order of 50 times the width as considered here – the width Y is modified by the following

$$Y(x,t)=\frac{Y_0}{2}\left\{erf\left[\frac{\ell}{4\sqrt{Gt}}\left(\frac{2x}{\ell}+1\right)\right]-erf\left[\frac{\ell}{4\sqrt{Gt}}\left(\frac{2x}{\ell}-1\right)\right]\right\} \tag{3}$$

the axis x being on the straight original shoreline, with zero in the central point. It is worth noting that, to obtain with the aforesaid relation appreciable differences between the values of Y and those of Y_0, it is necessary to consider large values of t. The simplified expression for Q_ℓ obtained under the hypothesis of shallow water (CERC formula, Shore Protection Manual [8]) is

$$Q_\ell = \frac{K_\ell H_{sb}^{5/2} \sqrt{g/\kappa}}{16(s-1)(1-P)} \sin(2\vartheta_b) \tag{4}$$

where s is the relative density of the sand, P the porosity, H_{sb} the significant wave height at breaking, ϑ_b the angle of attack relative to the shoreline normal, κ the ratio, assumed as constant, between the wave height and the depth h, and K_ℓ a dimensionless coefficient of experimental determination, dependent on the grain size. For G the following expression is obtained

$$G = \frac{K_\ell H_{sb}^{5/2} \sqrt{g/\kappa}}{8(s-1)(1-P)(h_* + B)} \cos(2\vartheta_b) \tag{5}$$

For small ϑ_b the last term is taken as 1.

In the aim of taking into account, as well as the wave height, also the peak period T_p, the mean bottom slope in the breaker zone S and the median grain size d_{50}, the following expression can be adopted

$$Q_\ell = K_\ell' H_{sb}^2 T_p^{1.5} S^{0.75} d_{50}^{-0.25} \sin^{0.6}(2\vartheta_b) \tag{6}$$

where K_ℓ' is a new dimensional coefficient of calibration. The above equation, proposed by Kamphuis [9], if compared with that of the CERC, leads to generally smaller volumetric transport rate, in coherence both with field observations and with the results that are obtained adopting more recent predictive formulas (Bayram et al. [10]; van Rijn [11]). Starting from eqn (6) the following expression for G is deduced

$$G = \frac{2K_\ell' H_{sb}^2 T_p^{1.5} S^{0.75} d_{50}^{-0.25}}{(h_* + B)} \frac{0.6}{\sin^{0.4}(2\vartheta_b)} \cos(2\vartheta_b) \tag{7}$$

An important indicator of the state of the nourishment is the percentage of platform surface remaining at time t. To obtain an analytical expression of such function, named $M(t)$, Dean [12] integrates the equation of diffusion, eqn (2), assuming a constant value for G, evaluated with reference to an averaged wave height. In the time interval included between the start of the process and the halving of the platform surface Dean suggests the asymptotic expression

$$M(t) = 1 - \frac{2}{\sqrt{\pi}} \frac{\sqrt{Gt}}{\ell} \tag{8}$$

Application of this equation with usual wave climates leads to half-lives of the order of ten years; this result once again highlights that the shoreline change due to longshore diffusion, as already observed, develops over long times.

2.2 Models for the evolution of the cross profile

The evolution of the beach profile associated with the cross-shore sand transport caused by single sea storms depends, besides on the offshore wave conditions, also on the characteristics of the sediments, native and filled, on the initial profile and on the actual storm sea level. Generally reference is made, with different applications, to Kriebel and Dean's model [13] which extends to the dynamic condition the static equilibrium profile concept. The cross-shore sediment transport rate per unit width of beach q results proportional to the difference between the actual energy dissipation per unit volume D defined by

$$D = \frac{1}{h}\frac{d(EC_g)}{dy}, \tag{9}$$

where (EC_g) is the energy flux per unit width, and the value D_{eq} occurring when the profile has achieved the equilibrium shape. It turns out

$$q = K(D - D_{eq}) \tag{10}$$

where K is a dimensional empirical coefficient difficult to determine. Through the equation of conservation of mass

$$\frac{\partial h}{\partial t} + \frac{\partial q}{\partial y} = 0 \tag{11}$$

the evolution of h from the trend of q is obtained.

The equilibrium profile generally adopted is the one proposed by Dean [14] starting from Brunn's model [15], which links the local depth h with the distance y from the shoreline through the dimensional parameter A, which in turn depends on the median grain size d_{50}

$$h = Ay^{2/3} \tag{12}$$

The value of A may be given by the empirical equation (Dean [12])

$$A = 0.067w^{0.44} \tag{13}$$

where A is in $m^{1/3}$ and w, fall velocity of the sediment in cm/s, is linked with d_{50} through the relation

$$w = 14d_{50}^{\ 1.1} \tag{14}$$

with d_{50} in mm. The Dean equilibrium profile is sometimes modified introducing a zone on the beach face with proper constant slope. The two profiles join at the depth at which their slopes are the same.

When this model is applied to the evolution of the cross profile, it is assumed that the volume of sand moved seaward is distributed until the depth of closure and that the shoreward contributions which may be found at the end of sea storms are negligible, due to the great extension of the depositing zone. Knowledge of the time evolution of the cross profile, which reduces the visible extent of the platform, especially in a short time, is of practical interest, but up to today consolidated indications on the time required to reach the equilibrium conditions are not available.

Recently Karasu *et al.* [4] proposed a model for control of beach nourishment that adopts, for the evolution of the cross profile, a new simplified scheme which, to be made operative, requires the specification of empirical coefficients. The scheme, used by the authors to interpret small-scale laboratory experiments with monochromatic waves on sand mounds with a high berm, is based on considerations of dimensional analysis which link the volume eroded with the energy of the incident wave, the geometry of the scarp and the characteristics of the sand. The equations proposed, which correlate the derivative in time of the volume removed V to the difference between the actual slope S of the scarp and the equilibrium slope S_o, are

$$\frac{dV}{dt} = K_c \frac{H^2 B}{(h_t + B)T}(S - S_o)^p \tag{15}$$

$$\frac{dy}{dt} = \frac{2h_t}{(h_t + B)^2}\frac{dV}{dt} \tag{16}$$

$$\frac{dz}{dt} = \frac{1}{\ell^*}\frac{dV}{dt} \tag{17}$$

where dy/dt is the moving back of the shoreline in time, H the incident wave height at the beach fill toe, B the berm height, h_t the depth at the toe, dz/dt the raising of the seabed, assumed uniform between the toe and the depth of closure h^* and ℓ^* the distance in between. The dimensionless coefficients K_c and p must be specified in the conditions under examination. From the results of laboratory tests Karasu *et al.* have deduced that the parameter K_c depends on the grain size and on the initial slope, but not on the wave period. Of this parameter they do not supply any value, not even for the experiments carried out. As regards p, they suggest assuming $p = 2$ as a reference value.

2.3 Applications of the models mentioned

Operative models aimed at quantifying the effect of the processes considered have been proposed by Larson and Kraus [16] and by Work and Dean [17]. These authors, despite recognising that the two processes are interacting in nature, apply the models in a separate way and introduce coefficients deduced from laboratory tests.

For what concerns diffusion, Larson and Kraus propose the 1-line model described in Section 2.1 and apply it to a triangular-shaped platform. The results are qualitatively similar to those obtained with a rectangular platform. For what concerns the evolution of the cross profile, the same authors suggest adopting different models for the four zones into which they divide the whole beach profile. In particular in the surf zone they adopt the equation of equilibrium relative to the balance of energy dissipated per unit of volume with reference to a Dean profile connected to a linear one on the beach face, assuming for K the value 8 $10^{-6} m^4 N^{-1}$. As an example the solution is given for a single sea storm described in analytical form.

Work and Dean [17] apply similar models to monitoring a beach fill. They introduce a model defined by them as 2-line for the longshore diffusion. For what concerns the evolution of the cross profile they still use the Dean's equation of equilibrium and the same beach profile, under the additional hypothesis of shallow water and assuming $K = 0.2\ 10^{-6} m^4 N^{-1}$. They conclude that the model adopted may also describe short-term behaviour, but state that it would not be correct to consider the two processes separately: " *...The next step will be a fully three-dimensional model that accounts for this interaction ...*".

With this in mind, Karasu *et al.* [4] recently proposed a model that links the two processes with one another. For the evolution of the cross profile they adopt their simplified scheme mentioned in Section 2.2. As regards longshore diffusion, the authors integrate directly, with an explicit first-order scheme, the equation of transport of the solid material coupled with Kamphuis' formulation for the volumetric transport rate Q_ℓ given by eqn (6), in which the slope changes in time depend on the progressive evolution of the profile. The authors apply the model step by step with monochromatic waves. They observe that the slope, which changes due to the cross transport, causes moving back of the berm crest and longshore diffusion. Furthermore, keeping the toe of the beach fill steady, they calculate the moving back of the shore line. The model results 3-line (berm, shoreline and toe). The results obtained by laboratory small-scale tests confirm the validity of the model. Despite not having been tested in field the model from Karasu *et al.* represents, from a conceptual point of view, an interesting alternative to the models previously mentioned. To make it operative it is necessary to specify the proper value of the coefficient K_c.

3 The proposed operative scheme

The operative scheme proposed here links the two processes of shoreline diffusion and evolution of the cross profile together assuming temporal steps corresponding to a proper succession of sea states.

3.1 The model adopted for the process of longshore diffusion

For the process of diffusion reference is made to the 1-line model adopting for G the expression obtained starting from Kamphuis' formula. Eqn (7) is simplified writing off the term in which ϑ_b appears, as the value of this term is slightly lower than one for the small values of θ_b here assumed

$$G = \frac{2K_\ell^{'}\, H_{sb}^{\;\;2}\, T_p^{\;1.5}\, S^{0.75} d_{50}^{\;\;-0.25}}{\left(h_* + B\right)} \tag{18}$$

The value $K_\ell^{'} = 0.0023\,\mathrm{m}^{1.25}\mathrm{s}^{-2.5}$ for the calibration coefficient is assumed according to Kamphuis' field and laboratory tests.

To assess the progressive loss of platform surface the simplified expression of *M(t)* proposed by Dean, given by eqn (8), is adopted, but considering for each step the value of G corresponding to the actual values of the wave characteristics and profile slope. In order to use the formula above, in which G is assumed as constant and the origin of the times corresponds to the initial geometry, a scheme is introduced for which, at each time step, a dummy value t^* is evaluated which corresponds to the time interval that would have been necessary to produce the effective diffusion of the platform assuming for G the actual value. The scheme is the following: given at the start of the *i-th* step the values of the remaining platform percentage M_{i-1} and of the slope of the cross profile S_{i-1} obtained in the step *i-1-th*, the value of G_i corresponding to the new breaking wave conditions is calculated; therefore the dummy time t^*, which would have lead to the value M_{i-1}, should G have kept a constant value equal to G_i is calculated. The value of M_i at the end of the *i-th* step is obtained adopting the value G_i associated with the time $t^*+\Delta t$, according to the equations

$$\sqrt{t_i^*} = \left(1 - M_{i-1}\right)\frac{\sqrt{\pi}}{2}\frac{\ell}{\sqrt{G_i}} \tag{19}$$

$$M_i = 1 - \frac{2}{\sqrt{\pi}}\frac{\sqrt{\left(t_i^* + \Delta t\right)G}}{\ell} \tag{20}$$

3.2 The model adopted for the process of evolution of the cross profile

To assess the evolution of the cross profile the volume eroded in the step under examination is evaluated utilizing eqns (15) and (16) proposed by Karasu *et al.*, here applied with the actual values of wave height at the toe of the fill and the slope obtained at the end of the previous step. The scheme requires expression of the equations in terms of finite differences and introduction of an appropriate value for the calibration coefficient K_c. To evaluate a proper value for this coefficient a scaling was carried out starting from the results of experiments by Dette *et al.* [18] to assess, on a large-scale physical model, the moving back and the modelling of beaches of different initial slopes S_v under the action of irregular waves with spectral wave height of the order of one metre. Among the slopes adopted by the authors the slope $S_v = 0.05$ is chosen, with $d_{50} = 0.33$ mm to which corresponds an equilibrium slope $S_o = 0.025$. The values are consistent with those suggested by Creed *et al.* [19] for free beach nourishments. Making reference to the equations proposed by Karasu *et al.* and adopting the value $p = 2$, for the case mentioned the value $K_c = 25$ is obtained. To test the variability of K_c the case $S_v = 0.066$ is also examined for which the value $K_c = 16.5$ was obtained. It can be seen that a decrease of K_c as the initial slope increases is consistent with the fact that an increase in the initial slope makes the evolution faster.

For each Δt, knowing the wave height at the toe of the fill, evaluated taking into account the set-up, the eroded volume and therefore the moving back of the shoreline, the reduction of the platform surface and the slope of the cross profile are all assessed. The scheme adopted is the following: with reference to the *i-th* step the equations are written in finite differences form and the moving back Δy_i and the slope S_i are obtained at the end of the step assuming the depth at the beach fill toe as fixed. The formulas adopted are

$$\Delta y_i = 2\Delta t\, K_c \frac{H^2 B h_t}{(h_* + B)^3 T} (S_{i-1} - S_o)^p \tag{21}$$

$$\frac{1}{S_i} = \frac{1}{S_v} + \frac{\Sigma \Delta y_i}{h_t} \tag{22}$$

where $S_{i\text{-}1}$ is the slope at the end of the previous step. The eroded sand volume, which is distributed up to the depth of closure, gives rise to a rising of the sea bottom that is assumed as negligible in the scheme adopted. The process of evolution of the cross profile is applied only when the wave attack corresponds to "erosion". For this assessment the rule of Larson and Krauss [16] is used as it adapts well to the field data; the erosion occurs when

$$\frac{H_o}{L_o} < 0.00070 \left(\frac{H_o}{w T_p} \right)^3 \tag{23}$$

where H_o is the mean wave height offshore, T_p the associated peak period, L_o the corresponding length and w the already mentioned sediment fall velocity.

4 An example of application of the procedure

The procedure proposed is applied to the hypothetical case of a free nourishment on a beach exposed mainly to head on waves. After having examined several recent sand fills, for the initial platform the following characteristics are assumed: length $\ell = 3{,}000$ m; width $Y = 60$ m, surface 180,000 m^2; berm height $B = 2$ m; depth at the toe of the fill $h_t = 1.50$ m, slope $S_v = 0.05$, resulting fill scarp length 30 m; diameter of the deposited sand $d_{50} = 0.33$ mm. In addition mean natural slope of the sea bottom $S_o = 0.025$ is considered.

For application of the model, reference will be made to a stretch of beach of the Tuscan coast exposed head on to SW seas (*libeccio).* The values H_o, T_p suitable as offshore data are those from the RON buoy of La Spezia. The ones recorded in the year 1998 are assumed because from January to December the data do not show any significant missing information. The values are known at three-hour intervals, thus the same interval is adopted for the time step in the application of the model. For the transfer to the coast of the deepwater data a scheme that takes into account the processes of shoaling and saturation is adopted (Rebaudengo Landò *et al.* [20]) up to the depth for which the effects of non linearity are not important (Gentile [21]); from this depth up to the breaking point only shoaling is considered.

To evaluate the longshore diffusion the values h_b and H_{sb} at breaking are evaluated utilizing Kamphuis' model [22]. To describe the evolution of the cross profile, the erosion conditions being verified by eqn (23), the following procedure is adopted: when $h_b<h_t$, i.e. the breaking takes place on the fill scarp, the wave height corresponding to the depth h_t is assumed; for $h_b>h_t$, the depth-dependent wave height is assumed corresponding to the depth at the toe of the fill increased by a set-up value assumed equal to 20% of the offshore significant wave height (Guza and Thornton [23]).

At the end of the first year after the filling the surface of the platform results 94,200 m^2, equal to about 52% of the original surface of 180,000 m^2, with a moving back of the central zone of about 22.5 m. The reduction is due for about 20% (17,400 m^2) to the diffusion - which induces the spreading of nourishment toward the lateral beaches – and for about 80% (68,400 m^2) by the evolution of the cross profile towards the equilibrium one.

Fig. 1 shows the offshore significant wave height H_0 and the corresponding remaining surface of the platform *RS* due to the evolution of the cross profile and that due to the two processes together. The figure highlights the fact that the most important transformations are found in the initial phase, mainly due to the strong effect of the evolution of the cross profile. At the end of the year the slope

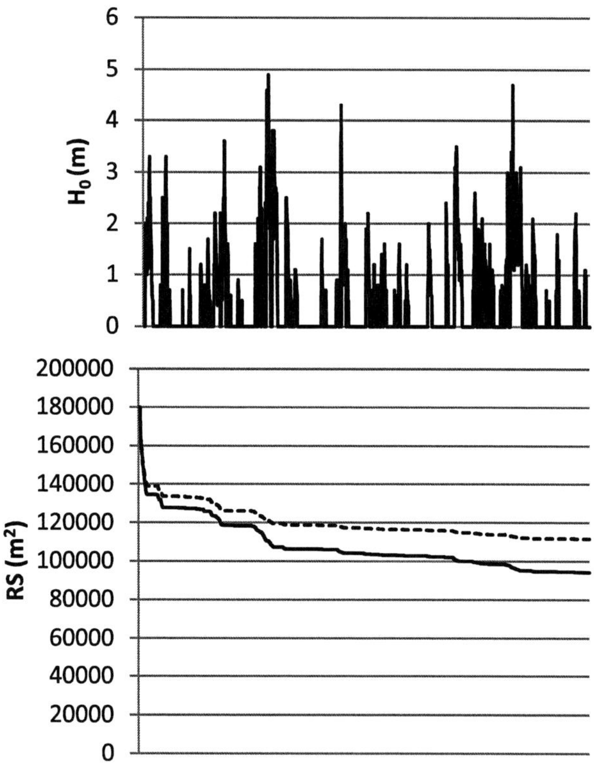

Figure 1: Three-hours values of H_0 from 01.01.98 to 31.12.98 and the relevant values of *RS* (--- only cross profile evolution).

of the profile is close to that of equilibrium (S = 0.028 in comparison with S_o= 0.025) therefore in the following years the evolution of the cross profile will be negligible and the modification due mainly to the process of diffusion.

To outline indicatively the final configuration of the platform, almost "bell shaped", the following first approach scheme is adopted: after having obtained the platform profile corresponding to the surface lost due to diffusion (17,400 m^2) by using eqn (8) with a proper constant value of G, the final plan profile is obtained by scaling the above profile in order to reach the final surface (94,200 m^2). From that procedure a length increased to about 4,500 m compared with the initial 3,000 m, a final width in the centre of about 37.5 m compared with the 60 m of initial fill and a width at the boundaries of about 18 m are obtained.

To highlight the importance of the initial slope of the front of the fill a small increase of this slope to S_v = 0.066, with K_c=16.5, is considered, all other conditions being the same. A faster evolution of the cross profile and a quite small variation in diffusion are obtained. The remaining surface at the end of the year considered, instead of 94,200 m^2 is reduced to 80,250 m^2 corresponding to 45% of the initial one; this result is due for 18% by longshore diffusion and for 82% by the evolution of the cross profile.

5 Conclusive considerations

The model proposed allows one to quantify the reduction that a sand platform built for coastal nourishment undergoes in the short term after the end of construction.

The model takes into account the two processes to which the front face of the fill is subjected due to wave attacks: longshore diffusion of the sand beyond the initial boundaries and evolution of the cross slope from the steep initial profile toward the equilibrium one. It has been highlighted that the most marked effects in the initial phase are produced by the evolution of the cross profile.

Fig.1, relevant to the example presented, shows that the sea storms of the first three months reduces the platform by 34%, of which 87% due to the evolution of the cross profile and 13% due to the longshore diffusion; after six months the overall platform reduction is of 45%, of which 83% due to the evolution of the cross profile and 17% due to the diffusion; at the end of the year the overall reduction is of 48%, of which 80% for the evolution of the cross profile and 20% for the diffusion. These results are in agreement with field observations of the rapid reduction of the platform surface that usually occurs after the sea storms of the first months. With the progressive reduction of the slope of the cross profile the process of diffusion becomes dominant and a slow trend of the shoreline toward the original straight line prevails.

If the calculation were repeated taking into account, as usually done, only longshore diffusion, with the same time history of wave attacks, less realistic results would be obtained because in the first year the moving back of the platform in the central part is not pointed out, thus the scheme, in which in addition the reduction due to diffusion alone is independent of the sequence of storm events but rather depends on their cumulative effects as observed by Dean [24], is not appropriate to follow the short time evolution of beach fill, that is the aim of the proposed model.

Acknowledgement

The study was made possible thanks to a research grant provided by MIUR (n° 2008YNPNT9_003)

References

[1] Hamm L., Stive M. (Ed), Shore nourishment in Europe. *Coastal Engineering., Special Issue*, **47 (2),** pp. 1-264, 2002.

[2] Kamphuis J.W., Coastal engineering - quo vadis?, *Coastal Engineering,* **53**, pp. 133-140, 2006.

[3] Elko N.A. & Wang P., Immediate profile and platform evolution of a beach nourishment project with hurricane influences, *Coastal Engineering,* **54**, pp. 49-66, 2007.

[4] Karasu S., Work P.A., Cambazoglu M.K. & Yuksek O., Coupled longshore and cross-shore models for beach evolution at laboratory scale, *J. of Waterway, Port, Coastal and Ocean Engineering,* **121(3)**, pp. 30-39, Jan./Feb. 30-39 2008.

[5] Pelnard-Considère R., Essai de Theorie de l'Evolution de Formes de Rivage en Plages de Sable et de Galets. IV Journees de l'Hydraulique. *Les Energie de la Mèr, Question III Rapport n 1*, 1956.

[6] Dean R.G., Principles of beach nourishment, *CRC Handbook of Coastal Processes and Erosion.* Komar P. D.(Ed), Boca Raton, pp. 217-232, 1983.

[7] Hanson K., Kraus N.C., GENESIS: Generalised Model for Simulating Shoreline Change, *Eng. Research Center Vicksburg Tech. Rep.,* CERC, 1989.

[8] *Shore Protection Manual*, CERC, Waterways Experiment Station, Vicksburg, USA, 1984.

[9] Kamphuis J.W., Alongshore transport rate of sand,. *Proc. of the 28th Int. Conf. on Coastal Engineering, ICCE 2002*, pp. 2478-2490, Cardiff, 2002.

[10] Bayram A. Larson M. & Hanson H., A new formula for the total longshore sediment transport rate, *Coastal Engineering,* **54**, pp. 700-710, 2007.

[11] van Rijn L.C., Longshore sand transport, *Proc. of the 28th Int. Conf. on Coastal Engineering, ICCE 2002*, pp. 2439-2459, Cardiff, 2002.

[12] Dean R.G., Coastal sediment processes: toward engineering solutions, *Proc. Costal Sediments,* ASCE, pp. 1-24, 1987.

[13] Kriebel D.L. & Dean R.G., Numerical simulation of time-dependent beach and dune erosion, Coastal Engineering, **119**, pp. 221-245, 1985.

[14] Dean R.G., Equilibrium beach profiles: characteristics and applications, *J. of Coastal Research.*, **7(1)**, pp. 53-84, 1991.

[15] Bruun P., Coast Erosion and Development of Beach Profiles, *U.S.Army Tech. Mem.,* **44,** Vicksburg, 1954.

[16] Larson M. & Kraus N., Mathematical modelling of the fate of beach fill, *Coastal Engineering,* **16**, pp.83-114, 1991.

[17] Work P.A. & Dean R.G., Assessment and prediction of beach nourishment evolution. *J. of Waterway, Port, Coastal and Ocean Engineering,* **121(3)**, pp. 182-189, May/June 1995.

[18] Dette H.H., Larson M., Murphy J., Newe J., Peters K. Reniers A. & Steetzel H., Applications of prototype flume tests for beach nourishment assessment, *Coastal Engineering*, **47,** pp. 137-177, 2002.

[19] Creed C.G., Bodge K.R. & Suter C.L., Construction slopes for beach nourishment projects, *J. of Waterway, Port, Coastal and Ocean Engineering,* **126(1)**, pp. 57-62, Jan./Febr, 2000.

[20] Rebaudengo Landò L., Gentile R. & Scarsi G., A nonlinear spectral model for directional random waves at decreasing depths, *J. of Offshore and Polar Engineering,* **9(2)**, pp 81-89, 1999.

[21] Gentile R., A procedure to evaluate the longshore sediment transport due to random waves, *Proc. of the 10th Int. Offshore and Polar Engineering Conference*, III, pp. 673-680, Seattle, 2000.

[22] Kamphuis, J.K., Incipient wave breaking, Coastal *Engineering*, **15**, pp 185-203, 1991.

[23] Guza R.T. & Thornton E.B., Wave set-up on a natural beach, *J. of Geoph. Research,* **86 C5**, pp. 4133-4137, 1981.

[24] Dean R. G., Beach nourishment projects as large scale experiments in nature. *Proc. of the 28th Int. Conf. on Coastal Engineering, ICCE 2002*, pp. 3649-3661, Cardiff, 2002.

www.witpress.com, ISSN 1743-3541 (on-line)

Section 8
Pollution and dispersion

Application of the MSB model to ballast water exchange in tidal embayments

S. Mecca[1], S. Morris[2] & R. Barber[3]
[1]*Department of Engineering-Physics-Systems, Providence College, Providence, USA*
[2]*General Dynamics, USA*
[3]*Computational Engineering Group, STFC Daresbury Laboratory, Warrington, UK*

Abstract

For safety reasons, at a captain's discretion, older vessels are often forced to exchange ballast water while in port. This paper examines questions of the optimal strategy for such ballast water exchange. The MSB (Mecca-Severino-Barber) model for well-mixed tidal basins is applied to this marine operation, parameterising the start of the ballast water release at different times in the tidal cycle. This enables the identification of the optimal release time for a typical single pump, single chamber vessel exchanging water in a hypothetical tidal embayment. In practical applications, the optimisation procedure is somewhat more complicated due to the fact that the problem can have multiple objectives: minimising the maximum pollutant concentration as a result of the ballast water exchange and/or minimising the relaxation time *i.e.* the time taken for the pollutant concentration to reduce to an acceptable level. This necessitates the application of multi-objective optimisation.
Keywords: tidal prism, pollution flushing, tidal embayment, ballast water exchange, multi-objective optimisation.

1 Introduction

The MSB (Mecca-Severino-Barber) single embayment model [1] enables a graphical and numerical representation of the process of pollution flushing from a well-mixed tidal embayment temporally loaded with a known quantity of pollutant. The model, which is written in Stella (Stella is a product of ISEE

www.witpress.com, ISSN 1743-3541 (on-line)
doi:10.2495/CP110221

Systems, Inc. of Hanover, New Hampshire, USA), allows for dynamic loading of pollutants [2] and realistic basin bathymetries. The essential rate equations for the pollutant concentration in the embayment have been discussed by Barber [3] and can be derived from the following mass flow equation:

$$\frac{d(CV)}{dt} = C\frac{dV}{dt} + V\frac{dC}{dt} = k + QC \tag{1}$$

where Q is the volumetric discharge across the entrance of the embayment, k is the pollution loading rate (measured as a mass per unit time) and C and V are the instantaneous concentration and volume of the embayment at time, t, respectively. During the ebb flow period, $dV/dt = Q$, allowing the QC term on the right-hand side of eqn. (1) to be cancelled with the CdV/dt term on the left.

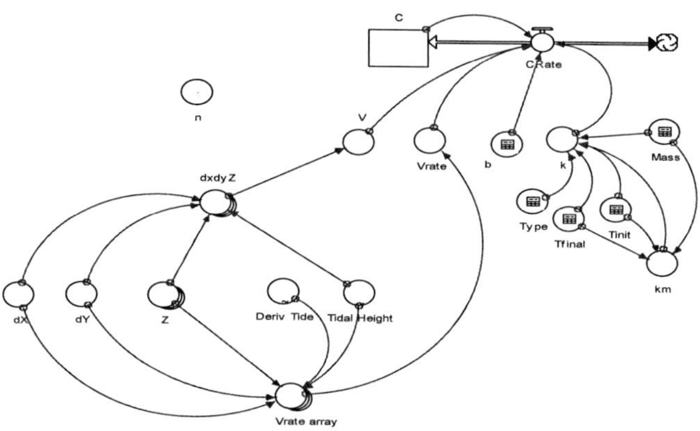

C(t) = Concentration of pollution
CrateFlood = Rate of change in pollution concentration during flood tide.
CrateEbb = Rate of change in pollution concentration during ebb tide.
b = Pollution return factor
k = Pollution loading rate (mass per unit time)
km= The highest value of k
n = Internal variable used to keep track of tidal cycles
Period = 12.42 hours in a tidal cycle
Tidal_Height =height above mean low tide (m)
V= Volume of the embayment (m3)
Vrate = Change in volume V over time based on tides
dX= Length of grid block
dxdyZ[i,j] = Volume of every block in the embayment grid at time (t)
dY = Width of grid block
max Tide = Max tidal swing (m)
Deriv Tide = Derivative of Tidal Height
Z[i,j] = Bathymetric data array for embayment
Mass=The total amount of pollutant to be dumped during a tidal cycle (kg)
Tinit=The time at which the polluting starts (hrs)
Tfinal=The time at which the polluting ends (hrs)
Type=This paramter decides the type of loading function to be used.
Type 1 is a constant loading
Type 2 is a user definable step function that is 0 until Tinit and is a constant value between Tinit and Tfinal and 0 again after Tfinal.
Type3 is an inclining ramp function with a definable start and end time.
Type 4 is an instaneous pulse.

Figure 1: Schematic representation of the single embayment MSB model.

During the flood flow period, the *QC* term on the right-hand side of the equation is zero provided there is no pollution return flow into the basin. This results in two rate equations for dC/dt, one for the ebb and another for the flood conditions. The MSB model employs the basic state variable approach used in *Systems Dynamics* to represent these rate equations, treating the time-varying pollutant concentration, *C*, in the embayment as a level. The single embayment model is shown graphically in Fig. 1 and described in references [1,2,4,5].

The return of polluted water during the flood tide is governed by an empirical factor, *b*, that is used to modify the volume of the tidal prism; this methodology has been discussed by Barber [3] and more theoretically by Sanford *et al.* [6]. The volume modifying factor is taken as (1-*b*), where *b* varies between 0, for no return flow, and 1, for total return flow. Recent studies using the MSB model have shown that the value of *b* can be mapped onto a physical interpretation of the ratio of the outer to inner basin volumes [2,4]. The MSB model has been validated against both physical tank tests [1] and rhodamine dye studies for single embayments [5] and more recently for concatenated basins [7].

The present study examines the scenario of a ship exchanging its ballast water in a tidal embayment using one, two, or three constant flow rate pumps operating for a sufficient length of time to exchange the ballast water, beginning at a particular time in the tidal cycle. Current regulations strongly recommend that ships exchange their ballast water outside of port. However, the regulations often leave the final decision to the discretion of the ship's captain who may decide for safety reasons to exchange ballast water in port to avoid the dangers associated with a change in the relative position of the centre of gravity and the centre of buoyancy of the ship [8]. Indeed, as ballast water is emptied from a single tank vessel, the centre of gravity and the centre of buoyancy move closer together increasing the potential for instability especially in rough sea conditions. Most modern vessels have multiple tanks/compartments and often multiple pumps. A brief summary of the ballast arrangements on a range of ships is shown in Table 1.

Table 1: Ballast arrangements on a range of ships – adapted from. Armstrong *et al.* [9].

Type Ship	Example	# of Pumps	Pumping Rates (m^3/hr)	# of Tanks	Total Capacity(m^3)
Bulk Carrier	Ormond (P&O)	2	2500	12	59568
Tanker	Similar to Bulk Carriers			3	
Containership	Jervis Bay (P&O)	1	500		16613
Cruise	Grand Princess (P&O)	1	250	29	4345

If a captain decides that it is best to exchange *water* whilst in port, the question arises as to when the pumping should be initiated. This decision will depend on many factors some of which relate to fees for having to wait in port longer than necessary, and other potentially more sensitive environmental issues such as consideration of both the maximum pollution concentration as well as the pollution relaxation time. It should be noted that often an overriding issue is that of invasive species taken on by ballast water at a source port and discharged at a destination port. Samples of ballast water reported in the work of Ivanov [10] show levels of eubacteria from 0.7 to 39.5%, enterobacteria from 0 to 2.5%, vibrio spp. from 0.2 to 35.8%, and E. coli. from 0 to 2.5%. In smaller embayments, the release of such ballast water can result in significant pollutant concentrations. The work of Drake *et al.* [11] concluded that ~10^{20} micro-organisms are discharged annually to the Port of Hampton Roads in the lower Chesapeake Bay.

2 Application of the MSB model

The model was applied to a typical tanker ballast exchange using constant rate pumping beginning at a particular time after high tide and continuing until the ballast water had been emptied. For the purposes of the study, the bathymetry of the embayment and the tidal variations were based on Great Salt Pond [5] which is situated on Block Island at the entrance to Long Island Sound. The release time was varied and the pollution concentration was tracked using the tidal variation shown in Fig. 2.

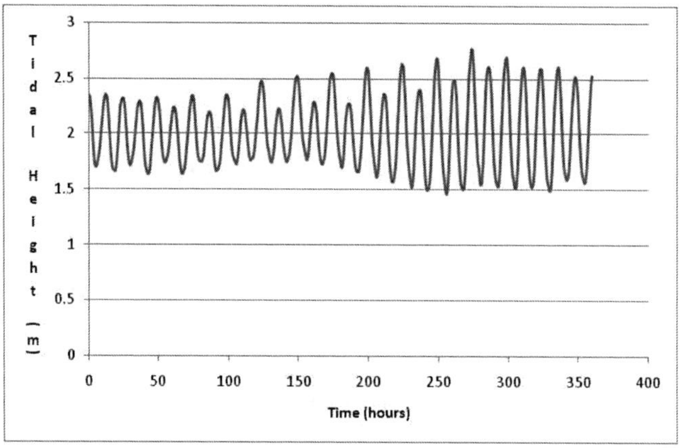

Figure 2: Tidal variation used in the simulations.

Both the maximum concentration and the time for the concentration to relax to 10% of its maximum value were tracked. The parametric simulations were carried out assuming one-, two-, and three-pump operations from a tanker. In addition, the pollution return parameter, *b*, was varied from 0 to 0.5 to determine the effects of the pollution return flow. In total, 192 simulations were performed to yield the results shown in Fig. 3.

The shape of the concentration profiles were found to be affected by the length of the release which is represented in the model as a pulse originating at the start of the release and terminating when the entire ballast contents have been emptied. Since the pumping rate is assumed constant, a square pulse is used for the pollution loading factor. While the conditions represented by these simulations are somewhat hypothetical (though we did use bathymetry and tidal

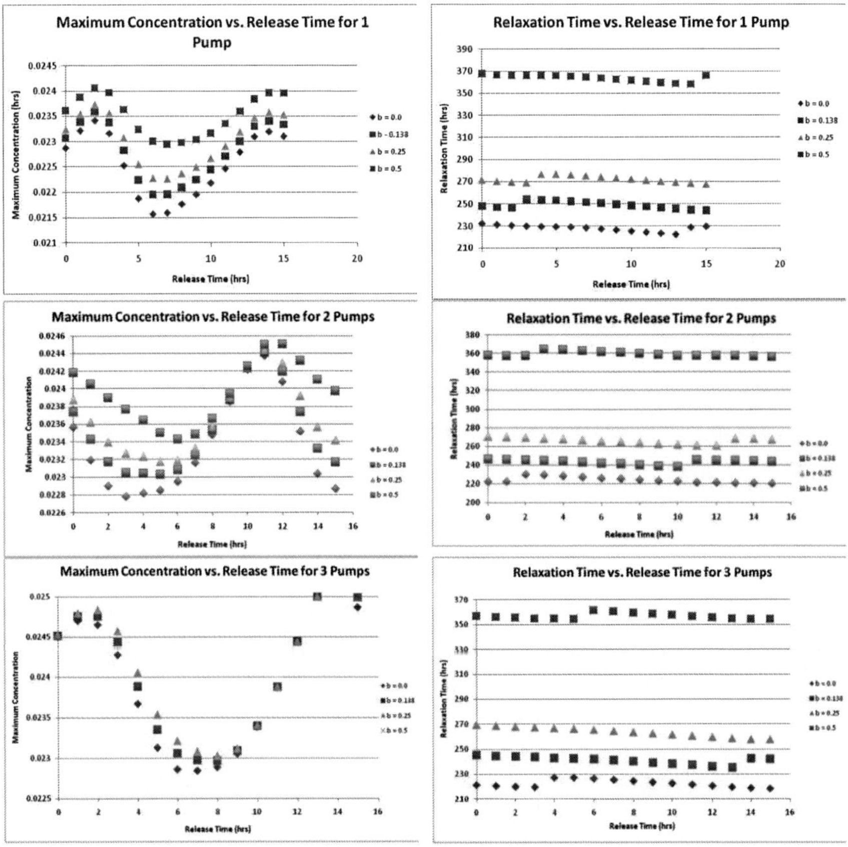

Figure 3: Maximum concentrations and relaxation times versus the release time for one-, two- and three-pump ballast exchange scenarios. See the text for details.

profiles consistent with an embayment used in a previous study [5]), there are nevertheless some interesting features in the results. The differences in maximum concentrations are rather small and the relaxation times span a week or more and are rather insensitive with respect to the release times. However, the concentration profiles have shapes similar to those observed in systematic studies of instantaneously released pollutants [2].

3 Optimisation considerations

It is informative to consider the results from the viewpoint of the optimal release times. Tables 2 and 3 give the release times corresponding to the smallest maximum concentrations and shortest relaxation times, respectively.

Referring to Table 2, if the objective is to minimise the pollution concentration in the embayment, then release times for a one-pump operation should be approximately 6-7 hours after high tide and this is relatively insensitive to the return flow parameter. For two-pump or three-pump operations, there is a similar insensitivity to b though the optimum release times vary with the number of pumps used. It should be noted that the table shows release times beyond 12.43 hours, the primary tidal period. If the tidal cycle was a pure sinusoidal function, the results for 12.43 hour release times would obviously be identical to those for 0 hour release times.

Referring to Table 3, if the objective is to minimise the relaxation time (here defined as the time for the pollutant concentration to return to 10% of its maximum value), then the release times vary quite markedly for different numbers of pumps with little sensitivity to b in the three-pump release times and modest sensitivity to b in the one- and two- pump scenarios.

Clearly, the results will depend on the bathymetry and the tidal variations in the embayment, and therefore a one-size fits all specification of optimal release time is not suggested.

Table 2: Optimal release times for smallest maximum concentration for four values of b and three different pumping scenarios.

No. of pumps	b			
	0.0	0.138	0.25	0.5
1	6	6	7	7
2	3	5	5	6
3	15	13	15	15

Table 3: Optimal release times for shortest relaxation times for four values of b and three different pumping scenarios.

No. of pumps	b			
	0.0	0.138	0.25	0.5
1	12	2	3	12
2	15	10	12	15
3	7	8	8	8

Finally, an obvious question is what is the optimum release time if both criteria, i.e. minimising the maximum concentration and shortening the relaxation time, are equally important? Indeed, when bacteria counts exceed a certain maximum, commercial and recreational activities may be curtailed. One also observes in many basins that shell-fishing and other activities may not be re-opened until pollution levels have relaxed to a certain value. So, both maximum levels of pollutant and times for return to pre-pollutant conditions may simultaneously be part of the objective function. This immediately suggests a multi-objective optimisation. If one can assign cost functions to each of these factors or otherwise assign weights to the multiple objectives, then usual optimisation schemes can be applied. Such assignments may not be too difficult as activities such as shell-fishing have measurable economic impact; recreational activities might be less tangible yet numerous studies of the impact of the BP Gulf of Mexico oil spill were able to assess these impacts. Lastly, the multi-objective optimisation problem can be simplified by a natural limiting of the domain of the problem by imposing constraints such as insisting that the maximum pollution concentration be less than some level related to a serious threat to health or to a threshold beyond which there would be an irreversible impact.

4 Conclusions

The MSB model for well-mixed basins has been applied to an important marine operation, namely, the discharge of ballast water from ships in port. The present study parameterises the optimum start time with respect to high tide. This has been carried out for one-, two- and three-pump operations for a hypothetical though not untypical ship in a somewhat hypothetical embayment to enable the question of when is the optimum time to commence the release of polluted ballast water. The answer(s) depend on whether the objective is to minimise the maximum pollutant concentration in the embayment or to minimise the time for the pollution to return to a particular fraction of the maximum value. More likely, the problem becomes a multi-objective optimisation question if one is trying to accomplish both. The results demonstrate the need to simulate basin-specific conditions including bathymetry, tidal variations at the time of the water ballast operations and, to a certain extent, the pollution return conditions of the tidal embayment. Finally, while the problem addressed in this paper is a real one for a number of vessels in existing shipping fleets, as retrofits and new ships come on line, there a number of ballast water management options, such as heat treatment using engine cooling water, filtration, uv irradiation, oxygen deprivation, shore-based discharge and treatment, and various electro-mechanical systems, that offer the promise that the problem will someday be eradicated. Such options have been in engineering and political discussions for some time [12].

References

[1] Mecca, S., Severino, C. & Barber, R., Pollution flushing models in Stella, *Development and Application of Computer Techniques to Environmental Studies X*, edited by G. Latini, G. Passerini & C.A. Brebbia, pp. 125-134, WIT Press: Southampton, UK, 2004.

[2] Mecca, S., Barber, R. & Tombarello, J., Alternative loading functions in the MSB pollution flushing model, *WIT Transactions on Ecology and the Environment*, **88**, pp. 89-97, edited by C.A. Brebbia, WIT Press: Southampton, UK, 2006.

[3] Barber, R.W., Analytical modeling of pollution flushing in well-mixed tidal embayments (Chapter 17), *Advances in Coastal Modeling*, edited by V.C. Lakhan, Elsevier Science B.V., 2003.

[4] Mecca, S., Severino, C. & Barber, R., Pollution flushing in cascading tidal basins, *WIT Transactions on the Built Environment*, **78**, pp. 93-100, edited by M. da Conçeicao Cunha & C.A. Brebbia, WIT Press: Southampton, UK, 2005.

[5] Mecca, S., Barber, R., Mellor, G. & Walsh, G., Application of the MSB pollution-flushing model to Great Salt Pond on Block Island, *WIT Transactions on Ecology and the Environment*, **104**, pp. 189-197, edited by C.A. Brebbia & K.L. Katsifarakis, WIT Press: Southampton, UK, 2007.

[6] Sanford, L.P., Boicourt, W.C. & Rives, S.R., Model for estimating tidal flushing of small embayments, *Journal of Waterway, Port, Coastal and Ocean Engineering*, **118 (6)**, pp. 635-654, 1992.

[7] Wegemer, C., Mecca, S. & Barber, R., Application of the MSB coupled embayment pollution-flushing model to Queenstown Creek, accepted for publication in spring 2011, *River Basin Management*, WIT Press.

[8] Abramowski, T. & Abramowski, E., Prevention of sea pollution from ships - The latest updates of standing regulations, *Marine Technology III, The Built Environment, Volume 45,* edited by: T. Graczyk, T. Jastrzebski & C.A. Brebbia, WIT Press, Southampton, UK, 1999.

[9] Armstrong, G. *et al.*, Ballast system design for flow-through exchange of ballast water, *Trans. I. Mar. E.*, **109 (3)**, pp. 257-269, 1997.

[10] Ivanov, V., Bacteriological monitoring of ships' ballast water in Singapore and its potential importance for the management of coastal ecosystems, *Environmental Toxicology*, edited by A.G. Kungolos, C.A. Brebbia, C.P. Samaras and V. Popov, WIT Press: Southampton, UK, 2006.

[11] Drake, L.A., Doblin, M.A. & Dobbs, F.C., Potential microbial bioinvasions via ships' ballast water, sediment, and biofilm, *Marine Pollution Bulletin*, **55**, pp. 333–341, 2007.

[12] Taylor, A.H. & Rigby, G., Suggested designs to facilitate improved management and treatment of ballast water on new and existing ships, *Discussion Paper prepared for Agriculture, Fisheries and Forestry Australia as part of the Research Advisory Group Ballast Water Research and Development Program,* 2001.

Environmental impact assessment based on a potential oil spill scenario: Eastern Black Sea coastal zone, Turkey

Ö. Acır[1], R. Aps[2], M. Fetissov[2] & M. Sert[3]
[1]*Petrako Oil & Gas Exploration Ltd., Ankara, Turkey*
[2]*Estonian Marine Institute, University of Tartu, Estonia*
[3]*Turkish State Meteorological Service, Ankara, Turkey*

Abstract

The coastal fishery made in the Eastern Black Sea coastal area satisfies the country's major demand for fish and seafood. The area has also attracted attention with the increasing international oil and gas exploration activities carried on recently. Large capacity tankers pass through this sea area to carry crude oil and petrochemical products while existing and planned hydrocarbon transportation pipelines increase the cargo traffic as well. Therefore the new environmental risks are threatening this region's ecosystem due to the potential events such as tanker accidents, well explosions or pipeline leaks. In this study, the off-shore oil spill scenario is modeled for the Eastern Black Sea area. Potential Incident Simulation Control and Evaluation System (PISCES 2) is used to simulate the spreading of the spilled oil. The following factors are taken into consideration in the mathematical model: 1) environmental parameters: coastline, field of currents, weather, wave height and water density, 2) physical properties of spilled oil: specific gravity, surface tension, viscosity, distillation curve and emulsification characteristics, 3) properties of spill sources. Results of the oil spill scenario modeling shows extremely negative environmental impact of the potential oil spill accident on the coastal zone of the Eastern Black Sea.
Keywords: environmental impact assessment, net environmental benefit analysis, Bayesian inference, oil spill simulation, Eastern Black Sea.

WIT Transactions on Ecology and the Environment, Vol 149, © 2011 WIT Press
www.witpress.com, ISSN 1743-3541 (on-line)
doi:10.2495/CP110231

1 Introduction

Recent oil spill occurred in Gulf of Mexico in the summer of 2010 has been the worst oil related accident in history. The event resulted in 11 people killed, more than 4.9 million barrels of crude oil have spilled to the sea (more than 60,000 barrels per day) whilst 42,000 people worked with 7,000 boats and more than 100 airplanes attended the response operations [1]. Operating company and US officials estimated the cost of overall damages and claims that they may reach to USD $20 billion in time as well. However this is not the only or unusual event that the world seas encountered. In US coasts only, there were 251 events reported in 1964–2009 where the amount of oil spilled to the sea is more than 260,000 barrels [2]. Besides US, such events also occur all around the world. Europe and Middle East regions are the other environmentally risk prone areas due to the heavy oil transportation traffic (Table 1).

Table 1: Major oil spills in Europe and the Middle East coasts [3].

Country	Dates	Bbbl
Spain	2002	441,000
United Kingdom	1996	504,000
United Kingdom	1993	595,000
Spain	1992	518,000
Italy	1991	1,008,000
Iraq	1991	5,740,000
Spain	1989	560,000
Iran	1985	490,000
Iran	1983	1,820,000
Greece	1980	700,000
Turkey	1979	715,000
Ireland	1979	448,000
France	1978	1,589,000
Spain	1976	700,000
Portugal	1975	616,000
Iran	1972	805,000

As seen in Table 1, the only spill accident in Turkey occurred at 1979 where the oil tanker "Independenta" carrying crude oil from Libya to Romania, crushed into another cargo ship at the Bosporus, which resulted a big explosion. According to the records, 43 people in the ship died in this tragic event whereas the fire could not be deflated through 27 days [4].

For the last couple of years, the load on Turkish coasts due to hydrocarbon and petrochemical transportation has been increasing sharply. One of the main reasons is that off-shore Black Sea coasts have been more attractive for international oil and gas exploration activities [5]. On the other hand Eastern

Black Sea coasts of Turkey provide the country's major fish and seafood needs [6]. Therefore new environmental risks are threatening this region's ecosystem due to the potential events such as tanker crashes, well explosions or pipeline leaks.

In this study, a scenario based oil spill is modeled for the off-shore Eastern Black Sea area while physical impact parameters of this event are simulated in special software. The long term wave statistics and oceanographic data sets were used for this purpose while the potential spill location is selected hypothetically (Figure 1). The scope of this study is to understand the nature and environmental effects of such a potential disaster and help local authorities to develop an action plan accordingly.

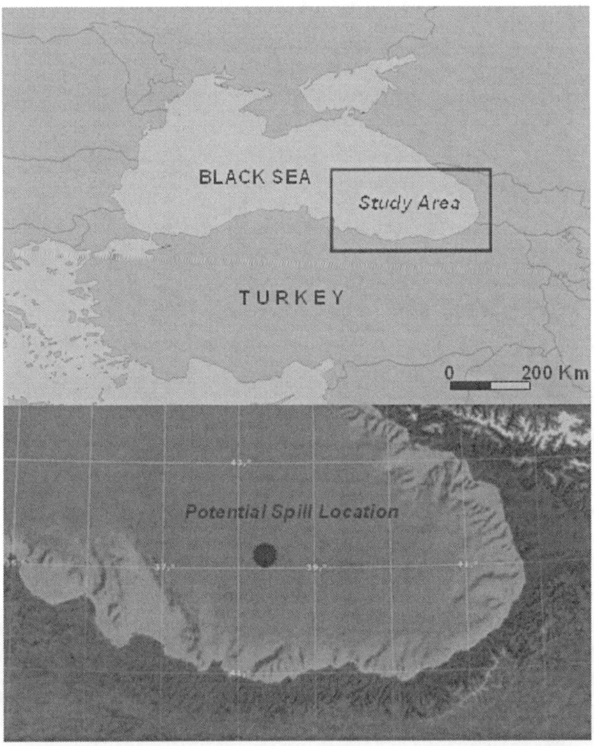

Figure 1: Study area with the position of the potential spill location.

2 Material and methods

PISCES 2 (Potential Incident Simulation Control and Evaluation System) of TRANSAS GROUP is used to simulate development of an oil spill incident scenario [7]. The PISCES 2 spill model simulates the processes in an oil spill on the water surface: transport by currents and wind, spreading, evaporation, dispersion, emulsification, viscosity variation, and interaction with the coastline (stranding or beaching).

In reality each accidental oil spill is unique in relation to the accident position, season and the accident time, the hydro-meteorological conditions, the type and the amount of the spilled oil. Therefore, the selection of scenarios is largely depending on the properties of the particular study. In our case two randomly selected scenarios were calculated with aim to get the general impression of the impact magnitude of the medium size oil spill under the most prevailing hydro-meteorological conditions.

Accordingly, two oil spill scenarios were simulated with incident position: Lat 42 03.934 N, Lon 38 22.110 E, spill rate: 4.9 t/h, and the duration 120 h and 240 h respectively. Hydro-meteorological conditions during the scenario calculations in September 2010 are imported from Turkish State Meteorological Service forecast for 5 days: sea surface temperature 24°C, air temperature 19–28°C, wind direction N, NE, NW and force 2–9 knots, sea surface current – Internal PISCES 2 Map of Surface Currents, direction SE, E, S and force 0.09–0.2 knots.

The Bayesian Belief Network (BBN) is constructed using the HUGIN RESEARCHER software. BBN is used to assess the general situation when answering the question: can the particular oil spill be combatted at sea using booms and skimmers? Current and wind speed in the oil incident sea area, predicted time interval of oil coming ashore are imported from PISCES 2 simulations. Mobilization time – the time needed for a ship/aircraft to arrive to an oil incident scene depends on the time the vehicle needs to get ready and the time needed to reach the location of the spill – is based on expert knowledge.

3 Results and discussion

3.1 What are the expected drift, behaviour and fate of the spilled oil?

Net Environmental Benefit Analysis (NEBA) is defined as a method to determine the most appropriate response option(s) in order to minimize the overall environmental impact of an oil spill [8, 9] and this method is often used for oil spill related risk assessment [10, 11]. Immediately after notification of a pollution incident at sea, the NEBA is to be performed, and a quick decision is to be taken on the most appropriate response option(s). This decision is based on the following information [8]: 1) what are the expected drift, behavior and fate of the spilled oil, 2) can the oil spill be combatted at sea, and 3) is the oil threatening a sensitive resource?

According to the calculation results of the Scenario 1 presented in Figure 2 the 120 hours of continuous oil spill resulted in the 4359 bbl (100%) of the spilled oil. Of that amount of the spilled oil the 2428 bbl (58%) was floating, 1882 bbl (39%) was evaporated and the rest of 50 bbl (3%) was dispersed and emulsified. The average speed of the spilled oil was 0.21 km/h.

When getting on the sea surface, oil spreads out generating an oil film, which drifts in response to wind and currents. The simulation takes into account the following factors: 1) shore description, 2) representation of field of currents, 3) data on meteorological conditions: water temperature, wind direction/force, 4) spill parameters: time, place and kind of spill; amount and properties of the oil product.

For computation of oil movement under action of wind and currents, the slick travels in the current direction at the current velocity, and in the wind direction at a speed equal to 3% of the wind velocity. The resulting movement direction is geometrical addition of wind direction and current direction

It is necessary to assess the dimensions of the entire pollution area which oil is getting on from the scenario start time up to scenario end time. Figure 2 is showing the area where oil was present at any moment of time from the scenario start to the current model time moment, in grey color.

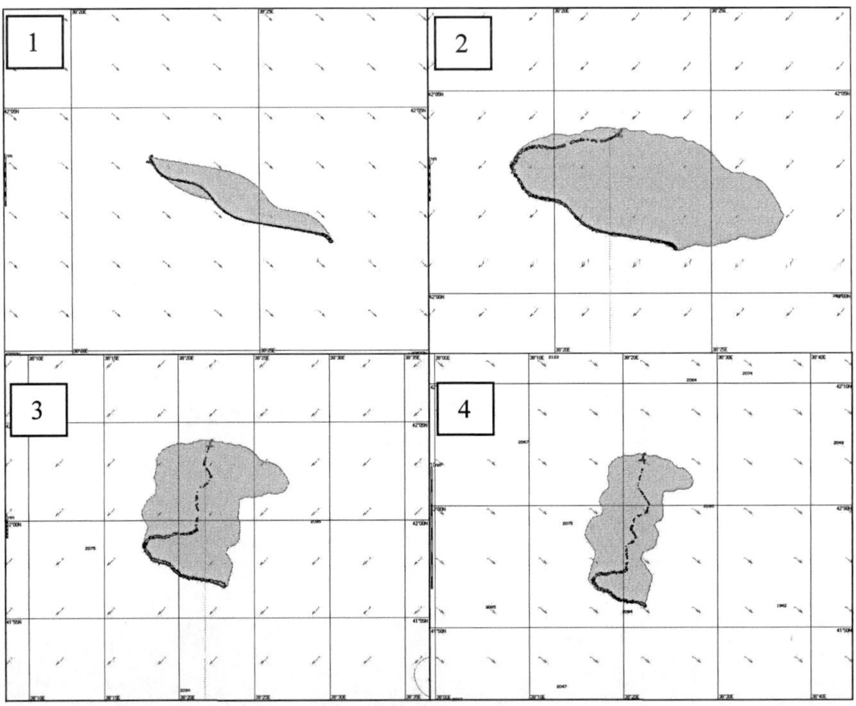

Figure 2: Display of pollution area for Scenario 1 after 1) 30 h, 2) 60 h, 3) 90 h and 4) 120 h of simulated accidental continuous oil spill.

Scenario 2 calculations showed that (Figure 3) during 240 hours of continuous oil spill the total amount of spill resulted in 8560 bbl (100%). Of that amount of the spilled oil the 3434 bbl (42%) was floating, 4482 bbl (48%) was evaporated, 149 bbl (4%) was dispersed and emulsified, and the 495 bbl (6%) was stranded. Calculated slick maximum thickness was 14.4 mm. Oil slick area (except footprint) was estimated 25 km^2 and the pollution footprint 1792 km^2.

The average speed of the spilled oil was estimated at 0.58 km/h and the total length of the polluted shoreline was estimated at 10 km.

For the description an oil slick, the PISCES model uses Lagrangian approach, which is an effective technique of oil spill numerical simulation. The oil spill is represented by an ensemble of particles moving under the effect of wind and current [7]. The distinctive feature of the model is the extension of Lagrangian approach by introducing interactions between oil particles. This allows some essential lacks of the traditional Lagrangian methods to be compensated, and oil interaction with different kinds of natural and artificial barriers to be described in a satisfactory manner.

As it was already mentioned above, the movement of an oil slick is assumed to be two-dimensional and is resulting from transport by currents and wind, spreading and stochastic diffusion. Diffusion adds a random component to this movement.

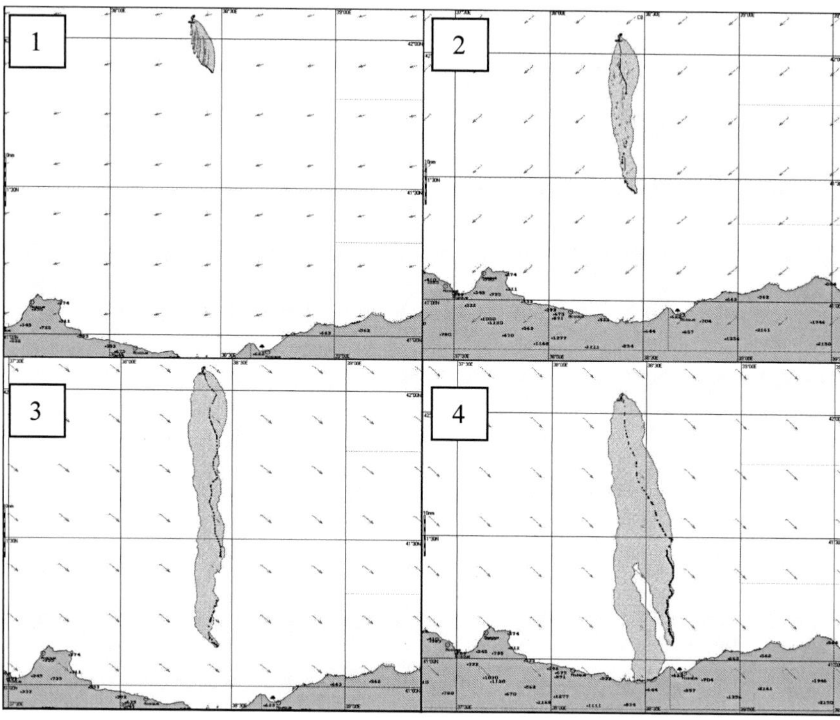

Figure 3: Display of pollution area for Scenario 2 after 1) 60 h, 2) 120 h, 3) 180 h and 4) 240 h of simulated accidental continuous oil spill.

3.2 Can the oil spill be combatted at sea?

According to [12] booms and skimmers are usually the first technique employed to remove the oil from marine environments but this technique can usually recover relatively small proportion of the spilled oil. At the same time, no boom is capable of containing oil against water velocities in excess of 0.58 m/s acting at right angles to it. It is estimated that the critical current velocity for many crude oils and refined products ranges from 0.7 (0.34 m/s) to 1.2 knots (0.58 m/s) while the 0.7 knots (0.34 m/s) is accepted as a conservative estimate. Skimmers are used to remove oil from water and put it into storage tanks but how well a skimmer works depends on the type of oil spilled, the thickness of the slick and, the weather conditions. It is also estimated that the wave height of about 1.5 m is critical for the most types of booms to work efficiently.

Monthly average wave height and the wind speed in the sea area concerned in 2009 are presented in Figure 4.

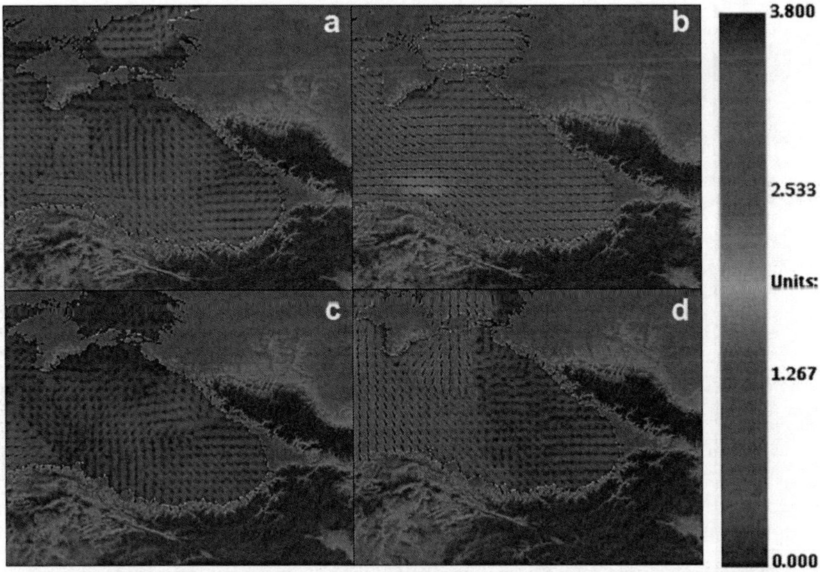

Figure 4: Monthly average wave height (m) in the study area in 2009: (a) January (b) March (c) June (d) September.

Referring to the basic hydro-meteorological characteristics of the sea area concerned (Figure 4 and 5) it is concluded that in general there are quite favorable conditions for the use of booms and skimmers in the case of accidental oil spill. The monthly average wave height in some parts of the study area was just about 1.5 m in March 2009 while in January, June and September the average wave height did not exceed 1 m. The monthly average wind speed was the highest in March 2009 reaching in some parts of the study area speed of 24 knots (12 m/s). At the same time the average wind speed in January, June and September was about 12 knots (6 m/s).

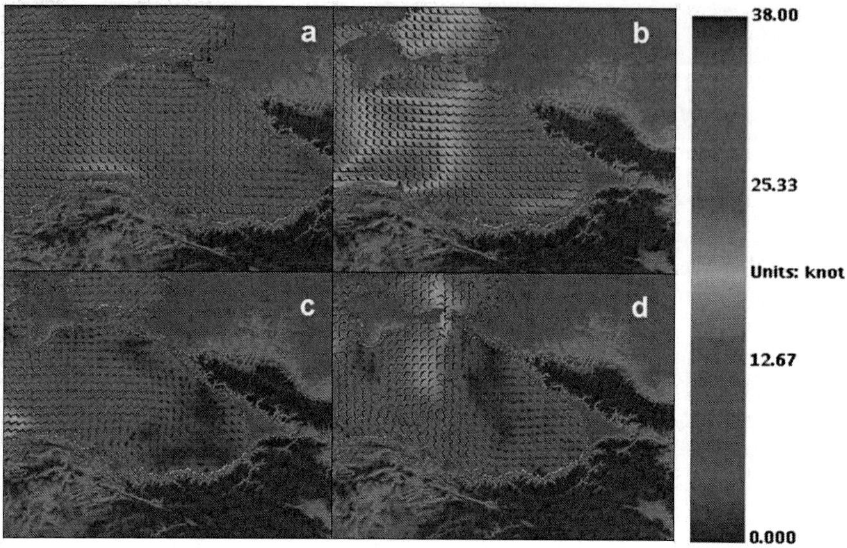

Figure 5: Monthly average wind speed (knots) in the study area in 2009: (a) January (b) March (c) June (d) September.

The BBN is constructed to assess general situation when answering the question: can the particular oil spill be combated at sea using booms and skimmers? Figure 6 communicates the basic conditions for efficient use of booms and skimmers.

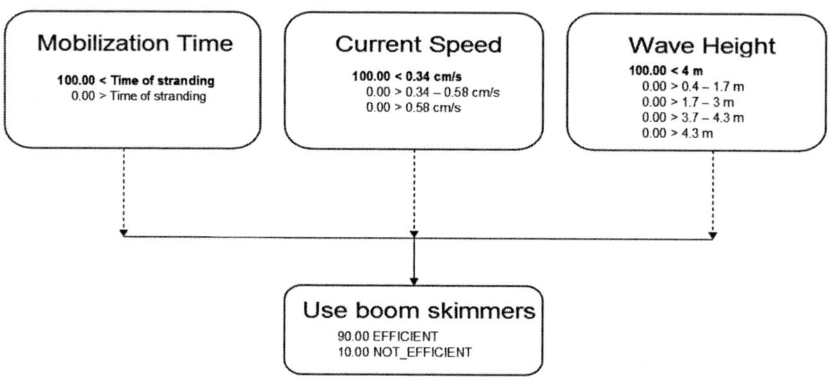

Figure 6: Low current < 0.34 m/s, and a wave height < 0.4 m. mobilization time is less than time for oil to wash ashore. Use of booms and skimmers is efficient with probability of 0.90, and inefficient with probability of 0.10.

However, if BBN simulations show that the use of the booms and skimmers is expected to be inefficient, it is almost impossible to prevent the oil from reaching ashore. In this case the advice on sensitive ecological resources likely to be impacted by the oil washing ashore is of critical importance in order to support decisions whether or not a response is necessary or what kind and extent of response is appropriate [11].

3.3 Is the oil threatening a sensitive resource?

Although both oil spill scenarios will have negative impact on the regional fishery activities, the second oil spill scenario will also have considerable impact on the coastline between Ordu and Giresun cities (Figure 7). This coastline is a center of tourism, sea food industry and transportation for about 1.5 million people [13].

Figure 7: Giresun – Ordu coastline is hosting 1.5 million people where fishery is the main means of living.

In order to carry out the accidental oil spill related integrated risk assessment the further studies are planned. The Environmental Sensitivity Index (ESI) maps will be further developed as an integral component for oil-spill related contingency planning and response. According to NOAA [14] shoreline geology is ranked according to its physical characters, including relative exposure to wave and tidal energy, shore slope, substrate type, grain size, tidal elevation. Biological resources are segmented into seven components: marine mammals, terrestrial mammals, reptiles and amphibians, invertebrates, habitats and plants, birds and fish. Human-use resources are divided into four components: high-use recreational access locations, management areas, resource extraction locations, and archaeological/historical resource locations. However, according to [15] the ESI values and impacted coastline extent based environmental risk assessment results are not the only determinants of oil spill response management decisions, with the oil spill response managers considering many other factors as well. Legal mandates and political, social, and economic considerations may lead oil spill response managers to make decisions that are more or less protective because the implementation of some of the potential alternative decisions may be too expensive or not technically feasible.

4 Conclusions

Referring to the monthly average wave height and the wind speed values of the Eastern Black Sea area in 2009, it is concluded that in general there should be a quite wide window of opportunity for efficient use of booms and skimmers in the case of accidental oil spill.

However, if due to bad weather conditions the use of booms and skimmers is expected to be inefficient, it is almost impossible to prevent the oil from reaching ashore. In this case the knowledge on sensitive environmental and human use resources likely to be impacted by the oil washing ashore is of critical importance in order to support the efficient oil pollution response decisions.

In order to carry out the accidental oil spill related integrated risk assessment the further studies are planned. In particular, the ESI maps will be further developed as an integral component for oil-spill related contingency planning and response.

Acknowledgements

Authors would like to thank PETRAKO LTD, Turkish General Directorate of Petroleum Affairs and Turkish State Meteorological Service for their kind support during this study. We also thank the TRANSAS GROUP for the kind permission to use the latest edition of the PISCES 2 software. The study was supported by the Estonian Science Foundation through grant No 7609 and target financing program SF0180104s08.

References

[1] BP. Internal investigation report for Deep Water Horizon, p.183, 2010.

[2] US Bureau of Ocean Energy Management, Regulation and Enforcement. *Spill summaries statistics 1970-2009*. 2010.

[3] Etkin, D.S. Analysis of oil spill trends in the US and Worldwide, *Proceedings of the International Oil Spill Conference*, pp. 1291–1300, 2001.

[4] Göktepe, B.G. and Gönençgil, B. Environmental Safety of Oil and Gas Transportation Risks through the Turkish Straits, *World Energy Council 18th Congress*, October 2001, Buenos Aires. 2001.

[5] GDPA. General Directorate of Petroleum Affairs. *Bulletin of Annual Activities,* 2000-2010.

[6] Yücel Ş. Social-Economic situation and fishery activities of the middle Black Sea region. *Journal of Fisheries & Aquatic Sciences*, pp. 529–532, 2006.

[7] Delgado, L., Kumzerova, E., Martynov, M. Simiulation of oil spill behavior and response operations in PISCES. *Environmental Problems in Coastal Regions VI including Oil Spill Studies.* Ed. C.A.Brebbia, Wessex Insitute of Technology, UK: WIT Press, pp. 279-292, 2006.

[8] IPIECA. Choosing spill response options to minimize damage – Net Environmental Benefit Analysis. *IPIECA Report Series Volume 10, International Petroleum Industry Environmental Conservation Association,* London, 20 p. 2000.

[9] Schallier, R., DiMarcantonio, M., Roose, P., Scory, S., Jacques, T.G., Merlin, F. X., Guyomarch, J., Le Guerroué, P., Duboscq, K., Melbye, A., Resby, J.L.M., Singsaas, I., Leirvik, F. NEBAJEX Pilot Project – Final Report. *Royal Belgian Institute of Natural Sciences.* 100 p. 2004.

[10] Aps, R., Fetissov, M., Herkül, K., Kotta, J., Leiger, R., Mander, Ü., Suursaar, Ü. Bayesian inference for predicting potential oil spill related ecological risk. In: M. Guarascio, C.A. Brebbia, F. Garzia (Eds). *Safety and Security Engineering.* UK: WIT Press, pp. 149–159. 2009.

[11] Aps, R., Herkül, K., Kotta, J., Kotta, I., Kopti, M., Leiger, R., Mander, Ü., Suursaar, Ü. Bayesian inference for oil spill related Net Environmental Benefit Analysis. In: C.A.Brebbia, G.Benassai, G.R.Rodriguez (Eds). *Coastal Processes* Southampton, Boston: WIT Press, pp. 235–246. 2009.

[12] Clark, R.B., Frid, C., Addrill, M. *Marine pollution.* Oxford UniversityPress, NY. 237 p. 2001.

[13] TUIK. Turkish Statistical Institute. *Turkey's Statistical Yearbook.* 2009.

[14] NOAA. Environmental Sensitivity Index Guidelines. Version 3.0. *NOAA Technical Memorandum NOS OR&R 11.* 192 p. 2002.

[15] Aps, R., Sawano, N., Hamada, S., Fetissov, M. Bayesian inference in oil spill response management. In: Ed. Brebbia, C. *Risk Analysis VII & Brownfields V.* UK: WIT Press, pp. 35–46. 2010.

Coastal and marine impact assessment for the development of an oil spill contingency plan: the case study of the east coast of Sicily

L. de Santoli, F. Cumo, D. Astiaso Garcia & D. Bruschi
CITERA (Interdisciplinary Centre for Housing, Heritage and Environment) Faculty of Architecture "Sapienza" University, Rome, Italy

Abstract

This project is one of the results of the multi-year collaboration in coastal and marine sustainable management, planning and policy between the General Directorate for Nature and Sea Protection of the Italian Ministry for the Environment, Land and Sea and the CITERA Research Centre (Centro Interdisciplinare Territorio, Edilizia, Restauro, Ambiente) of the Sapienza University of Rome. The main objective of the project was to analyze the parameters involved in preparedness, prevention and response actions required by the Italian Ministry for the setting up of an Oil Spill Contingency Plan. Due to its critical anthropic impacts and its huge environmental heritage in terms of biodiversity richness and coastal geomorphological variety, the Eastern Sicily coastline, bounded by the geographic vertices of Cape Peloro and Cape Passero, has been chosen as the study area. Moreover this area is characterized by significant economic activities such as fishing and tourist industries. One of the major aims of the study was to rank coastal areas on the basis of their sensitivity to an unforeseen oil spill. Physical, geomorphological, biological and environmental attributes have been gathered using remote sensing techniques and field works, and they have been subsequently elaborated for the valuation of oil spill environmental sensitivity indices (ESI). In addition to ESI values, the elaborated maps contain other important information like landscaping and environmental constraints, human-use resources, archaeological and historical sites, and marine benthic biocenosis. In order to achieve the tasks and draw up the maps, the whole data set has been reported in a Geographical Information System (GIS) database. The obtained results include operational maps

WIT Transactions on Ecology and the Environment, Vol 149, © 2011 WIT Press
www.witpress.com, ISSN 1743-3541 (on-line)
doi:10.2495/CP110241

representing useful tools for local stakeholders and practical guidelines for the elaboration of a regional oil spill contingency plan.

Keywords: oil spill contingency plan, marine impact assessment, risk assessment, GIS database.

1 Introduction

The progressive increase of sea traffic since the 1960s has caused growing concern for one of the fundamental aspects of pollution prevention: avoiding spills of hydrocarbons into the sea, which is particularly damaging to waters that are closed and subjected to heavy traffic such as the Mediterranean Sea [1].

Nowadays it is known that the majority of the hydrocarbons present in the sea comes from anthropogenic sources and represents a serious cause of marine pollution. As recently observed in the last environmental disasters in the Gulf of Mexico, oil spills can bring substantial damages to marine ecosystems, causing a loss of biodiversity with the consequent instability of fragile environmental balances. Moreover, as considered by the International Union for Conservation of Nature (IUCN), the Mediterranean basin is recognised as a biodiversity hotspot: its flora diversity is outstanding with 15,000 to 25,000 species, 60% of which are unique to the region and about one third of the Mediterranean fauna is endemic [2]. Furthermore, in recent years, a significant increase in the volume of crude oil and petroleum products transported in the Mediterranean Sea has been recorded: representing less than 1% of surface water in the world, about one third of the total traffic of oil was recorded in the Mediterranean Sea. Due to its central position, about half of all the crude oil that arrives each year in the Mediterranean ports, is directed to Italy [3].

This paper resulted from the multi-year collaboration between the General Directorate for Nature and Sea Protection of the Italian Ministry for the Environment, Land and Sea and the CITERA Research Centre (Interdisciplinary Centre for Housing, Heritage and Environment) of the Sapienza University of Rome. Starting from the case study of a specific area, Eastern Sicily, as well as offering a powerful instrument for reducing pollution in coastal waters, the aim is to provide protection for the Italian coastline, which is an important resource in terms of the natural, economic, social, historical and cultural phenomena. It is necessary to implement an integrated management and control of the whole area to preserve or recover its environmental integrity, optimising the cost/benefit ratio and continually improving the quality of human life and local communities in the area.

Ultimately, compared to the high probability of an oil spill in the Mediterranean Sea, this research wishes to provide useful tools and practical guidelines for planning the actions of prevention, preparedness and response required for the creation of a Contingency Plan in case of oil spills, through the classification of coastal areas on the basis of their sensitivity to an unforeseen oil spill and the creation of environmental sensitivity maps. To achieve these objectives, this research, after a short description of the legislative framework on sea pollution, provides some oil spill sensitivity maps of Eastern Sicily, bordered by the geographic vertices of Cape Peloro and Cape Passero.

2 Regulatory framework

The legislative framework for environmental issues and in particular for sea protection issues is constantly developing. Therefore, to avoid engaging into a misleading legal disquisition, but at the same time to get an idea on the subject, below is offered a simple overview of the normative levels that govern this matter.

At global level, there are several international conventions that deal with maritime safety and prevention of marine pollution. Among these, the most significant are:

- International Convention relating to Intervention on the High Seas in Cases of Oil Pollution Casualties (IHSOPC 1969), as amended by the related Protocol of 1973: it is the first International Convention that deals with marine oil pollution and it affirms the right of a coastal State to take such measures on the high seas as may be necessary to prevent, mitigate or eliminate danger to its coastline or related interests from pollution by oil or the threat thereof, following upon a maritime casualty;
- International Convention on Civil Liability for Oil Pollution Damage, as amended by the Protocols of 1976 and 1992 relating thereto (CLC 1969): it ensures that adequate compensation is available to persons who suffer oil pollution damage resulting from maritime casualties involving oil-carrying ships;
- International Convention for the Prevention of Pollution from Ships, as amended by the Protocols of 1978 and 1997 relating thereto (MARPOL 73/78): it aims at preventing and minimizing pollution from ships, both accidental pollution and that from routine operations. It currently includes six technical Annexes;
- International Convention on Oil Pollution Preparedness, Response and Co-operation, as amended by the Protocol of 2000 relating thereto (OPRC 1990): it provides for international co-operation in combating major incidents or threats of marine pollution. Parties are required to establish measures for dealing with pollution incidents, either nationally or in co-operation with other countries. Ships and operators of offshore units are required to have an oil pollution emergency plan, to be developed by IMO [4]; and
- Third United Nations Conference on the Law of the Sea (UNCLOS III, 1973-1982).

As regards the specific case study of the Mediterranean Sea, a convention that extends the international legislative framework related to marine pollution is the Convention for the Protection Of The Mediterranean Sea Against Pollution (Barcelona Convention).

Indeed, after the creation of the United Nations Environment Programme (UNEP) Regional Seas Programme in 1974, the Mediterranean became the first region to adopt an Action Plan (Mediterranean Action Plan - MAP) in 1975. This was quickly followed by the adoption of the Convention for the Protection of the Mediterranean Sea against Pollution (Barcelona Convention) in 1976, which entered into force in 1978, and a succession of seven landmark Protocols. In 1976, a Conference of Plenipotentiaries representing sixteen Mediterranean coastal States and the European Communities adopted the Convention for the Protection of the Mediterranean Sea Against Pollution (Barcelona Convention), which aims at protecting the marine environment and coastal zones through prevention and reduction of pollution, and as far as possible, elimination of pollution, whether land or sea-based. Finally, in 1995, the Action Plan for the Protection of the Marine Environment and the Sustainable Development of the Coastal Areas of the Mediterranean (MAP Phase II) was adopted, replacing the Mediterranean Action Plan of 1975.

3 Study area

The study area (Fig.1) forms part of Sicilian coastline which starts from Cape Peloro (15°39’09.13”E and 38°16’08.30”N) and covers the entire east coast up to Cape Passero (15°04’46.69”E and 36°38’51.20”N) (WGS 84 coordinate system).

This specific area was chosen as case study due to its critical anthropic impacts and its huge environmental heritage in terms of biodiversity richness and coastal geomorphological variety. Moreover this area is characterized by significant economic activities like fishing and tourist industries.

From the geomorphological point of view, these coastlines include bays, estuaries, rocky headlands, sandy beaches, creeks and cliffs. In order to simplify the study area description, it was divided into 8 segments (stretches), resulting in about 416 kilometres of shoreline classification.

Moreover, this specific Italian coast area was chosen for its real risk of an oil spill given the strong oil activities and for its particularly critical anthropic-environmental features. Indeed, in addition to the strong urban concentration and the great service industry presence along the coast, the stretch of sea in front of is characterized by one of the most intense maritime transit of the Mediterranean seas. Furthermore, the whole study area is characterized by a high seismic instability, factor that can potentially contribute to rage natural disasters.

Therefore, the Eastern Sicily coastline appears to be a highly sensitive zone, not only for its habitats and ecosystems uniqueness and diversity, but also and for its shipping traffic and its hydrographical and meteorological factors (in particular, with regard to the strong currents that affect the east coast of Sicily sea) that considerably increase the probability of maritime accidents and consequently the risk of marine and coastal pollution.

Due to its high environmental value, in this zone there are a significant number of protected areas that have important natural values. Among these there are 18 Sites of Community Importance (SCI areas), 8 Special Protection Areas (SPA areas) and one Ramsar area (wetlands of international importance). This area supports also many endemic plants and animals. On the marine side, besides the presence of two MPAs (Marine Protected Areas) (Cyclops Islands and Plemmirio), the relevance of this area should be stressed as important resting and nesting sites for many migratory birds species.

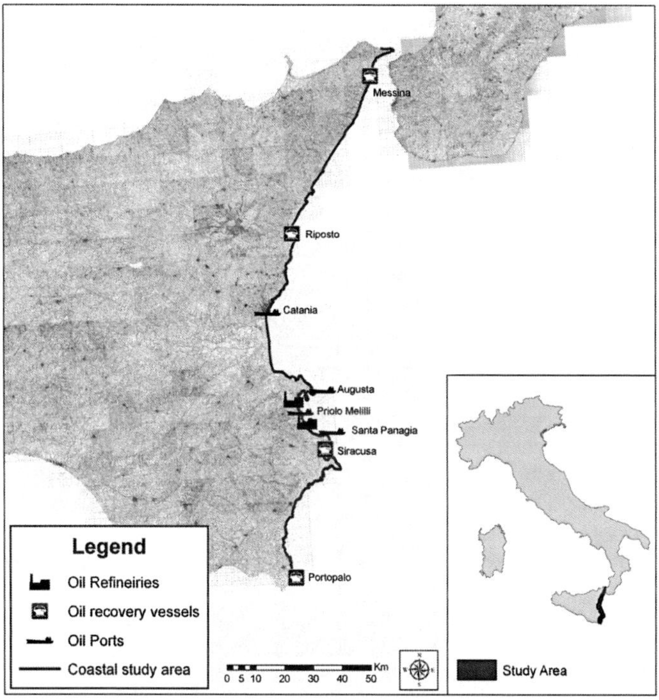

Figure 1: Oil ports, refineries and facilities along the study area.

With regard to the strong oil activities, this area is marked by heavy petroleum traffic. In fact, this zone is crossed by thousands of ships, especially ships carrying dangerous or polluting materials such as oil tankers, chemical tankers and gas carriers. Many of these are now obsolete or without the double hull or equivalent technologies. Besides the heavy commercial vessel traffic, along the coast there are 4 oil ports - Augusta, Priolo Melilli, Catania e Santa Panagia - in which more than 23 million tonnes of crude oil are landed each year (about 20% of total oil unloaded in Italy) and 2 oil refineries, Priolo and Augusta (figure 1); particularly the Priolo refinery is one of the biggest in the whole Mediterranean sea).

In addition, the port of Messina, located at the north of our study area, holds the record for carried passenger numbers between all the Italian ports: considering boarding and disembarkation, it records more than 10 millions of passengers every year [5].

Finally, to make a complete risk assessment of oil spills, it should be considered the huge amount of oil and hydrocarbon components carried through the sea in the petrol tank of ferries, pleasure crafts, fishing boats and merchant ships.

4 Methods

Between all the coastal hazards and pollutants, oil spills are certainly one of the most damaging because of they may bring long term impacts on coastal and marine habitats. According to the International Tanker Owners Pollution Federation (ITOPF), the oil spills impacts depend on a range of factors, from spill size and physicochemical characteristics of the oil, to weather conditions (e.g. waves intensity and frequency) [6].

Even if it is not possible to predict the impacts of an oil spill with any certainty, it's possible to evaluate the vulnerability of a specific area to a defined spill scenario based on the economics, environmental and cultural resources present in that marine zone.

This paper describes the design and development of the coastal prioritisation index taking into account the vulnerability of natural-cultural and socio-economic resources. The creation of maps, elaborated considering all the criticality criteria, represents an important key activity in the production of contingency plans about the oil spill at sea. It is fundamental, that these maps can show which areas need prior action, thus providing to decision makers and stakeholders essential information for the protection of the entire coastline.

Therefore, to classify the impacts related to an oil spill, it was considered the appropriate methodology published by the International Maritime Organization (IMO) and by the International Petroleum Industry Environmental Conservation Association (IPIECA) [7], which, in its turn, is built on the work proposed by the Federal U.S. NOAA (National Oceanic and Atmospheric Administration) [8]. However, in order to consider carefully the whole archaeological or historical sites that typically marks the Italian and generally the Mediterranean coastlines, some changes have been made. In fact, in the Mediterranean context, the historical elements have such an inestimable value that should be considered in a balanced evaluation process with the ecological and socio-economical aspects.

Physical, geomorphological, biological and environmental attributes have been gathered using remote sensing techniques and field works that, subsequently, have been elaborated by the evaluation of oil spills environmental sensitivity indices (ESI). Remote sensing and Geographical Information System (GIS) software played a very significant role in the mapping of different classes of shoreline and for preparing different thematic layers. Moreover, in order to implement these gathered data, the Sicily's shoreline coastal habitats mapping was performed also with the support of WebGis informative systems like the Italian National Cartographic Portal (PCN) and the network of the Italian National Environmental Information System (Sinanet). In the end, to draw up the maps, the whole data set has been reported in a GIS database containing a comparative spatial analysis in order to classify on nationwide, the criticalities of Italian coast and seawaters.

4.1 Shoreline classification

As mentioned before, the Eastern Sicily shoreline classification and its consequential analysis in this study are based on the ESI methodology created by NOAA, which was modified and adapted in order to properly map these Mediterranean coastlines. According to this methodology, habitats would be more or less sensitive to oil spills according to their degree of exposure to hydrodynamic action and substrate characteristics [9]. The different types of habitat identified and their associated main characteristics are listed in Table 1.

Table 1: Values of coastal environmental sensitivity indices to oil spill impacts.

Vulnerability Index	GIS Code	Coast Features	Comments on environmental rehabilitation
1.A Exposed high rocky shores 1.B Exposed artificial coastal structures	CRAE ACE	Rocky shores with steep slope exposed to the action of waves, which produce a natural and fast oil removal. The substrate is waterproof	It is not required or recommended any type of intervention
2. Exposed low waterproof coasts	CBIE	As for the index 1, the action of waves and the natural forces of sea remove most of the oil in a few weeks. The substrate is waterproof (rocky or clay lithology)	In most cases there is no need of cleaning. The only problem could be the possible presence of sediment because of the morphology of the coast
3. Compact clay or muddy sand beaches	SSC	Grain size of sand <0.06 mm. The compactness of the grains of the beach prevents the deep penetration of oil. The use of specific machinery is often required for clean-up interventions. The substrate is semi-waterproof	Intervention measures recommended: 1. wait until all the oil has been deposited on the beach; 2. not repeatedly switch on the areas where there is oil; 3. remove only the amount of contaminated sand. It is recommended manually collect of the sediments.
4. Sandy beaches	SS	Grain size of sand: 0.06-1.0 mm. The substrate is permeable and the oil penetrating average 25 cm in depth	The oil can penetrate deeply in the sand. Difficult clean-up operations. At the beginning, the action of the waves makes the burial of oil, while in the medium long term it will remove most of the oil (several months). It is recommended to use specific machines.
5. Mixed sand and gravel beaches	SM	The substrate has a medium-high permeability. The oil easily penetrates up to 50cm deep, and its burial can happen in a few days.	The oil in this type of beaches may remain for a long time. Moreover, its removal can be very difficult.

Table 1. Continued.

6.A Gravel and pebbles beaches 6.B Natural coasts with blocks of stone melted 6.C Artificial reef	SG CNP AS	Grain size: 1-256 mm and above. The substrate is highly permeable and it has a high absorption capacity. The oil can penetrate up to 1 meter deep	The natural rate of removal is very low, so oil removal from these beaches is very laborious. When a beach was highly oil contaminated, it is practically impossible to clean without removing a large amount of sediment. This could be very harmful to the biocenosis present in the area. In the case of artificial reefs the only way to clean them is to remove and replace them.
7. Exposed tidal flats coasts	CEM	The substrate is permeable. Generally, the oil is not stagnant and does not penetrate into the soil	Interventions are necessary only to prevent contamination of surrounding areas. In fact, the tidal action involves a natural cleaning action. The use of the equipment should be used only if the presence of oil is very intense
8.A Sheltered rocky shores 8.B Sheltered artificial coastal structures 8.C Sheltered low waterproof coasts	CRR ACR CBIR	The substrate is hard and mostly composed by rock, artificial material or clay	These areas are sheltered from waves and ocean currents. Oil pollution can remain for years. Both in the presence of impermeable rock or more permeable materials, due to the strong presence of biological activity and the low rate of natural removal, it is often necessary a cleaning action even if it will be difficult and invasive. For this reason are not recommended cleaning operations, unless oil is not present in large quantities
9. Sheltered tidal flats coasts	CRM	The substrate is semi-waterproof	These areas are characterized by small waves, but with strong organic production. Interventions are not recommended except in the case of major pollution. The removal of oil is in fact impossible without compromising the conservation of the biological communities.
10. Marshes and brackish coasts	CP	The substrate typology varies depending on the particular characteristics of each considered area	The protection of these areas should generally have the highest priority. In fact, in terms of biological activity, these areas are the richest and the most productive marine environments. Moreover, given the very low rate of natural removal, oil may remain in these areas for more than 10 years. The environmental restoration of these habitats requires about 20 years.

The mapping process is achieved using coloured lines to represent the different shoreline types, and a standardized colour code is associated to each sensitivity value. The ESI classification is a convenient form to summarize information relating to the coast and does not take into account some possible uses of the considered coasts. For example, an exposed high rocky shore ranked with ESI 1 (low sensitivity) could accommodate an internal colony of seabirds with high sensitivity, at least during the breeding season. Or a sandy coast, with ESI 3 (relatively low sensitivity), could be an important site for tourism or for spawning by sea turtles in a given year (*Caretta caretta* in the Sicilian coast). Therefore in this study are presented ESI maps that implement the ESI classification with other biological and human-use information.

5 Results

In addition to ESI shoreline classification, the elaboration of an oil spill contingency plan is also related to the analysis of biological and human-use resources like: landscaping and environmental constraints, archaeological and historical sites, marine benthic biocenosis, etc.

Therefore, given the amount of information contained in these sensitivity maps, in order to make them more user friendly, it was decided to divide these information, thus obtaining two types of maps for each area analyzed (figures 2 and 3).

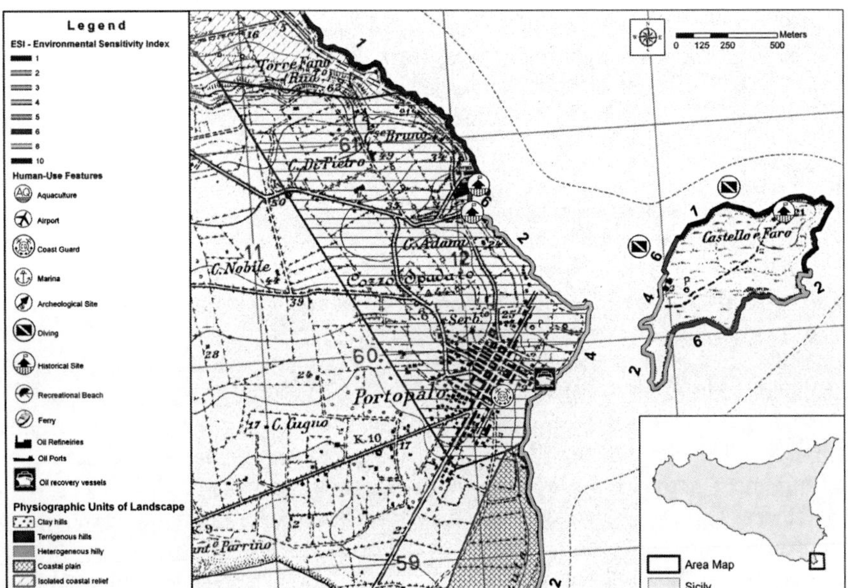

Figure 2: Environmental sensitive indices, human-use features and physiographic units of landscape in a coastal zone of the study area.

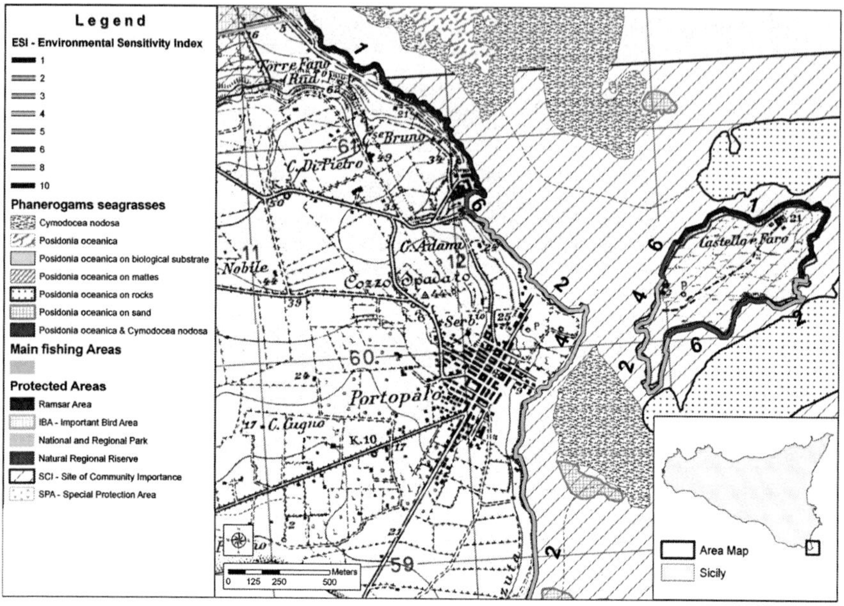

Figure 3: Environmental sensitive indices, marine phanerogams grasslands, main fishing and protected areas in a coastal zone of the study area.

Thus, these maps consider three main types of information:

1. the ESI classification, according to a scale which takes into account the different response capacities of each type of coast against a possible impact on oil-spill, considering the natural behaviour of oil and the ecological resilience of the area;
2. the presence of biological resources, defined as rare and protected animals, plants and entire habitats;
3. the presence of human activities, such as the existence of parks and marine reserves, archaeological sites, areas of aquaculture, etc.

The total length of the analyzed shoreline is about 416 km, and specifically 308 km of natural coastal habitats and 108 km of artificial coastline. In the whole zone the most present ESI value recorded is 6, while there are no coastal areas characterized by 6 or 7 ESI value.

Therefore, in order to plan oil spill response actions the obtained results emphasize the importance of identify coastal areas with higher ESI values along a shoreline generally represented by low or medium sensitivity.

Otherwise, the presence of fragile ecosystems and habitats, not to mention the significant human-use coastal resources and historical/archaeological sites makes Eastern Sicily shorelines an area of high priority in the management of a national or regional contingency plan.

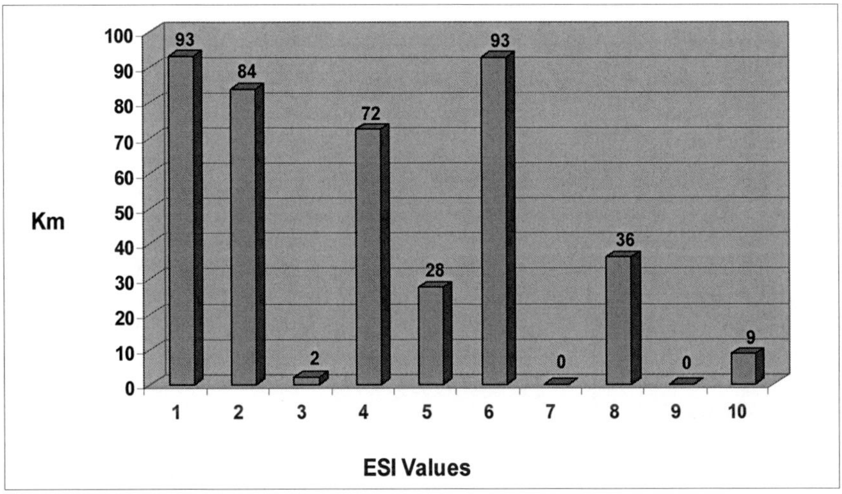

Figure 4: Coastal length of each ESI values in the Eastern Sicily shoreline.

6 Conclusions

The paper offers the opportunity to validate the tested methods and obtain an overview of the sensitivity to hydrocarbon spills in a particularly complex area where coexist important historical and archaeological resources, protected marine species and intense human activity, especially related to the tourism.

In addition, the results obtained in the pilot project confirmed that the used methodology provides a clear system for localising environmental, landscape and cultural priorities for the protection of those areas that will results most vulnerable.

The GIS database containing the collected data and their subsequently elaboration is a valuable instrument for the application of the principles of Integrated Coastal Zone Management by local communities and public administrations, as well as for the prevention of major environmental disasters from unforeseen hydrocarbons spill.

Ultimately, this project contains a series of technical information valuable for the development and the implementation of a specific Contingency Plan for the Ionian Sea, identifying high priority areas for the planning of preparedness and response actions aimed to reduce the risks and the impacts of marine pollution related to maritime hydrocarbon traffic.

References

[1] Cinquepalmi F., *Traffico petrolifero nel Mediterraneo: il quadro normativo e le politiche ambientali.* in "Biodiversità marina e strategie di riduzione del rischio connesso al trasporto di Idrocarburi", study day minutes, Rome, October 2005

[2] Biliardo U., Mureddu G., *Traffico petrolifero e sostenibilità ambientale. Unione petrolifera*, 2005

[3] Cuttelod A. et al., *The Mediterranean: a biodiversity hotspot under threat*. Published by IUCN, Switzerland, 2008

[4] Slim H., Scovazzi T., *Study of the current status of ratification, implementation and compliance with maritime agreements and conventions applicable to the Mediterranean Sea Basin – Part 1*. AGRECO Consortium, 2009

[5] Ministero delle Infrastrutture e dei Trasporti, *Relazione sull'attività delle autorità portuali*. 2009

[6] ITOPF, *Fate of Marine Oil Spills*. ITOPF Technical Information Paper N° 2, 2002

[7] IMO/IPIECA Report Series – Volume One: *Sensitivity Mapping for Oil Spill Response*. 1990-2005

[8] NOAA, *Environmental sensitivity index guidelines*. Version 3.0, 2002

[9] Wieczorek A., Dias-Brito D., Carvalho Milanelli J.C., *Mapping oil spill environmental sensitivity in Cardoso Island State Park and surroundings areas, Sao Paulo, Brazil*. Ocean & Coastal Management, Volume 50, 2007

Relation between natural cleansing effect and angularity of coastal gravel

S. Hamada[1], N. Sawano[2] & R. Aps[3]
[1]*Geological Survey of Hokkaido, Department of Marine Geosciences, Otaru City, Japan*
[2]*Inaoki Educational Institute & Seiryo Women's Junior College, Kanazawa City, Japan*
[3]*University of Tartu, Estonian Marine Institute, Tallinn, Estonia*

Abstract

Serious environmental contamination will arise once spilled oil hits the coastal area. A gravel beach is commonly hard for oil recovery because these areas are hard to approach in order to conduct cleanup work. Also, stranded oil will easily be penetrated into the coastal sediment of gravel. In this study, energy of wave exposure was analyzed using the method of image analysis to gravel shapes. As a result, the shape of gravel on a beach is affected by the energy of wave and then provides natural cleansing ability. These effects were examined by an actual oil spill by "Nakhodka" which happened in 1997. In the field survey, projected images of cobble-sized andesitic gravel were taken by a digital camera. Then shapes of them were analyzed by the software "ImageJ" and a common spreadsheet. All the images were changed to 720-angled polygon then "bending angles" were measured to decide "Angularity Indexes" of gravel. Oil residual time of monitoring points of Nakhodka and the indexes were examined and then a strong correlation coefficient (R^2=0.858) has been observed. Also, the natural cleansing ability for oil can be estimated by image analyses of coastal gravel.

Keywords: oil spill, gravel beach, ESI mapping, risk assessment.

1 Introduction

In the continental shelf of the Island of Sakhalin, a consortium led by Royal Dutch Shell is operating the Sakhalin II oil and gas project that is the largest

www.witpress.com, ISSN 1743-3541 (on-line)
doi:10.2495/CP110251

foreign investment of Russia [1]. Considering the risk associated with oil spills, contingency plans will have to be established for Soya strait.

In January 1997, the Russian tanker *Nakhodka* spilled more than 8,000 tons of heavy-C oil. The northern part of the Noto Peninsula (Figure 1) was heavily damaged. Shoreline cleanup and recovery works were hard and cost much for oil events. Recovery works were difficult in gravel beaches because access for traffic to the oiled sites is dangerous and stranded oil is penetrated easily into the sediment [2]. Based on the lessons of Nakhodka, the natural cleansing ability depending on the feature of gravel will have to be evaluated then be reflected in contingency planning.

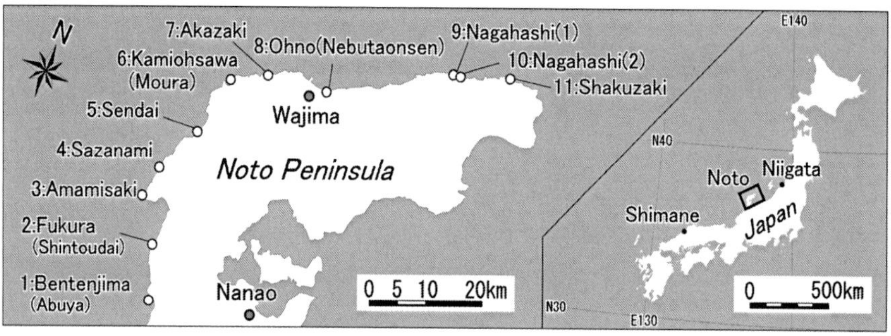

Figure 1: Location of field survey area.

Sawano [2] reported a relation between oil residue and the type of coast (sandy, rocky, etc) by his on site monitoring of Nakhodka. Also, Hamada [3] made a report on the oil impacted beach of Nakhodka from the point of beach topography and sediments and showed the relation between gravel shape and oil residue of the gravel beaches.

The relation between “angularity” of coastal gravel and oil residue was discussed, based on the case study of Exxon Valdez by Hayes [4]. His evaluation of angularity, however, was basically based on human impression using the “roundness chart” proposed by Powers [5]. Visual inspection is a convenient way for field research, but an inevitable difference occurs even between well trained personnel. Our study presents the relation between image-analyzed angularity and length of oil residue after a big scale oil spill. In this paper, a quantitative approach will be made to evaluate the angularity of coastal gravel.

2 Evaluation strategies for gravel angularity

It is known that “roundness of particle” was defined by Wadell [6]. Generally, evaluation of gravel roundness has been measured by visual impression using the “roundness chart” proposed by Powers [5] or “Krumbein’s roundness chart” [7].

Then another approach was used to evaluate angularity using image analyses without eye bias.

Some of the oiled shorelines of this study are hard to access even by specially equipped automobiles, and then correctly weighted equipment was selected for logistic purposes. A common digital camera and a portable stand (Figure 2) were used in the study sites. Images of coastal gravel are taken using the equipment. After a field survey, a public domain software “ImageJ” and spreadsheet were used to calculate the value of gravel angularity.

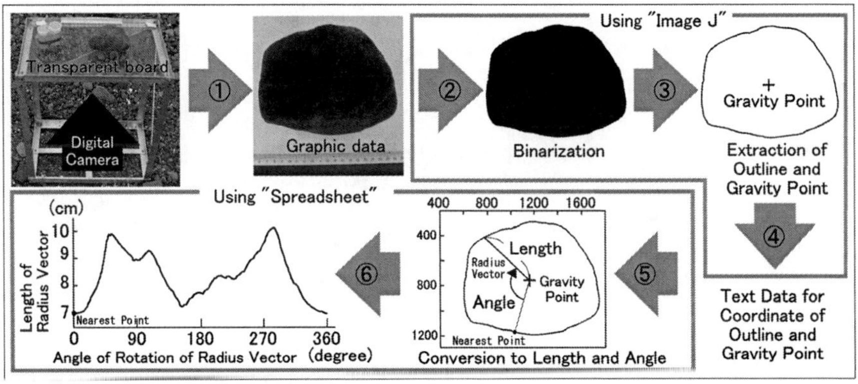

Figure 2: Photographing of gravel image and extraction of outline data.

2.1 Imaging method and profile extraction

2.1.1 Gravel sampling

Gravel samples for this study were andesitic large-cobble-sized (the diameter was from 12.8 to 25.6 cm). Fifty cobbles were taken from the beach where oil remained. In some study sites where the oil was removed completely, samples were collected from the part of the berm located in the upper tidal zone of the gravel beach.

2.1.2 Images of gravel shape taken by digital camera

The maximum projected plain of gravel was used as the shape of gravel. Gravel was put on the transparent board, and taken by the camera fixed under the board (Figure 2). Shadows did not appear around the gravel and clear-cut of projected shapes can be taken on this process. With our equipment, 1 pixel of the image is equivalent to 0.15 mm of actual length.

2.1.3 Binarization and extracting profile

The photo images of the gravel were binarized into black (for gravel) and white (for background) by ImageJ (Figure 2) and the circumference of shape was extracted from the image (Figure 2). Subsequently, the coordinate of the center of gravity and the length from the center of gravity to circumference (length of radius vector) were calculated (Figure 2). Based on the length and the angle of radius vector, the gravel shape was expanded to a bent line on the graph

(Figure 2). If the sample is rounded, the bent line becomes smooth, because the length between the centre of gravity and the peaks are always stable. On the other hand, if the sample was angular, the graph becomes irregular and wavy in accordance with the surface roughness of the sample.

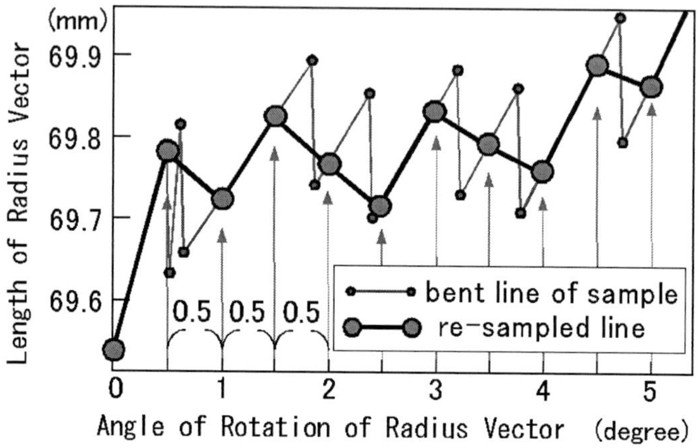

Figure 3: Re-sampling of bent line at intervals of 0.5 degrees.

2.1.4 Re-sampling

The bent line acquired from the photo image of gravel has many bending points and interval of those are irregular (Figure 3). For the purpose of measuring the angularity of the line, new lines were sampled again. The bending points of the new lines were measured every 0.5 degrees of angle rotation of radius vector (Figure 3). In these processes, each bent line has transformed to a 720-angled polyline.

2.2 Angularity index based on bending angle

In order to determine the best index for the evaluation of gravel angularity, we have developed 5 kinds of index [8] and compared them. As a result, we conclude that the angularity index based on the bending angle is the best.

The index was calculated from the bending angle at each 720 bending point of the 720-angled polyline (Figure 4). For example, the bending angle at Pn in Figure 4 is defined as an angle between the linear line from Pn-1 to Pn and the line from Pn to Pn+1. The horizontal distance from Pn-1 to Pn or Pn to Pn+1 has 5 degrees. At every 720 points, the bending angles are measured in the same way. If the angle is convex, the value of bending angle becomes plus. On the other way, concave angle shows minus. If the gravel shape is complete round, every bending angle becomes 0 degrees.

The 95 percentile value of all bending angles at the 720 points was used as the angularity index. The more “angular” the gravel is, the “larger” the index becomes.

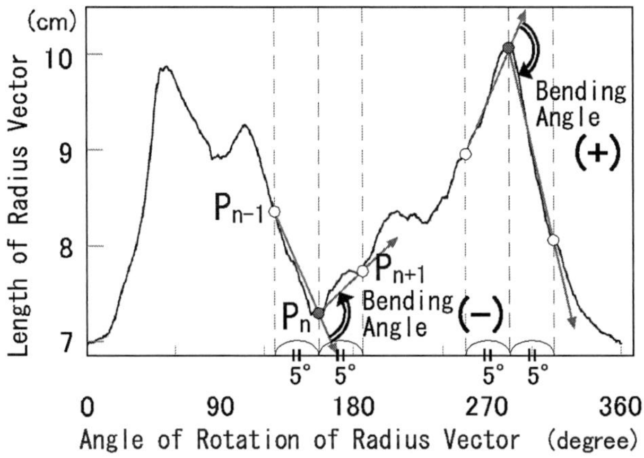

Figure 4: The bending angle at each bending point of the line.

3 Gravel angularity and oil residual time in *Nakhodka* oil spill

More than 8000 tons of heavy crude oil was discharged from the wreck of the tanker Nakhodka on January 2nd 1997. More than 1300 km of shoreline was contaminated. The oiled shoreline along the Japan Sea was extended from Shimane to Niigata prefecture. Especially heavily contaminated sites in Noto Peninsula of Ishikawa prefecture were rich in gravel beach and cliff, then the oiled shoreline was hard to access, and the recovery works on it were seriously difficult. Subsequently, a considerable amount of stranded oil has been untouched on the beaches after the oil spill. The above mentioned quantitative estimations of angularities are applied to check the relation between the oil residual period and the index of gravel angularity (Figure 1). The survey points are located from Bentenjima to Shakuzaki on the Noto peninsula.

3.1 Oil residual time

Oil residual time at 74 beaches on the Noto Peninsula after the Nakhodka oil spill have been monitored by Sawano [2]. His study showed the degree of oil residue as “Heavy”, “Light” or “None”, and made it clear how it had changed at each survey point. Oil residual time in this paper uses his results, and defined as the time from “Heavy” to “Light” or “Light” to “None”.

3.2 Comparison between the angularity and oil residual time

An angularity index was compared with the oil residual time at each survey point (Table 1, Figure 5). The shape of coastal gravel at No.4 Sazanami, No.5 Sendai, and No.8 Ohno are relatively rounded, and the stranded oil has remained for 2 to 4 years. On the other hand, at No.7 Akazaki, No.9 and 10 Nagahashi (1) (2), No.11 Shakuzaki, the shapes are relatively angular, and the oil has remained up to 8 to 9 years or more. As a result of the comparison, a high correlation coefficient was obtained between the angularity index defined by this study and oil residual time (table 1). The index defined by this study shows the more angular in shape the gravel is, the longer the oil stays. And the index obtained from the bending angle of convex has the highest correlation coefficient (R^2 = 0.858) with the oil residual time. Figure 5 shows the regression. The actual oil residual time on the gravel beach is observed within 1.5 years from the regression line. The equation of the regression line is as follows:

$$\text{Residual time at gravel beach in Nakhodka oil spill (year)} = (\text{Index} - 3.41) / 2.89$$

Table 1: Correlation coefficient between the angularity index of coastal gravel and residual time of stranded oil at each survey point. The index of each survey point is the average of 50 pieces of gravel.

No.	Survey point	Residence time of oil (years) *1	Angulality index of coastal gravel (deg.)
1	Bentenjima (Abuya)	> 8.6 *2	30.5
2	Fukura (Shintoudai)	6	20.7
3	Amamisaki	5	20.6
4	Sazanami	2	11.7
5	Sendai	4	11.1
6	Kamiohsawa	5	19.8
7	Akazaki	> 8.9	27.9
8	Ohno (Nebutaonsen)	4	11.4
9	Nagahashi 1 (Ohtani)	> 8.9	33.1
10	Nagahashi 2 (Gas Station)	> 8.9	24.4
11	Shakuzaki	> 8.9	29.3
Correlation coefficient between shape index of gravel and residence time of oil after *Nakhodka* oil spill (R^2)			**0.858**

*1: after Sawano,2007 *2: Oil residue was find at field survey in July 2005

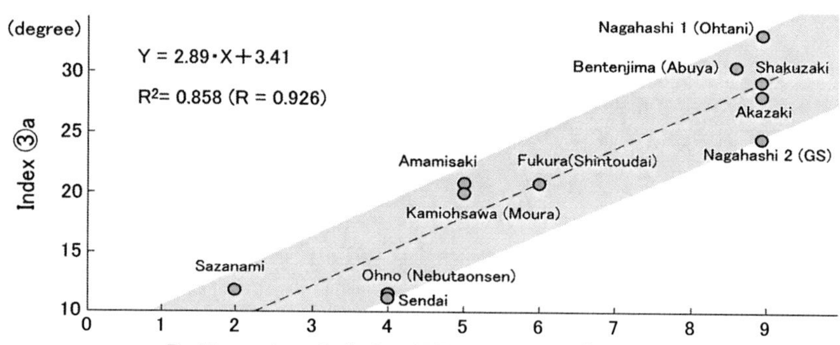

Figure 5: Relation between oil residual time at each survey point and gravel angularity index.

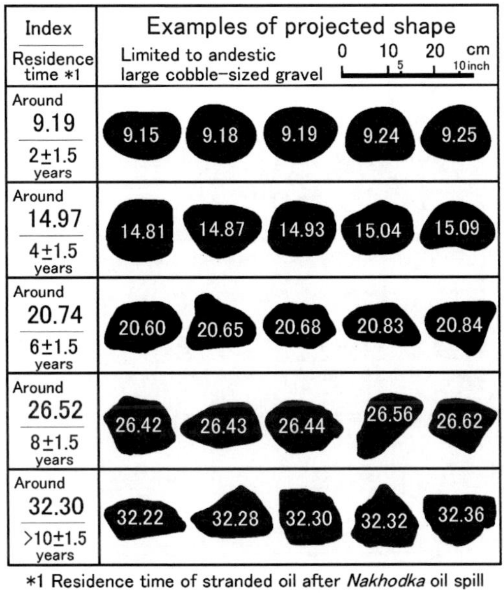

Figure 6: Examples of projected shapes of gravel angularity index.

Figure 6 shows the examples of the shape of Index (around 9.19, 14.97, 20.74, 26.52 and 32.30). In total, each classification of the shape is equivalent to the oil residual time of 2, 4, 6, 8 and 10 years after the Nakhodka incident.

3.3 Relation between angularity of gravel and natural cleansing effect

Natural cleansing effects on a gravel beach are thought to be providing a wave washing effect. Based on the field observations, wave strength at the beach is different from coastal topography. For example, on a sheltered gravel beach where wave energy was decreased by wave cut bench, the waves could not run up to the beach (Figure 7 lower). On the other hand, on the exposed gravel beach where strong waves directly hit, the gravel is rolling up and down (Figure 7 upper).

The highest Backshore Height ("BH" in Figure 7) shows the maximum elevation of waves run up in the shoreline. It was observed as a vegetation limit or storm berm in the beach. These are showing "maximum wave strength" on a beach. In this study, we have measured and compared the elevation of highest Backshore Height (BH) with the angularity index of coastal gravel.

As a result, the angularity index of gravel highly correlated with the highest BH of the gravel beach (R^2=0.79) (Figure 8). On a sheltered beach, the elevation of BH becomes lower in association with the shape of gravel as it becomes angular. On the other hand, on an exposed gravel beach, the level of BH becomes higher and the shape of gravel becomes rounded.

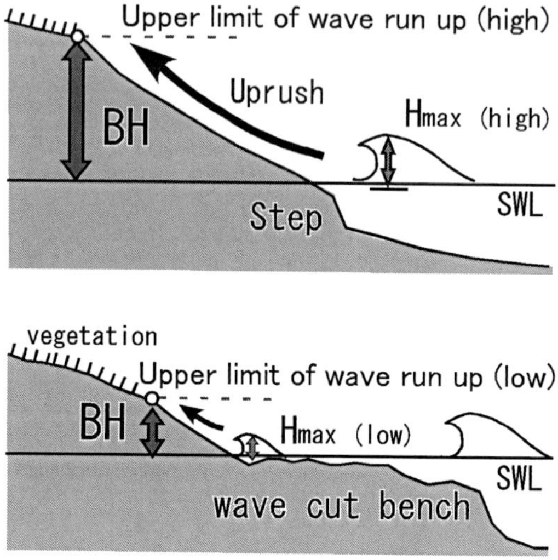

Figure 7: Relation between wave strength at beach and upper limit of wave run up height.

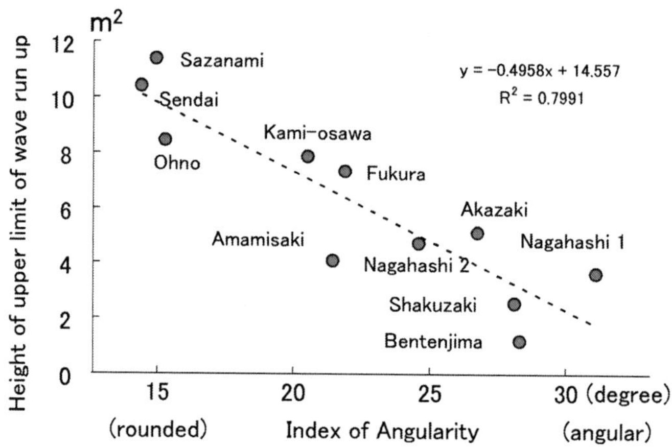

Figure 8: Relation between the index of angularity and height of upper limit of wave run up.

4 Conclusion

In this study, we have tried a quantitative evaluation of gravel angularity then compared oil remaining time. As a result, the angularity index of the gravel has high correlation with the oil residual time on a gravel beach (R^2=0.878). And we propose to examine the "convex of the shape" to evaluate the natural cleansing effect on a gravel beach.

The angularity index of gravel shape has a high correlation with the BH of a gravel beach. It means that the wave energy has a great influence on gravel shape.

It is concluded that the angularity of the gravel shows the natural washing effect on a gravel beach, and quantitative angularity will be used for the index of natural persistence of oil. As of the moment, it is difficult to make a "correct" prediction of the rate of oil removal from the oiled gravel beaches because so many conditions may affect it. However, a "relative" comparison of the self-cleansing effect on the gravel beach, for example, which part of shoreline should be cleaned faster, will be predicted. This result should be reflected for developing a shoreline information map for oil spill counter measures.

References

[1] Murakami, T., Oil and natural gas developing projects in the Sakhalin continental shelf, *Sakhalin Offshore Oil and Gas Development and Environmental Protection*, ed. T. Murakami, Hokkaido University Press: Sapporo, pp.250, 2003.

[2] Sawano, N. Analysis of Oil Residue based on Long term Shoreline Monitoring after Nakhodka Oil Spill, Papers on Environmental Information Science, 20, pp. 303-308, 2007.

[3] Hamada, S. Remaining position of stranded oil in beach profiles. Report of the Geological Survey of Hokkaido, 72, pp. 73-84 (In Japanese), 2001.

[4] Hayes, M. O. An Exposure Index for Oiled Shorelines, Spill Science & Technology Bulletin, 3(3), pp. 139-147, 1996.

[5] Powers, M. C. A new roundness scale for sedimentary particles, Journal of Sedimentary Petrology, 23, pp. 117-119, 1953.

[6] Wadell, H. Sphericity and roundness of rock particles, Journal of Geology, 41, pp. 310-331, 1933.

[7] Krumbein, W. C. Measurement and geologic significance of shape and roundness of sedimentary particles, Journal of Sedimentary Petrology, 11, pp. 64-72, 1941.

[8] Hamada, S. Relation between oil residues and angularity of coastal gravel. Proceedings of PACON 2007, pp. 108-115, 2007.

Section 9
Hydrodynamics

Influence of seasonal sea ice conditions on hydrodynamic processes and oil spill drift

R. Aps & Ü. Suursaar
Estonian Marine Institute, University of Tartu, Estonia

Abstract

Over the last hundred years, the annual maximum ice extent in the Baltic Sea has varied between 12 and 100% of the surface area of the sea, and the number of days with ice has fluctuated between zero and about 160 along the Estonian coast of the Gulf of Finland. Apparently, variations in ice conditions have a strong and diverse influence on hydrodynamic processes. The aim of the paper is to analyse the influence of sea ice on the waves, currents and coastal processes. Ice conditions are also closely related to the issues of shipping safety and pollution spread in oil spills. The study is based on meteorological and ice data gathered by the EMHI, measurements of waves and currents using the RDCP complex, and wave hindcasts. The hydrodynamic observations performed in the mouth section of the Gulf of Finland in winter 2009/10 showed somewhat weaker currents during ice cover, but the density-driven estuarine circulation was still rather efficient. The waves were absent when the whole gulf (measuring site) froze over, but when the sea was only partially covered with ice, the waves just decreased as a result of decreased fetches from certain wind directions. In addition to the solely meteorologically forced wave hindcasts, versions of the calculations with "ice mask" were produced for the period of 1966–2008. The ice correction considerably influenced the trend of annual mean ice statistics. Detecting, tracking and combating of oil spills are more difficult in ice conditions.
Keywords: sea ice, currents, wave modelling, oil spill, Baltic Sea.

1 Introduction

In the Baltic Sea, icy conditions vary greatly within the seasonal cycle as well as interannually [1]. Over the last hundred years, the annual maximum ice extent

WIT Transactions on Ecology and the Environment, Vol 149, © 2011 WIT Press
www.witpress.com, ISSN 1743-3541 (on-line)
doi:10.2495/CP110261

has varied between 52 000 and 405 000 km^2 (i.e., 12–100% of the surface area of the sea) [2], and the number of days with ice has fluctuated between zero and about 160 along the Estonian coast of the Gulf of Finland [3] (Fig. 1). In the 20th century, the Baltic Sea was almost entirely ice-bound in 1939/40, 1941/42, 1946/47 and 1986/87 [2]. On the other hand, there was very little or practically no ice in the Estonian coastal waters in 1948/49, 1960/61, 1988/89, 1989/90, 1991/92, 1992/93, 1994/95, 2007/08 and 2008/09, and such winters have become twice as frequent as in the middle of the 20th century [3]. Apparently, the variations in ice conditions have a strong influence both on marine hydrodynamic and water exchange processes, as well as on economic behaviour of coastal communities, including fisheries and navigation.

Ice conditions are also closely related to the issues of shipping safety and possible spread of pollution in case of accidental spills [4, 5]. The Gulf of Finland is a sensitive brackish water area, which has some of the busiest cargo and oil shipping routes in the world. In 2010, the amount of transported oil could be estimated at nearly 200 million tons in the Gulf of Finland. Despite improving navigation measures, there is a growing risk for incidental spills [5].

Figure 1: Map of the study area with locations of *in situ* measurements and wave hindcast marked by "RDCP(1)" for the Neugrund-Pakri site and "RDCP(2)" for the Letipea-Kunda site. The background shows the ice conditions on 12 January 2010 according to the ice charts by the SMHI. Dashed arrows exemplify restricted by ice cover fetch directions and the solid arrow represents a non-restricted fetch.

The aim of the paper is to analyse the influence of sea ice on the waves, currents and coastal processes on the basis of oceanographic measurements and modelling performed in the practically tideless Baltic Sea. We also discuss the issues of coastal zone management and pollution control in the changing ice conditions.

2 Material and methods

2.1 In situ measurements of waves and currents

The study is based on measurements of waves and currents using the Doppler-effect based current profiler (RDCP) in the two locations of the Gulf of Finland in 2006-2010, wave hindcasts, and some hydrodynamic modelling experiments performed and reported by ourselves earlier [5, 6]. Most importantly, extensive data were obtained in the Neugrund Bank area (59° 19′47″N; 23° 29′33″E, Fig. 1) during 20 November 2009–24 April 2010. The bottom mounted RDCP-600 (by AADI Aanderaa) was moored to the depth of approximately 15 m. Over 155 days and under various ice conditions, it provided hourly data on currents at seven depth layers (without cell overlap), wave parameters, sea level variations and water column properties. Similar measurements at Letipea–Kunda area (Fig. 1) in 2006–2009 were more thoroughly described in [7]. In this work, we only used data on long-term wave hindcast from that site.

2.2 Meteorological and ice data

For describing the meteorological conditions during the *in situ* measurements at Neugrund and to force wave model, we used data from the meteorological station at Pakri (59° 23′22″N; 24° 02′24″E). Being just about 30 km to east, it is the closest station to the mooring site. For the Letipea site, we used the wind data from Kunda station (59° 31′04″N; 26° 32′43″E), about 10 km to the west. Both stations are operated by the Estonian Meteorological and Hydrological Institute (EMHI). The digitized wind data are available since 1966. Winds have been measured by “weathercocks” (wind vanes of Wild’s design) during 1966–1976, automatic anemorhumbometers during 1976–2003, and MILOS-520 automatic weather complexes since September 2003. The older data in the database have been slightly corrected for homogeneity following the procedure described in [8].

For ice conditions, we used observational ice data gathered by the EMHI, as well as daily (during ice season) ice charts (see e.g. the excerpt in Fig. 1), which are available from the websites of the Swedish Meteorological and Hydrological Institute (SMHI) [9], Finnish Meteorological Institute (FMI), and the EMHI.

2.3 Wave hindcast

To study the influence of sea ice conditions on waves we performed a wave hindcast with a semi-empirical SMB-type wave model. The SMB-model, also called the significant wave method, is based on the fetch-dependent equations of

Sverdrup, Munk, and Bretschneider [10]. Based on wind data, it calculates the significant wave height (H_s), wave period and wavelength for a chosen location. As the role of remotely generated waves (swell) is small and the memory time of the wave fields in the Baltic Sea is relatively short [1], this relatively simple method can deliver reasonably good and fast results on the study area [7, 8]. A comparison of hindcasts between the SMB model and 3rd generation WAM model [11] showed that such SMB-type models can be used for site-specific long-term hindcasts, despite their limitations.

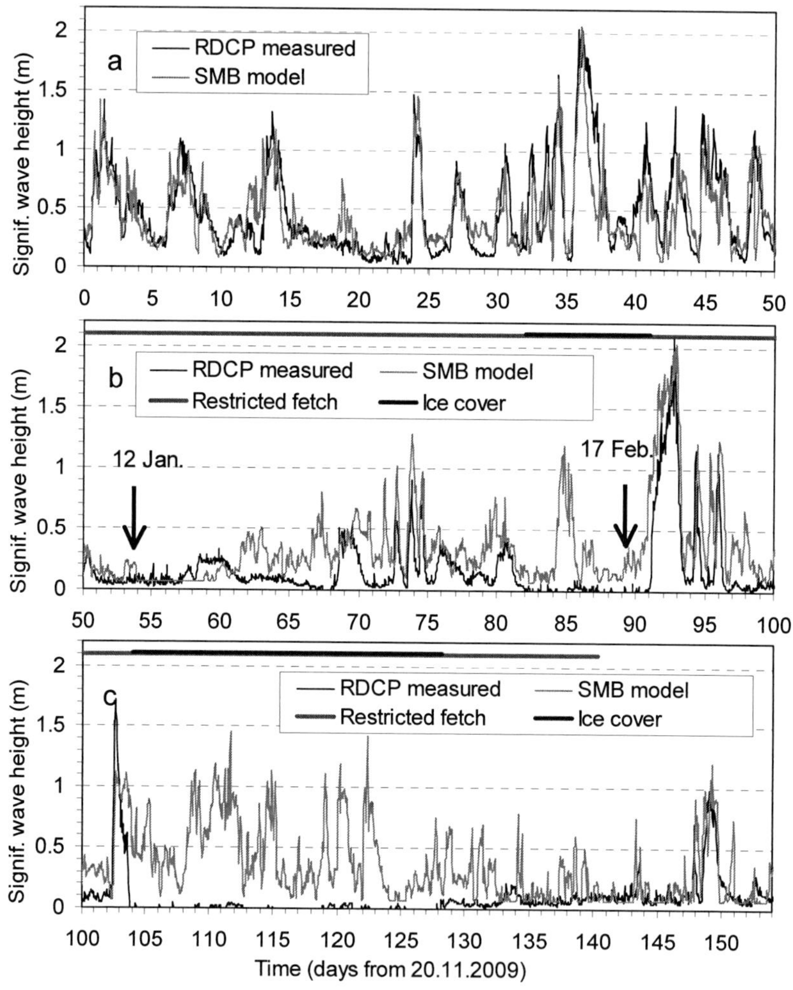

Figure 2: Variations in measured and modelled wave heights from 20 Nov. 2009 to 23 Apr. 2010. Periods of restricted fetches and ice cover are marked in the upper parts of the figures (b, c). The first 40 days (a) served as the calibration period for the SMB wave model.

The first 40 days of the full mooring (Fig. 2a) served as a calibration period for the model. The calibration was done by prescribing (i.e., modifying geographical "true") fetch distances for different wind directions and by attempting to minimize the differences between the modelled and measured series. For more details on the procedure, also see the paper by Suursaar *et al.* in this volume [12].

In ice conditions, we compared the actual waves to the modelled ones, which ought to happen under the same forcing conditions, but without ice. At the Letipea site, the model was calibrated separately using the Kunda wind data and Letipea wave measurements [7], and was further used in multi-year (1966–2008) wave hindcast at a 3h interval. Afterwards, a time-dependent "ice mask" was applied on the hindcast and a new set of annual wave statistics was computed, showing the influence of inter-annually varying ice conditions on wave climatology and coastal processes.

3 Results and discussion

3.1 Influence of ice on waves

After the first 40 day calibration period, the model performed adequately in ice-free conditions (in days 40–50 and 145–155), but the more the gulf froze over, the larger the differences between measured and modelled series grew (Figs. 2, 3). By February 2010, when the ice covered the measuring site (Fig. 4), the waves were practically absent. But when the sea was only partially covered with ice, the waves just decreased as a result of shortened fetches from certain wind directions (see also the graphical explanation in Fig. 1). The influence was larger in NE and N directions (Fig. 3), where the fetch reduction was the largest in that winter. It was smaller for westerly directions, until the site was fully covered up.

The influence of wave height reduction on long-term wave statistics, coastal processes and pollution combat issues is discussed in chapters 3.3 and 3.4.

3.2 Influence on currents

The average measured sub-surface (1–3 m) current speed (module) was 9.8 cm/s, and the maximum reached 36 cm/s in this location. The speed decreased with depth, being just 6.4 cm/s as an average in the near-bottom (12–14 m) layer. As expected, both the hydrodynamic observations performed with the RDCP in winter 2009/10, as well as hydrodynamic simulations showed somewhat weaker currents during ice cover. The ice eliminates wind-induced momentum input from the sea surface, but mainly density and sea level difference driven components remain. Surprisingly, at the Neugrund site, the partial coverage of the Gulf of Finland by ice virtually did not affect the flow magnitudes (Fig. 5). Just the dependence between wind forcing and current was weaker than in entirely ice-free conditions. The scatter was larger in ice conditions (for both full and limited coverage). The so-called wind coefficient was about 20% smaller with full ice cover (Fig. 6). This means that the current is indeed somewhat weaker below the ice cover, but not that much as one might expect.

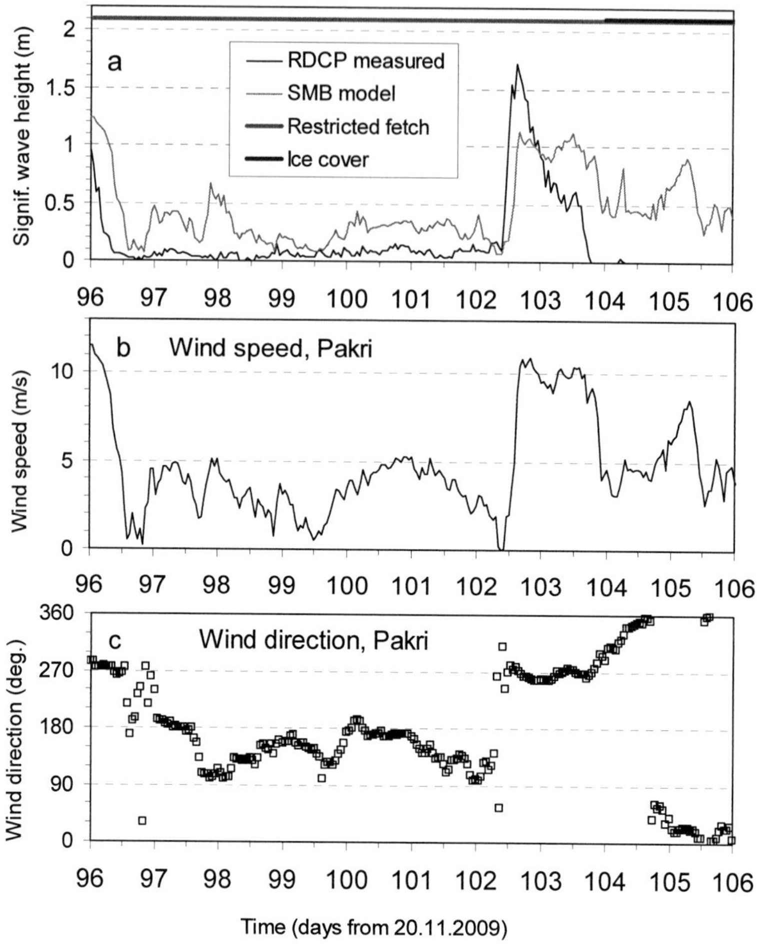

Figure 3: Variations in measured and modelled wave heights (a), wind speeds (b) and directions (c).

Apparently, the estuarine, density and sea level difference driven circulation is still rather important in the mouth section of the Gulf of Finland [1]. Somewhat differently, our previous flow measurements in the Suur Strait (near Virtsu, Fig. 1) in 1993–96 showed considerably weaker currents below the ice cover. The velocities were typically around 10 cm/s with the maximum value of 36 cm/s, whereas velocities up to 80–100 cm/s were frequently measured in ice-free, stormy conditions [13]. Over a winter, the modelled integral volume transport through the strait constituted only a half, when the Gulf of Riga and the Archipelago Sea were covered by ice [13]. In fact, the currents in the funnel-shaped straits are partly generated by wind stress and partly by the sea level difference between the strait's ends (which in turn, is also generated by wind). In icy conditions, both the direct wind stress and the sea level fluctuations are suppressed.

3.3 Influence on coastal geomorphic processes

In the practically tideless Baltic Sea, the main hydrodynamic forces acting on seashores are waves and (infrequent) storm surges [14]. Ice can directly be a geomorphic agent as well when wind-driven and broken ice masses attack coasts and form ice ridges [15].

However, more often the ice cover protects coasts from waves and hampers possible surges during cold and stable winters. Recent vitalization of coastal processes in the West Estonian Archipelago is supposed to occur as a result of increased storm surge and wave heights in one hand [16, 17], and more frequent absence of protective ice cover in the last decades in other hand [3]. While near Vilsandi (Fig. 1) both high sea level and high wave events are increasing [16], the wave heights are probably decreasing along the northerly exposed North Estonian coast as a result of northward shift of atmospheric pressure systems [7].

Figure 4: Excerpt from the ice chart by the SMHI [9] on 17 February 2010 close to the maximum ice extent for the ice season 2009/10 (see also Figs. 1, 2). The distance from the Neugrund measuring and modelling site to the ice edge was approximately 90 km.

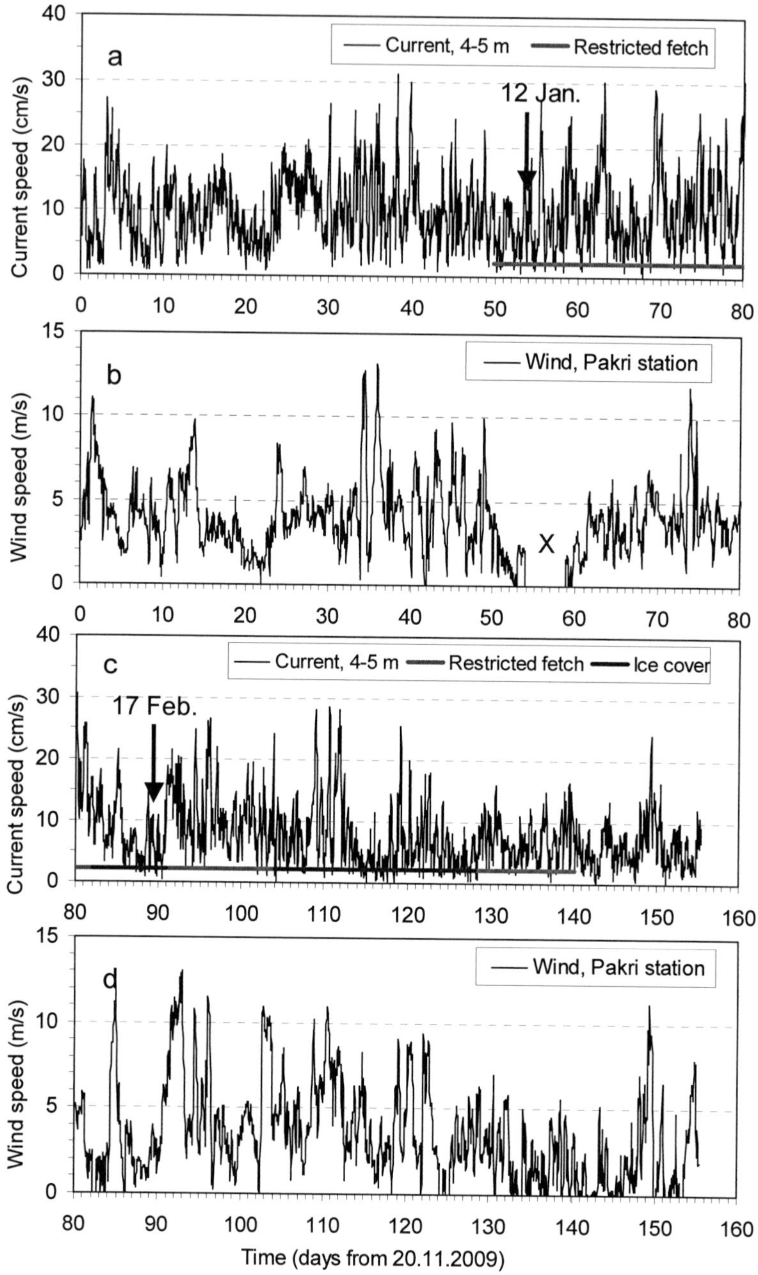

Figure 5: Variations in measured current speeds at the 3–5 m depth in 20 Nov. 2009–7 Feb. 2010 (a)–23 Apr. (c) and corresponding wind speeds (b, d). Periods of partial ice cover ("restricted fetches") and full ice cover are marked (on a, c).

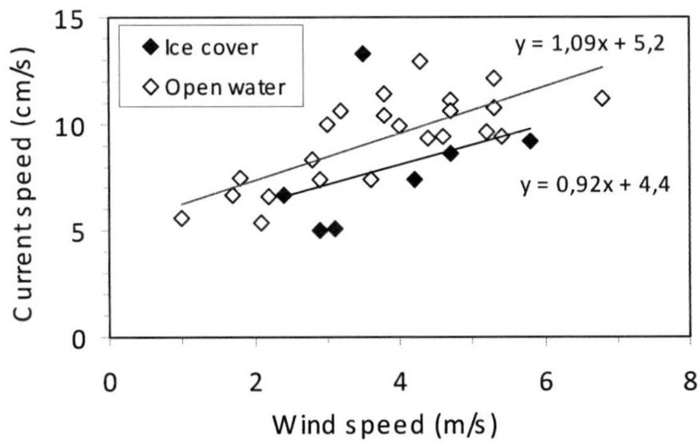

Figure 6: Relationships between wind and current speeds (averaged over 5 days periods) for icy and (entirely or partially) open water conditions.

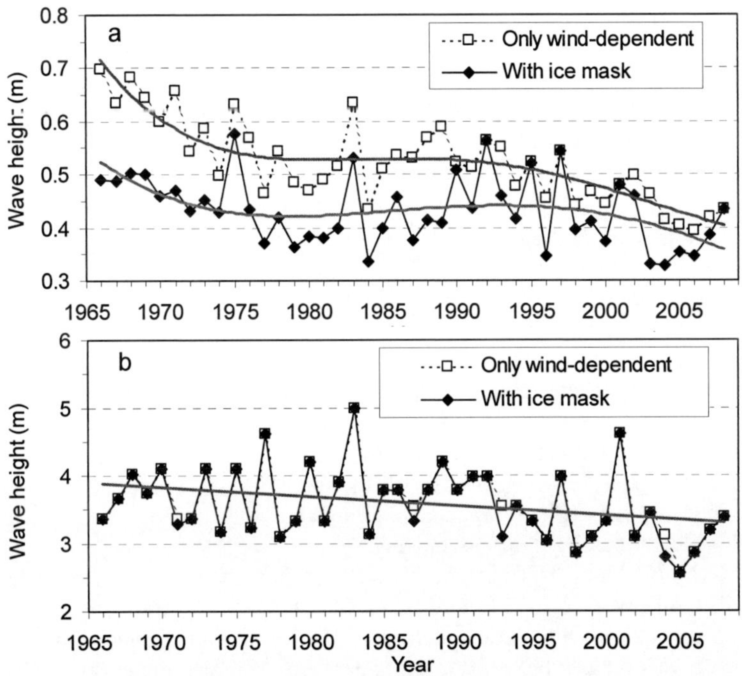

Figure 7: Decadal variations in average (a) and maximum (b) values of annual significant wave height data as a comparison between meteorologically forced hindcast and the hindcast with ice mask applied in the periods with real ice cover in the Letipea-Kunda region of the Gulf of Finland.

Compared to the West-Estonian coastal sea, the Letipea-Kunda coastal area is also more influenced by variations in ice conditions, and particularly by the 20–40 days (according to linear trend over 1949–2008) decrease in number of ice days [3]. In addition to the solely meteorologically forced wave hindcast, a version of the calculations with "ice mask" was produced for the period of 1966–2008. The ice correction influenced the annual mean ice statistics and the trend became less steep (Fig. 7a). But the variations in annual wave maxima remained practically the same (Fig. 7b). Evidently, strongest winds (and waves) mainly occur during cyclonically active and ice free winters, whereas ice cover may develop only during anticyclonic, calm and cold winter weather conditions.

3.4 Influence on oil spill drift

According to the analyses compiled for the Baltic Sea by HELCOM [4], the relative frequency of wintertime shipping incidents where oil spill is involved may be about 10 times higher than in summer (ice-free) conditions. The last major shipping accident in the Gulf of Finland occurred in March 2006, when the cargo ship Runner 4 sank after colliding in the caravan. Due to severe ice conditions, the spill was difficult to detect, track and combat [18].

DeCola *et al.* [19] suggest that the nature of the ice tends to dominate other factors in impacting the behaviour of oil after a spill: ice coverage below 30% is not believed to significantly impact oil behaviour, although it has been observed to impact oil spill recovery activities. However, the 30% ice coverage or greater will significantly impact the behaviour of spilled oil. Authors argue that due to the density difference between oil and water, spilled oil will likely rise to the surface of slushy oil and ice mix while the oil trapped under ice may freeze and remain there as it cannot evaporate. Consequently, the oil is moving with the ice until the spring melt and may ultimately be released some distance from the spill site. It has been reported [18, 20] that oil released under fast or pack ice will not spread as evenly as it might on the water surface: the absence of direct wind drag and the rough underside of the ice will cause the oil to pool in some places, unless the current is strong enough to keep the oil moving. It has been shown that the sea current must exceed 0.20 m/s, under even the smoothest ice, to produce substantial relative motion of the oil. Under-ice sea current speed reported by this study (Fig. 5) exceeds 0.20 m/s only occasionally and would not cause significant under-ice oil movement.

A key finding from the latest field-tests was that although different types of crude oil would behave in different ways following an oil spill, in general oil breaks up and mixes with water much more slowly in Arctic conditions and the presence of pack ice can provide a natural barrier to oil movement, thus acting as a system of natural booms that can prevent the oil from spreading over an excessively wide area [21]. On the other hand, it has been stated that the slow-weathering phenomenon, a consequence of relatively low water temperatures and low levels of wave action, would extend the time window during which some oil recovery techniques could be applied. Thus, once the spill is detected, it will give responders more time to plan their actions and move any necessary equipment to the response site.

4 Conclusions

Ice can influence coastal processes directly (mechanically), but more important are the indirect influences by reducing or eliminating wave heights, hampering storm surges and reducing current speeds. The influence on currents and water exchange processes may be larger in the straits of the West Estonian Archipelago Sea and it is smaller in the mouth section of the Gulf of Finland, where estuarine circulation can be rather efficient even below ice cover. As ice conditions in the Baltic Sea have become milder than about 50 years ago, these gradual changes have also influenced long-term wave statistics.

In icy conditions, shipping safety decreases and oil spills are more difficult to detect and combat. The rough underside of the ice usually causes the oil to pool in some places, unless the current is strong enough to keep the oil moving. Under-ice sea current reported by this study exceeds 0.20 m/s only occasionally and would not cause significant under-ice oil movement.

Acknowledgements

The study was supported by Estonian Science Foundation grant No. 7609, Estonian target financed theme SF0180104s08, and the EU-funded Central Baltic INTERREG IVA Programme Project "OILRISK – Applications of ecological knowledge in managing oil spill risk".

References

[1] Soomere, T., Myrberg, K., Leppäranta, M. & Nekrasov, A., The progress in knowledge of physical oceanography of the Gulf of Finland: a review for 1997–2007. *Oceanologia*, **50**, pp. 287–362, 2008.

[2] Seinä, A. & Palusouo, E., The classification of the maximum extent of ice cover in the Baltic Sea 1720–1992. *Meri*, **20**, pp. 5–20, 1993.

[3] Sooäär, J. & Jaagus, J., Long-term variability and changes in the sea ice regime in the Baltic Sea near the Estonian coast. *Proc. Estonian Acad. Sci. Eng.*, **13**, pp. 189–200, 2007.

[4] HELCOM. Maritime Activities in the Baltic Sea – An integrated thematic assessment on maritime activities and response to pollution at sea in the Baltic Sea region. *Balt. Sea Environ. Proc.*, **123**, 65 pp., 2010.

[5] Leiger, R., Aps, R., Fetissov, M., Herkül, K., Kopti, M., Kotta, J., Mander, Ü. & Suursaar, Ü., Oil accident response simulation: allocation of potential places of refuge. Coastal Processes. Book Series: *WIT Transactions on Ecology and the Environment,* **126,** WIT Press: Southampton, Boston, pp. 247–258, 2009.

[6] Suursaar, Ü. & Kullas, T., Influence of wind climate changes on the mean sea level and current regime in the coastal waters of west Estonia, Baltic Sea. *Oceanologia*, **48**, pp. 361–383, 2006.

[7] Suursaar, Ü., Waves, currents and sea level variations along the Letipea – Sillamäe coastal section of the southern Gulf of Finland. *Oceanologia*, **52** pp. 391–416, 2010.
[8] Suursaar, Ü. & Kullas, T., Decadal variations in wave heights off Cape Kelba, Saaremaa Island, and their relationships with changes in wind climate. *Oceanologia,* **51**, pp. 39–61, 2009.
[9] SMHI Ice Service. Web-page: http://www.smhi.se/oceanografi /iceservice/is_prod_en.php
[10] Seymour, R.J., Estimating wave generation in restricted fetches. *J. ASME WW2, May 1977*, pp. 251–263, 1977.
[11] Räämet, A., Suursaar, Ü., Kullas, T. & Soomere, T., Reconsidering uncertainties of wave conditions in the coastal areas of the northern Baltic Sea. *Journal of Coastal Research, SI* **56,** pp. 257–261, 2009.
[12] Suursaar, Ü., Szava-Kovats, R. & Tõnisson, H., Wave climate and coastal processes at the Osmussaar – Neugrund region, Baltic Sea. *Coastal Processes II.* Book Series: WIT Transactions on Ecology and the Environment, **145**, WIT Press: Southampton, Boston, pp. xx–xx, 2011.
[13] Astok, V., Otsmann, M. & Suursaar, Ü., Water exchange as the main physical process in semi-enclosed marine systems: the Gulf of Riga case. *Hydrobiologia*, **393,** pp. 11–18, 1999.
[14] Suursaar, Ü., Jaagus, J., Kont, A., Rivis, R. & Tõnisson H., Field observations on hydrodynamic and coastal geomorphic processes off Harilaid Peninsula (Baltic Sea) in winter and spring 2006–2007. *Estuarine Coastal and Shelf Science,* **80,** pp. 31–41, 2008.
[15] Orviku, K., *Estonian seacoasts*. Valgus: Tallinn, 112 pp., 1974. [in Russian].
[16] Suursaar, Ü. & Kullas, T., Decadal changes in wave climate and sea level regime: the main causes of the recent intensification of coastal geomorphic processes along the coasts of Western Estonia? *Coastal Processes*. Book Series: *WIT Transactions on Ecology and the Environment*, **126,** WIT Press: Southampton, Boston, pp. 105–116, 2009.
[17] Tõnisson, H., Orviku, K., Jaagus, J., Suursaar, Ü., Kont, A. & Rivis, R., Coastal Damages on Saaremaa Island, Estonia, Caused by the Extreme Storm and Flooding on January 9, 2005. *Journal of Coastal Research*, **24**, pp. 602–614, 2008.
[18] Wang, K., Leppäranta, M., Gästgifvars, M., Vainio, J. & Wang, C., The drift and spreading of the Runner 4 oil spill and the ice conditions in the Gulf of Finland, winter 2006. *Estonian Journal of Earth Sciences*, **57**, pp. 181–191, 2008.
[19] DeCola, E., Robertson, T., Fletcher, S. & Harvey, S., *Offshore Oil Spill Response in Dynamic Ice Conditions: A Report to WWF on Considerations for the Sakhalin II Project*. Nuka Research: 74 pp., 2006.
[20] AMAP. *AMAP Assessment Report: Arctic Pollution Issues. Arctic Monitoring and Assessment Programme.* Oslo, Norway: 859 pp., 1998.
[21] Bailey, A., Arctic Directory: JIP publishes Arctic oil spill results. *Petroleum News*, **15** (38), pp. 9–11, 2010.

Hydrodynamic forces acting on an oscillating structure

D. Mirauda[1], A. Volpe Plantamura[1] & S. Malavasi[2]
[1]*D.I.F.A. Basilicata University, Italy*
[2]*D.I.I.A.R. Politecnico di Milano, Italy*

Abstract

The submerged and semi-submerged structures (off-shore platforms, wind farms, floating buildings, marine turbines, oil conduits, buoys) in sea currents are often subjected to strong vibrations that condition the structural stability and also give rise to serious damage to the mechanical parts of the structures, with consequent efficiency loss of the system and above all, repercussions in the environment in which they are located. These vibrations increase when the structures are immersed in bounded flows (presence of the bed and/or the free surface). In order to attenuate and eliminate these problems, analytical and experimental means are used to describe the dynamic behaviour of structures for different flow field configurations and boundary conditions. Recent research has demonstrated that integrated numerical modelling (considering both structure and fluid characteristics) is the more suitable way to value the dynamic response of a system. To check on the reliability of the resolution methods experimental controls are needed that, for the complexity of structures, requires models with simplified geometries (spheres, cylinders, rectangles). In this context, the present work aims to provide some experimental results obtained analyzing both the oscillations and the forces acting on a structure to simple geometry (sphere) characterized by low values of mass and damping and invested by a free surface steady flow.
Keywords: sphere, free surface, bed, hydrodynamic forces.

1 Introduction

The study of flow-structure interactions is important in many scientific fields as in hydraulics, in civil and environmental engineering, in river, maritime and in

www.witpress.com, ISSN 1743-3541 (on-line)
doi:10.2495/CP110271

ocean constructions. In particular, the analysis of the hydrodynamic forces and the evaluation of vibrate modes of structures in water are fundamental for the correct design of off-shore structures like wind farms, marine turbines, oil conduits and platforms, buoys, river constructions, bridges, embankments and also new plants producing electric energy by the sea currents.

The complex phenomena generating by the fluid-structure interaction depend by different elements: the physic characteristics of the fluid (density and dynamic viscosity), the flow cinematic conditions, the location of the structure respect to the depth of the current (level of submergence of the system in the current) and also the mechanical characteristics of the structure itself.

All these elements can cause conditions of structural instability, with consequent efficiency loss and repercussions in the sea environment.

At the moment, these problems are exceeded through an increasing of the off-shore structures dimensions, assuring in this way high security coefficients to the detriment to a general reduction of costs, but above all amplifying the environmental impact, causing visual externalities, when the installation is not completely submerged and located near the coast.

In order to eliminate these problems, recent research (see Jonkman and Sclavounos [1]) have demonstrated as the integrated numerical modelling, considering the characteristics of both structure and fluid, is the more suitable way to value correctly the dynamic response of such systems and thus to give more information for a better design. The difficulty of this type of approach consists in the check on the reliability of the resolution methods. Because of the complexity of the structures, these checks require models with simplified geometries (spheres, cylinders, rectangles, etc.). For this purpose, it is important to underline as the study of the dynamic response of a structure at simple geometry is fundamental to make deeper the analysis of these complex phenomena. Moreover, simple geometries allow an easier comparison between the experimental and numerical approach which, also if with approximate solutions, gives certain advantages in the modelling of complex structures.

In literature there is a large number of fundamental studies on the fluid-structure interactions of submerged cylinders (two-dimensional bodies) (Sarpkaya [2], Griffin and Ramberg [3], Bearman [4], Parkinson [5], Williamson and Govardhan [6], Blevins [7], Naudascher and Rockwell [8], Sumer and Fredsøe [9], and Anagnostopoulos [10]), but there are fewer more recent studies on three-dimensional structures (spheres), because of the difficult interpretation and analysis of the flow fields generating around these bodies (Williamson and Govardhan [11]; Govardhan and Williamson [12, 13]; Jauvtis *et al.* [14], Mirauda and Greco [15, 16]; Franzetti *et al.* [17]; Greco *et al.* [18]; Mirauda *et al.* [19–21], and Volpe Plantamura *et al.* [22]).

In order to characterize the phenomena involving the interaction between fluid and three-dimensional structures, this paper analyzes in detailed way the response of a spherical body in a free surface steady flow to change of the location of the structure inside the current. The knowledge of the behaviour of the body in these conditions could be interesting because could avoid the structural over-dimensioning of the same.

2 Experimental details

Experiments have been performed in a non-tilted Plexiglas open water channel in the Hydraulics Laboratory of the Politecnico di Milano, having a rectangular cross section (0.6m high, 0.5m wide and 7.0m long) just used and characterized in previous works (Malavasi and Guadagnini [23]). A water filled sphere, with a diameter D=0.087m and a mass m=0.46Kg, has been used as the obstacle. The sphere surface is made of PVC and covered with an episodic paint to reduce the surface roughness. A rod, made of stainless steel and Delrin, has been used to connect the sphere to a fixed structure mounted on the channel. The stainless steel part, which is connected to the sphere, is 0.080m long and 0.003m in diameter, while the Delrin part completing the rod has a diameter of 0.012m.

The sphere-rod system is sustained by a robust support realized with aluminium bars, fixed to the channel which permits at the sphere to move in the both directions (stream wise, x, and transversal respect to the current flow, y).

In all the experiments the depth of current was equal to 5 times the sphere diameter (H=0.435m), in order to maintain the blockage coefficient constant (γ_b=0.03, ratio between the body frontal area and the transverse section of the channel), while the mean flow velocity, U, changed between 0.2 and 0.5m/s. The sphere, instead, was located in three different submergence conditions respect to the flow, h^* (ratio between the distance from free surface and upper surface of the sphere and the body diameter): the first at an elevation of 2D above the bottom wall of the channel and from the free surface, in order to assume the mimicking of the symmetrical flow condition for the sphere (h^*=2) and to neglect the wall and free surface influence (fig. 1(a)); the second at a distance of 3mm (h^*=3.97) from the bottom (fig. 1(b)), with the aim to value if the wall closeness have relevant effects on the dynamic response of the body; finally, near the free surface ($h^*\approx$0) in asymmetric bounded flow (fig. 1(c)), in order to analyze the effects of the free surface distortion on transverse and stream wise oscillation amplitudes.

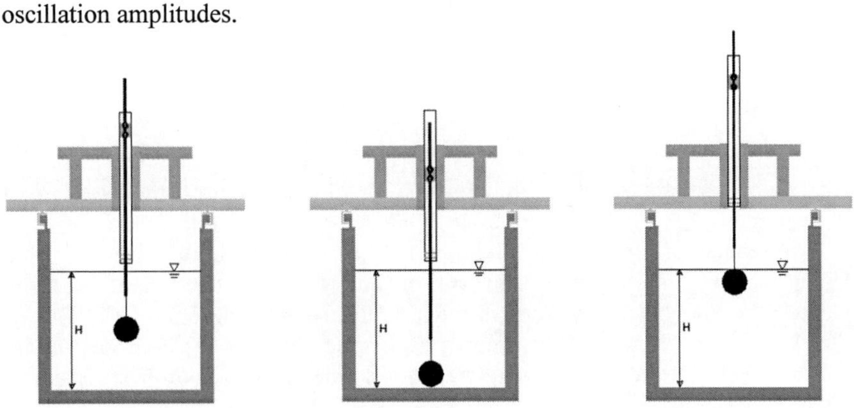

Figure 1: Transverse section of the channel with sphere at: (a) h^*=2; (b) h^*=3.97, (c) h^*=0.

The sphere movements, for the different flow velocities U (or likewise for the normalized velocity $U^*=U/f_{nw}D$, where f_{nw} is the body natural frequency in water), have been measured by an image analysis through a CCD camera (Charge Coupled Device) set on acquisition frequencies from 30Hz to 50Hz. The camera, located below the water channel, allows us to memorize the 2D displacements, thanks to the presence of a marker placed below the centre of the sphere in line with its centre of gravity. Table 1 summarizes the experimental tests of the present work.

Table 1: Experimental tests.

TEST	1	2	3	4	5	6	7	8	9	10	11	12
h^*	2	3.97	0	2	3.97	0	2	3.97	0	2	3.97	0
U (m/s)		0.2			0.3			0.4			0.5	
U^*		1.90			2.84			3.79			4.74	
Re		17400			26100			34800			43500	

3 Analytical approach

The forces acting on the sphere in the flow principal direction, F_x, and in transverse that, F_y, have been valued by means of the system of two uncoupled dynamic equilibrium equations:

$$\begin{aligned} m\ddot{x} + c\dot{x} + kx &= F_x \\ m\ddot{y} + c\dot{y} + ky &= F_y \end{aligned} \tag{1}$$

in which the instantaneous velocity ($\dot{x}$, $\dot{y}$) and acceleration ($\ddot{x}$, $\ddot{y}$) of the sphere have been obtained by a Centred Finite Difference Method (CFDM) with error of 4th order for both the first and the second derivative, thanks to the knowledge of the displacements (x and y). Because the scheme of the equations of the used CFDM does not change for the two directions, we report below the system only for the stream wise case:

$$\begin{aligned} \dot{x}_i &= \frac{x_{i-2} - 8x_{i-1} + 8x_{i+1} - x_{i+2}}{12\Delta T} + \frac{1}{30}\overset{.....}{x}(\xi)\Delta T^4 \\ \ddot{x}_i &= \frac{-x_{i-2} + 16x_{i-1} - 30x_i + 16x_{i+1} - x_{i+2}}{12\Delta T^2} + \frac{1}{90}\overset{......}{x}(\xi)\Delta T^4 \end{aligned} \tag{2}$$

where ΔT represents the sampling time or the inverse of the acquisition frequency.

The numerical resolution of the eqn. (1) has required the knowledge of the modal parameters of spring constant, k, and damping coefficient, c. Such parameters have been calculated through the analysis of the free oscillations of the sphere in air (fig. 2). In particular, the spring constant has been derived by the expression of the natural frequency in air, valued by the power spectrum of the free oscillation (fig. 3):

$$f_n = \frac{1}{2\pi}\sqrt{\frac{k}{m}} \tag{3}$$

The value obtained of the spring constant was k=36.72Kg/s^2.
Instead, the damping parameter, c, has been deduced by the following equation:

$$\zeta = \frac{1}{2\pi}\sqrt{\frac{c}{km}} \quad (4)$$

in which the damping ratio, ζ, has been valued through the classical method of logarithmic decrement applied to the successive displacement peaks of the same free oscillation signal shown in fig. 2. The value of the damping ratio was ζ=0.0044, while the structural damping was c=0.036Kg/s. Finally, the values of f_{nw} and ζ_w in water equal to 1.21Hz and 0.004, have been obtained multiplying the eqns. (3) and (4) for the coefficient $[m^*/(m^*+C_A)]^{1/2}$, where C_A is the added mass coefficient taking into account the water presence and is equal to 0.5 for spheres, while m^* is the mass ratio equal to the ratio between the mass of the sphere and the displaced fluid mass ($m^*=m/\pi\rho D^3/6=1.34$).

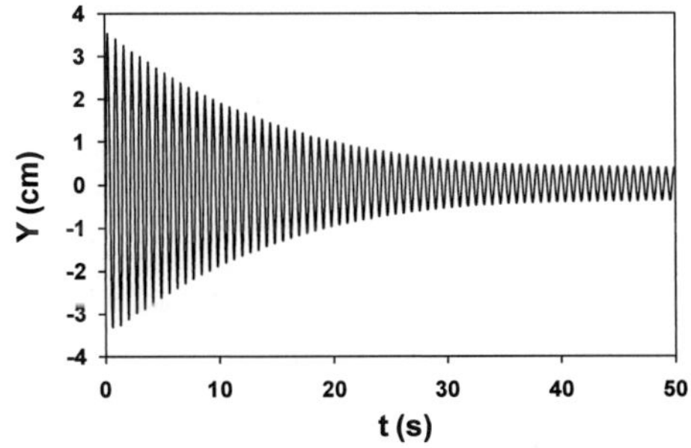

Figure 2: Evolution in the time of the free oscillation of the sphere.

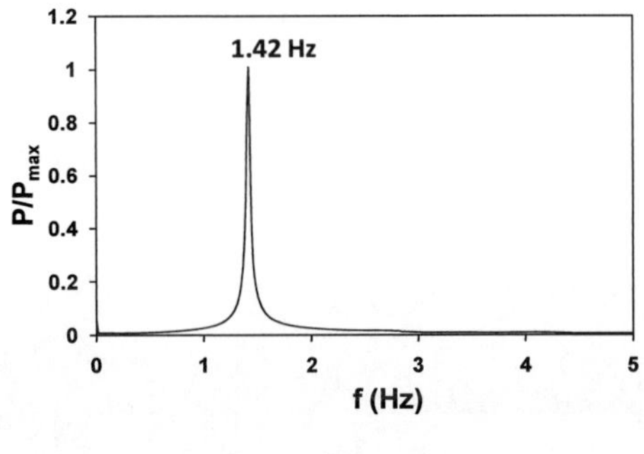

Figure 3: Power spectrum of the free oscillation of the sphere.

4 Forces analysis

In order to study the behaviour of the structure at the different positions inside the current, h^*=2, h^*=3.97 and h^*=0, the displacements in the flow principal direction and transverse to the current and the respective acting hydrodynamic forces have been evaluated.

In particular the following figures report the stream wise and transverse oscillations normalized through the characteristic dimension of the body, that in this case is represented by the sphere diameter, for the lower (U^*=1.90) and higher normalized velocity (U^*=4.74) of the investigated currents and the relative forces normalized respect to the observed maximum value.

In the case of h^*=2, representing the more studied condition in literature, an increase of stream wise and transverse displacements is observed to increasing of the mean flow velocity (fig. 4(a)). In particular, the stream wise oscillations for U^*=4.74 increase of 15 times respect to the case of U^*=1.90, while the transverse those of around 75 times. The major increase of transverse displacements implies a trajectory at shape of crescent topology, observed also in literature for spheres with similar mass ratio and completely submerged. Obviously, the mean displacement of the sphere in the main flow direction respect to the static position (zero position) increases with the mean flow velocity. The increase of the displacements is confirmed by an increasing of the forces. In fact, the stream wise forces in fig. 4(b) show an increase of the mean component with the increasing of the velocity, while the fluctuating component seems to be lightly growing with U^*. On the contrary, in fig. 4(c), the transverse forces show a more

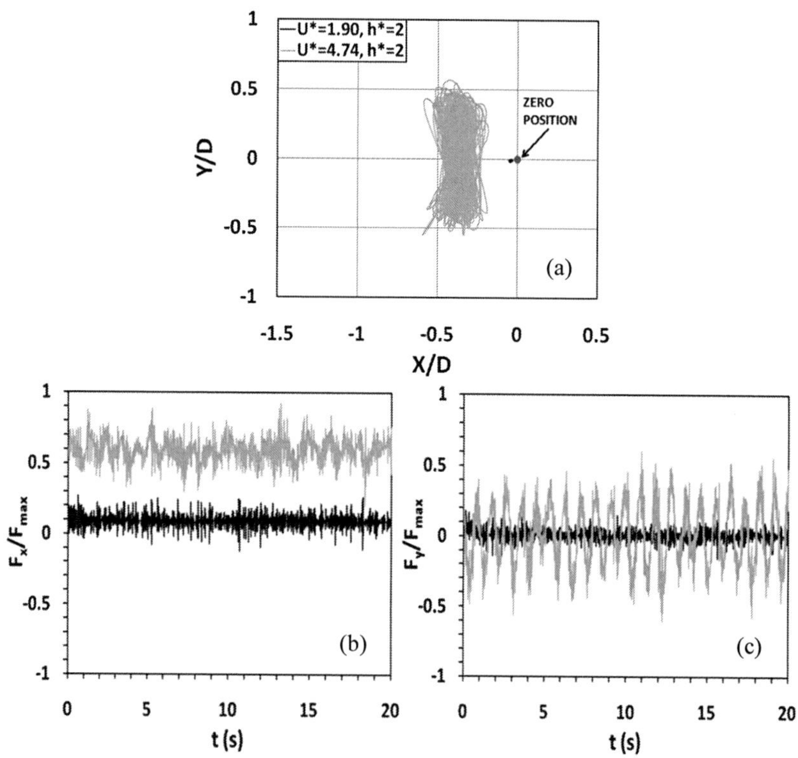

Figure 4: (a) Trajectory of the sphere at h^*=2 for U^*=1.90 and 4.74; (b) and (c) evolution in the time of stream wise F_x and transverse forces F_y.

significant increasing of the fluctuating component with the increasing of the velocity, while the mean component remains constant around the zero value for the body symmetry condition (fig. 4(c)).

Similar results to the case of h^*=2 can be observed for the sphere placed near the bottom of the channel, h^*=3.97 (fig. 5). In this case, the asymmetrical boundary condition due to the closeness of the wall lightly influences the sphere movements in respect to the previous case. In particular, as it is possible to note from fig. 5(a), the presence of the channel floor significantly reduces the stream wise and transverse mean displacements respecting the condition of h^*=2. For the higher velocity, U^*=4.74, we note a reduction of about 20% in stream wise and of about 10% in the transverse mean displacements. Also in this case the shape of the trajectory for high velocity is that at crescent topology. The slight differences in the displacements are however less evident in terms of mean and fluctuating values of the acting forces (fig. 5(b) and 5(c)).

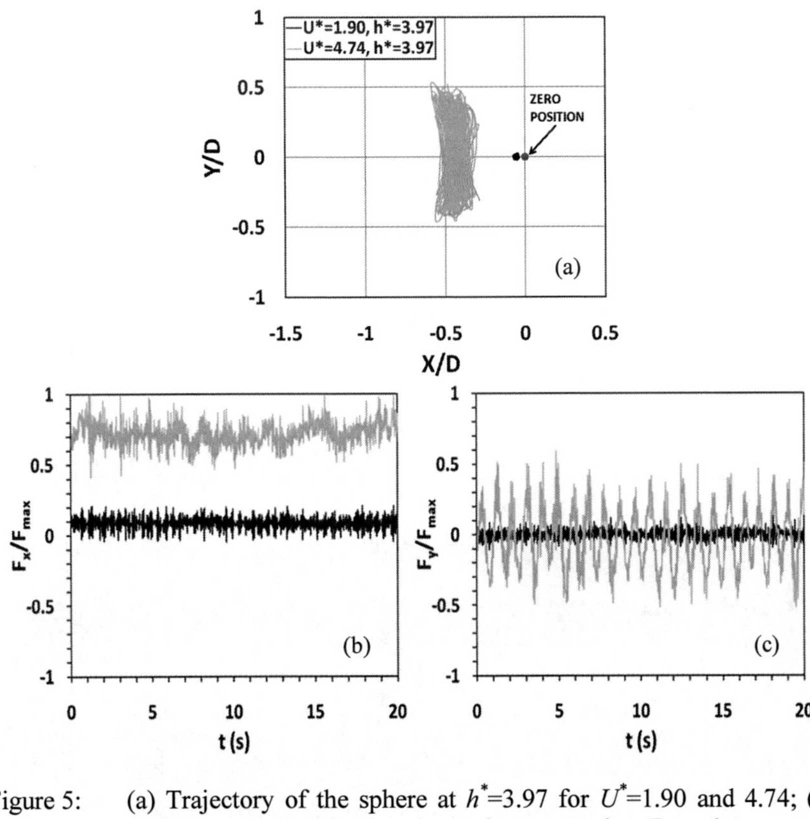

Figure 5: (a) Trajectory of the sphere at h^*=3.97 for U^*=1.90 and 4.74; (b) and (c) evolution in the time of stream wise F_x and transverse forces F_y.

Comparing the results of h^*=2 and h^*=3.97 with the case of body close to the free surface (h^*=0), we observe as the latter presents a very different behaviour both in terms of the sphere movements and acting forces. Fig. 6(a) shows as for the two different flow cinematic conditions the sphere trajectories are similar, and present stream wise and transverse displacements comparable. This experimental result confirms that reported by Mirauda *et al.* [21] and Volpe Plantamura *et al.* [22], putting in evidence displacements ranges around to values of about 0.1*D*, and showing a chaotic trajectory. This behaviour could be explained with the strong distortion of the free surface that reduces the transverse movements of the sphere for higher flow velocities and increases the body displacements in the both directions for lower normalized velocities. Instead, as observed in the previous cases, correctly at U^*=4.74 the sphere results more displaced in the principal flow direction respect to the static position. These results appear confirmed from the analysis of the forces acting on the structure. Fig. 6(b) shows as the mean value of the stream wise force, F_x, for U^*=4.74 is translated less towards the high respect to the other cases of relative submergence, and the fluctuations are less growing with the increase of the current velocity. From the fig. 6(c) it is possible to note as, for the two analyzed conditions of current flow velocity, also in this case, the mean value of F_y is

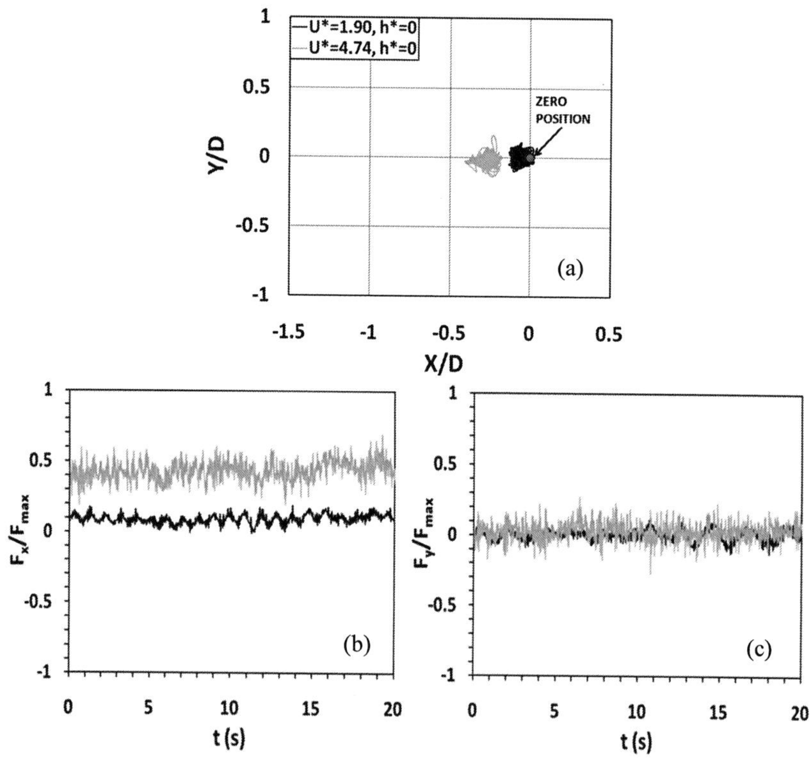

Figure 6: (a) Trajectory of the sphere at h^*=0 for U^*=1.90 and 4.74; (b) and (c) evolution in the time of stream wise F_x and transverse forces F_y.

correctly equal to zero because of the symmetry of the sphere, while, instead, in different way respect to the other submergence conditions, the force fluctuations respect to the mean value for U^*=4.74 are strongly reduced.

Fig. 7 summarizes and compares the behaviour of the cinematic and dynamic parameters characteristic of the structure response for the two different conditions of normalized velocity (U^*=1.90 and 4.74) and for the three considered relative submergences (h^*=2, 3.97 and 0). In particular, the root mean square values of stream wise and transverse components of normalized displacements and forces are reported. In the stream wise direction, it is evident as the displacements and forces values are always crescent with the current velocity; while for the transverse direction it is shown clearly the different trend of the fluctuating values of both displacements and forces, when the sphere is close to the free surface at h^*=0.

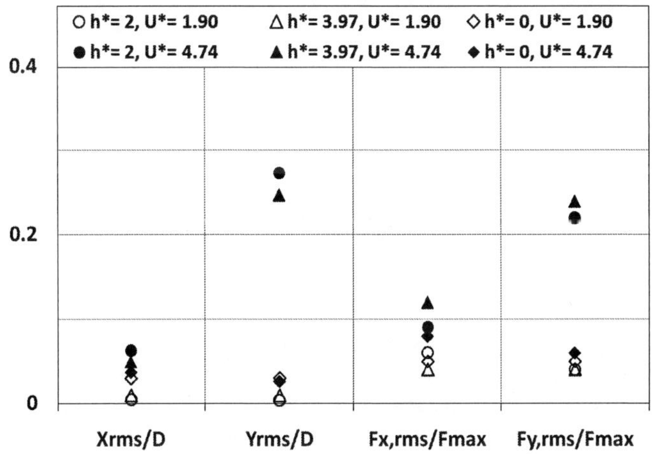

Figure 7: Root mean square values of normalized displacements and forces in stream wise and transverse direction, for h^*=2, 3.97 and 0 at low (U^*=1.90) and high normalized velocity (U^*=4.74).

5 Conclusions

In the present work we have analyzed the forces acting on a spherical structure characterized by a low mass-damping parameter $m^*\zeta$=0.005 and immersed in a steady flow at constant blockage coefficient and at three different relative submergences: two asymmetrical bounded condition, with sphere close to the free surface (h^*=0) and to the bottom (h^*=3.97) and a symmetrical position (h^*=2), which represents the more analyzed condition in literature.

The results evidence the strong influence of the closeness of the free surface (h^*=0) both on the sphere movements and acting hydrodynamic forces, respect to the symmetrical condition of complete submergence (h^*=2) and to the asymmetrical condition to the bottom (h^*=3.97). In particular, the free surface distortion induces two different effects: a reduction of displacements and force fluctuations in the transverse direction for high current flow velocity, U^{*}=4.74, and an increase of stream wise and transverse movements and forces, even if not much significant, in the case of low normalized velocity, U^{*}=1.90.

Acknowledgements

The authors express their sincere thanks to Dr. Luca Martinelli, of the Department of Structural Engineering, Politecnico di Milano, for his precious assistance in the choice and resolution of the analytical method for the evaluation of the forces acting on the body.

References

[1] Jonkman, J.M. & Sclavounos, P.D., Development of Fully Coupled Aeroelastic and Hydrodynamic Models for Offshore Wind Turbines: Preprint. *ASME Wind Energy Symposium*, Reno, Nevada, January 10-12, 2006.

[2] Sarpkaya, T., Vortex-induced oscillations. *ASME J. Appl. Mech.*, **46**, pp. 241-258, 1979.

[3] Griffin, O.M. & Ramberg, S.E., Some recent studies of vortex shedding with application to marine tubulars and risers. *ASME Journal of Energy Resour. Tech.*, **104**, pp. 2-13, 1982.

[4] Bearman, P. W., Vortex shedding from oscillating bluff bodies. *Annual Review of Fluid Mechanics*, **16**, pp. 195-222, 1984.

[5] Parkinson, G., Phenomena and modelling of flow-induced vibrations of bluff bodies. *Prog. In Aerospace Sci.*, **26**, pp. 169-224, 1989.

[6] Williamson, C. H. K. & Govardhan, R., Vortex-induced vibrations. *Annual Review of Fluid Mechanics*, **36**, pp. 413-455, 2004.

[7] Blevins, R. D., *Flow-Induced Vibrations*. Van Nostrand Reinhold, 1990.

[8] Naudascher, E. & Rockwell, R., *Flow-Induced Vibrations: An Engineering Guide*. Balkema, 1994.

[9] Sumer, B. M. & Fredsøe, J., *Hydrodynamics around Cylindrical Structures*. World Scientific, 1997.

[10] Anagnostopoulos, P., Flow-Induced Vibrations in Engineering Practice. WIT Press, 2002.

[11] Williamson, C. H. K. & Govardhan, R., Dynamics and forcing of a tethered sphere in a fluid flow. *Journal of Fluids and Structures*, **11**, pp. 293-305, 1997.

[12] Govardhan, R. & Williamson, C. H. K., Vortex induced motions of a tethered sphere. *Journal of Wind Engineering Industrial Aerodynamics*, **69-71**, pp. 375-385, 1997.

[13] Govardhan, R. & Williamson, C. H. K., Vortex-induced vibrations of a sphere. Journal *of Fluid Mechanics*, **531**, pp. 11-47, 2005.

[14] Jauvtis, N., Govardhan, R. & Williamson, C. H. K., Multiple modes of vortex-induced vibration of a sphere. *Journal of Fluids and Structures*, **15**, pp. 555-563, 2001.

[15] Mirauda, D. & Greco, M., Transverse vibrations of an sphere at high combined mass-damping parameter. In *Shallow Flows*, eds. Jirka & Uijttewaal, Balkema Publisher, Taylor & Francis Group, London, pp. 111-115, 2003.

[16] Mirauda, D & Greco, M., Flow-induced vibration of an elastically sphere at high combined mass-damping parameter. *Journal of IASME Transactions*, **1**, pp. 486-491, 2004.

[17] Franzetti, S., Greco, M., Malavasi, S. & Mirauda, D., Flow Induced Excitation on basic shape structures. In *Vorticity and turbulence effects in fluid structure* interaction, eds. Brocchini M., Trivellato F., Collana AFM - Advances in Fluid Mechanics, Wessex Institute of Technology Press, Southampton, United Kingdom, pp. 131-156, 2005.

[18] Greco, M., Malavasi, S., & Mirauda, D., Interaction between a tethered sphere and a free surface flow. In *Fluid Structure Interaction and Moving Boundary Problems*, eds. Chakrabarti S.K., Hernandez S., Brebbia C.A., Wit PRESS, Southampton, Boston, **84**, pp. 205-213, 2005.

[19] Mirauda, D., Malavasi, S., Greco, M. & Volpe Plantamura, A., Kinematic analysis of the movement of a tethered sphere immersed in a free surface flow. *In 9th International Conference on Fluid Control, Measurement and Visualization*, Flucome 2007, Tallahassee, Florida, USA, September 16-19, Paper 151, 2007.

[20] Mirauda, D., Malavasi, S. & Volpe Plantamura, A., Analysis of effects of the free surface on the movement of a tethered sphere immersed in a steady flow. *In 9th International conference on flow-induced vibrations*, FIV 2008, eds.: Zolotarev I., Horacek J., Praga, 30 June - 3 July, pp. 835-840, 2008.

[21] Mirauda, D., Volpe Plantamura A., Malavasi, S. & Martinelli, L., Forze idrodinamiche agenti su una struttura oscillante. *In XXXII Convegno di Idraulica e Costruzioni Idrauliche*, ed. Walter Farina, Palermo, Italy, September 14-17, 2010.

[22] Volpe Plantamura, A., Mirauda, D. & Malavasi, S., Analisi dell'influenza della superficie libera sulle vibrazioni di una sfera investita da una corrente stazionaria ed uniforme. *L'Acqua*, **4**, pp. 71-76, 2010.

[23] Malavasi, S. & Guadagnini, A., Interactions between a rectangular cylinder and a free surface flow. *Journal of Fluids and Structures*, **23**, pp. 1137-1148, 2007.

Detrended fluctuation analysis of significant wave height time series

L. Cabrera & G. Rodríguez
Departamento de Física, Universidad de Las Palmas de G. C., Spain

Abstract

Mean daily significant wave height time series are analysed by means of the detrended fluctuation analysis (DFA) method to examine the existence of long-range correlations, or long memory. The results indicate that the scaling behavior of significant wave height fluctuations is not constant over the considered time scales but there is a decrease of the scaling exponent with increasing time scale. Two crossover times have been identified, indicating three different scaling behaviors. Fluctuations associated to time scales between 10 and 100 days, approximately, show long-range correlation while fluctuations above and below this range present white and Brownian noise-like behavior, respectively.

1 Introduction

The knowledge of sea state conditions is critical not only for studying many oceanic and atmospheric processes but also for many offshore and nearshore operations. The evolution of sea state conditions is also of utmost importance in coastal erosion, several questions concerning safety, reliability and feasibility of offshore activities, as well as in maritime transport.

It has been highlighted by many researchers [1, 2] that the design and construction of marine structures and the coastal and offshore development activities require a relatively accurate knowledge of wave conditions to be expected during those activities and the expected lifetime of the structures. However, acquiring accurate knowledge of wave conditions evolution proves to be a difficult task. Ocean wind waves are generated in the interface connecting the turbulent boundary layers of air and water. The coupling of these two different viscous fluids is strong and nonlinear. Waves are inherently forced by the wind, break intermittently and interact strongly with surrounding turbulence. Moreover, in

WIT Transactions on Ecology and the Environment, Vol 149, © 2011 WIT Press
www.witpress.com, ISSN 1743-3541 (on-line)
doi:10.2495/CP110281

a given sea surface area, apart from an actively wind driven wave field, there may be one or more wave fields that have been generated elsewhere in the past and propagated toward this zone. These different wave fields interact through nonlinear interactions. As a result, in general, wind-generated waves are non-stationary, irregular and highly nonlinear, representing a process with a highly complex dynamics [3].

Many natural phenomena can be described as stochastic processes. A key feature of this type of process is the correlation time or scale, characterized by the autocorrelation function, involving either long range or short range correlation characteristics. The autocorrelation function decay for a long range correlated process lacks a characteristic time. In fact, the de-correlation time is infinite. On the other hand, short range correlation processes decay is described by an exponential function and therefore they exhibit finite time scales. Although, in recent years, several studies have shown that many complex natural systems, including hydro-meteorological processes, are characterized by long-range correlations [4, 5], a standard assumption of time series analysis is that observations separated by a large time span are roughly independent. However, for processes with long-range dependence, these observations are not independent.

Experimental data are often affected by non-stationarities, i.e. varying mean and standard deviation. These effects have to be well distinguished from the intrinsic fluctuations and correlations of the system in order to find the correct scaling behavior. Sometimes, type and reasons for underlying non-stationarities in collected data are not known [6]. Furthermore, it is difficult to identify and quantify long-range correlations using traditional tools like the autocorrelation function or power spectrum, partly because of the nonstationarity of the signal.

In this context, long-term time series of significant wave height (the parameter most frequently used to describe wave conditions) exhibit a number of features, namely random variability, serial correlation, seasonal periodicity, nonstationarity, and, possibly, long-term trends, evolving over different time scales. If there is long range dependence in significant wave height time series then forecasts in the long term, as well as the knowledge on the physics underlying the observed dynamical processes, could improve.

The key objective of this paper is to analyze significant wave height time series over a range of temporal scales to identify the possible existence of long-range correlations. For this, the detrended fluctuation analysis(DFA) technique is used, because it permits the detection of long-range correlations embedded in a seemingly nonstationary time series and also avoids the spurious results that are possibly caused by nonstationarity.

The rest of the paper is structured as follows. Section 2 describes some characteristics of the wave data used and introduces the detrended fluctuation methodology applied for detecting scaling properties in wave time series. Section 3 deals with a summary and discussion of results obtained. Conclusions are presented in Section 4.

2 Data and analysis methods

2.1 Significant wave height data

Wave conditions during a sea state can be properly characterised by the directional spectral density. However, the parameter most frequently used to describe a sea state is the significant wave height, defined as the average of the highest one-third of wave heights in a record or in terms of the spectral density function, as four times the square root of the zero order moment of the wave spectrum. These definitions are equivalent for seas with narrow band spectra.

Significant wave height data used in this paper have been provided by Puertos del Estado. Wave data were recorded during 12 years (from 1998 to 2009) using a SeaWatch directional waverider buoy placed at 382 m water depth, off Estaca de Bares Cape (Coruña), north-west of Spain. Wave data were recorded with 2 Hz sampling frequency during a period of 30 minutes, approximately, for each hour. These data were used to compute the wave spectrum and the significant wave height, as well as other spectral wave parameters. Estimated significant wave heights have been averaged to obtain a mean daily value.

One of the main obstacles in analysing long time series of significant wave height, as in almost any environmental application, is that usually datasets measured in nature are incomplete due to a number of reasons, such as interruption of measurements due to instrument failure or maintenance, accidental loss of data, or discard of erroneous measurements. A logical approach to alleviate this inconvenient is to use some of the various methods suggested for filling gaps in time series. Unfortunately, this type of approach is not free of problems. Chen et al. [6] observed that removing segments from a signal and stitching together the remaining parts does not affect the scaling behavior of positively correlated signals, even when up to 50% of the points are removed. Thus, stitching together segments of data obtained from discontinuous experimental recordings, or removing some noisy and unreliable parts from continuous recordings and stitching together the remaining parts has become a procedure commonly used in preparing data for scaling analysis and is the approach used in this study.

2.2 Detrended fluctuation analysis

The Detrended Fluctuation Analysis (DFA) technique was introduced by Peng et al. [7, 8] to investigate long-range power-law correlations along DNA sequences. The key advantages of the DFA over conventional methods are its ability to detect long-range correlations embedded in a seemingly non stationary time series, and also to avoid the spurious detection of apparent long-range correlations that are an artifact of nonstationarity. Thus, DFA has been established as an important method to reliably detect long-range correlations in data affected by trends.

Briefly, the detrended fluctuation analysis of a time series can be described as follows. Given a time series $\{x(i)\}$ with N values taken at equidistant intervals, a new time series $\{y(k)\}$ is obtained by estimating its average value,

$$\bar{x} = \sum_{i=1}^{N} x(i) \tag{1}$$

and integrating the original time series,

$$y(k) = \sum_{i=1}^{k} (x(i) - \bar{x}) \tag{2}$$

Next, the integrated time series of length N is divided into $N_b = N/n$ nonoverlapping boxes, each containing n data points. Then, the local trend within each box is estimated by fitting an adequate function, $y_p(k)$, to the segment. The local trend in each box is commonly defined as the ordinate of a linear least-square fit of the data points in that box,

$$y_p(k) = ak + b \tag{3}$$

Thus, the possible influence of linear trends on scales larger than the segment are eliminated by subtracting $y_p(k)$ from the integrated time series in each box. However, it is interesting to remark that the use of other detrending functions may improve the accuracy of the DFA technique by removing higher order trends [9]. In this sense, the DFA method is conventionally denoted as DFA$-q$, with the index q denoting the order of the polynomial function used to characterise the local trend (DFA1, DFA2,...). In the DFA$-q$ approach, the influence of possible $(q-1)$th-order trends on scales larger than the segment size are removed by subtracting the best qth-order polynomial fit from the integrated time series, $y(k) - y_p(k)$, in each box.

The root-mean-square values of the integrated and detrended signal are calculated for each segment of length n to yield the detrended fluctuation function,

$$F_r(n) = \left[\frac{1}{n} \sum_{k=(r-1)n+1}^{rn} |y(k) - y_p(k)|^2 \right]^{1/2}, \qquad r = 1, 2, \ldots, N_b \tag{4}$$

Averaging $F_r(n)$ over the N_b intervals gives the mean value of the fluctuation function for a given subseries length n,

$$\langle F(n) \rangle = \frac{1}{N_b} \sum_{r=1}^{N_b} F_r(n) \tag{5}$$

This computation is repeated over many time scales (box sizes) to provide a relationship between $\langle F(n) \rangle$, and the segment length n. Typically $\langle F(n) \rangle$ will increase with box size n.

A linear relation between $\log F(n)$ versus $\log n$ indicates the presence of scaling, which can be characterised by the slope a of the regression line fit

$$\langle F(n) \rangle \sim n^{\alpha} \tag{6}$$

A fluctuation plot is a graph of $\log F(n)$ versus $\log n$ and the slope of the fluctuation plot is the so called α scaling exponent which characterizes the correlation in a time series. A slope $\alpha = 0.5$ corresponds to a process with no long-term correlations (random uncorrelated process, i.e white noise, or short-term memory) whereas $\alpha = 1.5$ corresponds to integrated random walk, i.e. Brownian noise. If $\alpha = 1$, temporal fluctuations are of flicker-noise type ($1/f$ noise), typical of systems in a self-organized critical state. The cases $\alpha > 1/2$ and $\alpha < 1/2$ should be physically distinguished. For $\alpha > 1/2$, there is persistence. In this case, if in the immediate past the signal has a positive increment, then on average an increase of the signal in the immediate future is expected. An exponent $\alpha < 1/2$ means antipersistence. In this case, an increasing value in the immediate past implies a decreasing signal in the immediate future, while a decreasing signal in the immediate past makes an increasing signal in the future probable (i.e. large and small values of the time series are more likely to alternate). Thus, data records with $\alpha < 1/2$ appear very noisy. They have a local noise of the same order of magnitude as the total excursion of the record.

Fluctuation analysis has been undertaken using custom-written IDL software. The computer program was tested by using synthetic signals with a known fractal dimension generated with the Weierstrass–Mandelbrot function [10].

3 Results and discussion

Significant wave height time series are obviously affected by seasonality and the concept of long-memory refers to non-periodic processes. Thus the seasonal variation of the sample mean, should be removed before analyzing the scaling behavior of the time series. Seasonality in each value of H_{s_i} has been removed by subtracting the mean of the corresponding day of the year, obtained by averaging over all years in the record. This approach has been also used by many authors (see e.g., [11]). Consequently, analysis of fluctuations is based on the departures of Hs_i from their mean daily value $\overline{H}_{s_i}$ for each calendar date i,

$$\Delta H_{s_i} = H_{s_i} - \overline{H}_{s_i} \tag{7}$$

The significant wave height time series is shown in Fig. 1a in which the seasonal behavior becomes evident. Time series of ΔH_s departures after removing the annual cycle is shown in Fig. 1b.

The integrated deviations function, or profile, of the mean daily significant wave height time series obtained by integration of the series of mean daily significant wave height departures from the corresponding calendar day mean is presented in Fig. 2. The complex behavior of H_s time series becomes apparent in this figure, since there seems to be structures over a variety of scales. Note that if during a

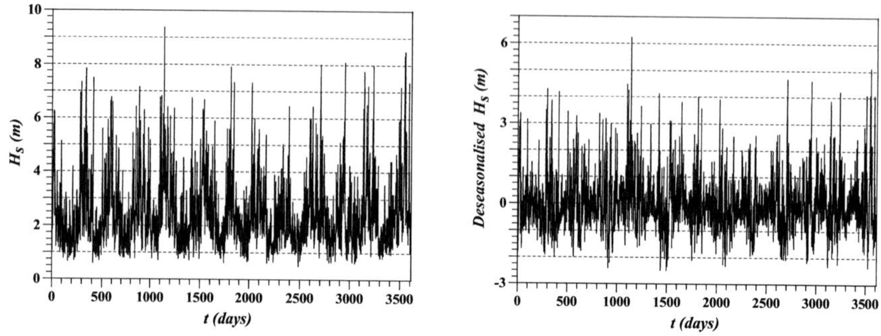

(a) Significant wave height time series

(b) Deseasonalised time series

Figure 1: Time series of daily averaged significant wave heights measured during the period (1998–2009) off Estaca de Bares, Spain, in the Cantabric Sea, and Significant wave height time series deviations, ΔH_s, after removing the seasonal component

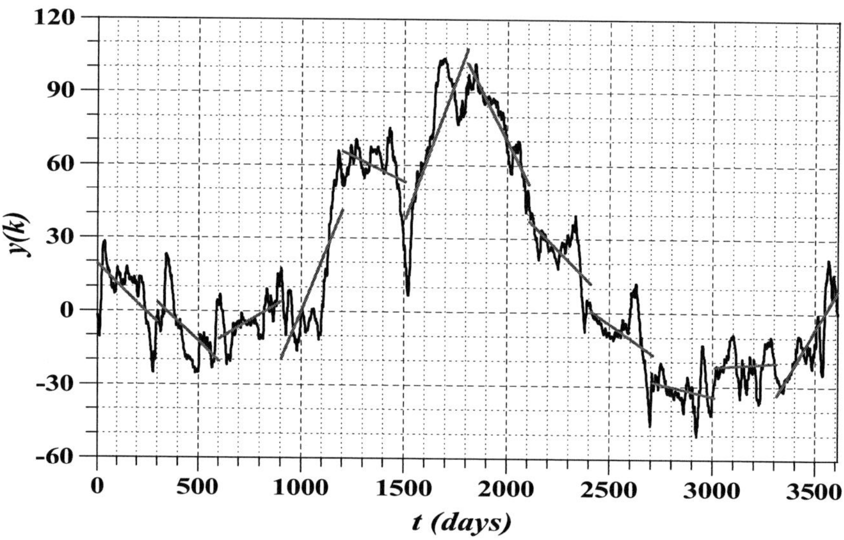

Figure 2: Mean-centred and integrated deviations of significant wave height. The straight lines are an example of the fitted local trend within windows of length $n = 301$ after partitioning the profile in $N_b = 12$ segments.

certain time period H_s is higher than the mean (stormy period), the slope of the integrated time series $y(n)$ for the specific period will be positive, while for periods with H_s lower than the mean (periods of calm), the slope will be negative.

The plot of $\log F(n)$ vs $\log n$ is shown in Fig. 3. Superposed straight lines are the best fit in the least-squares sense and indicate scaling with α exponent. To obtain statistically reliable estimates of the fluctuation function, the largest box size used is $n_{max} = N/4$, such as suggested in [12]. It can be noticed that with increasing the box size the fluctuation function increases as well. Nevertheless, it is interesting to observe that the plot of $\log F(n)$ vs $\log n$ does not fit a single line over the whole range of time scales. The plot, in fact, does not define a single long range correlation exponent but instead two or more correlation ranges.

Examination of Fig. 3 reveals the existence of two crossover and provide evidence of three distinct scaling exponents over an equal number of time scales. While the data resemble Brownian noise, $\alpha = 1.36$, dominance over short time scales, up to 10–11 days, persistent long-range correlations, $\alpha = 0.79$, are identified over time scales between 10 days and 115 days, approximately. This persistence crosses over to an almost independent random process scaling behavior, $\alpha = 0.47$, for larger time scales. This value of the DFA scaling exponent, lower than 0.5 may indicate a crossover from persistent to anti-persistent random behavior. However, this value is almost 0.5 and probably represents a random behavior for larger time scales.

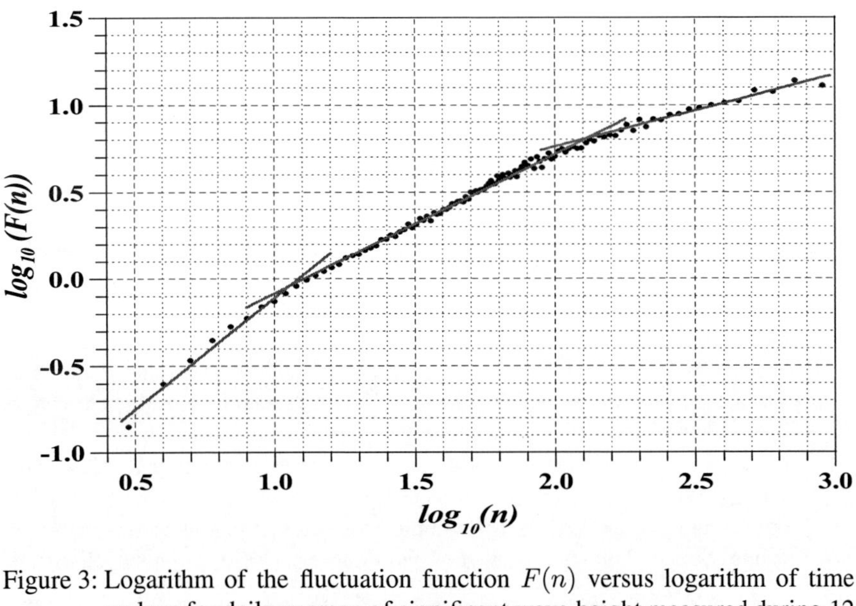

Figure 3: Logarithm of the fluctuation function $F(n)$ versus logarithm of time scale n for daily average of significant wave height measured during 12 years off Estaca de Bares, Spain, in the Cantabric Sea.

4 Conclusions

Detrended fluctuation analysis of mean daily significant wave height fluctuations estimated from wave data taken in the Cantabric Sea, off Estaca de Bares Cape, Coruña, in deep waters, during the period 1998–2008 is presented.

As is common in the fluctuation analysis of natural phenomena, the observed DFA plot is not a single straight line but presents three different scaling regions separated one from another by a crossover time.

It is shown that the associated time series display long-range dependence for time scales from 10–11 days up to time scales close to 100 days, characterized by a scaling exponent $\alpha = 0.79$. For smaller time scales fluctuation analysis reveals a Brownian noise-like behavior while for time scales larger fluctuations it tends to depict a fully random behavior.

The observed difference in the scaling behavior may be due to variabilities in the underlying physics governing the process. In particular, it may be related to changes in the external atmospheric and oceanic forcing mechanisms over various time scales but the origin of this diversity in scaling behavior should be investigated in depth.

Acknowledgements

The authors are grateful to Puertos del Estado (Spain) for kindly provide the wave data used in this study. This research has been undertaken within scientific project TEC2009-14217 financed by the Spanish Ministry of Science and Innovation.

References

[1] Goda, Y., Random Seas and Design of Maritime Structures (3rd Edition), World Scientific Publishing, 2010.

[2] Ochi, M.K., Ocean Waves: The Stochastic Approach, Cambridge University Press, 2005.

[3] Komen, G.J., L. Cavaleri, M. Donelan, K. Hasselmann, S. Hasselmann, and P. Janssen, Dynamics and Modelling of Ocean Waves, Cambridge University Press, Cambridge, 1994.

[4] Beran, J., Statistics for Long-Memory Processes, Chapman & Hall, New York, 1994.

[5] Fox R. and Taqqu M.S., Large-sample properties of parameter estimates for strongly dependent stationary gaussian time series, The Annals of Statistics, 14, 517–532.

[6] Chen, Z., P.C. Ivanov, K. Hu, and H.E. Stanley, Effect of nonstationarities on detrended fluctuation analysis, Physical Review E, 65(4), 15 pp, 2002.

[7] Peng, C.K., S.V. Buldyrev, S. Havlin, M. Simmons, H.E. Stanley and A.L.Goldberger, Mosaic organization of DNA nucleotides, Physical Review E, 49(2), 1685–1689, 1994.

[8] Peng, C.K., Havlin, S., Stanley, H.E., and Goldberger, A.L., Quantification of scaling exponents and crossover phenomena in non stationary heartbeat time series, Chaos, 5(1), 82–87, 1995.

[9] Bunde, A., S. Havlin, J.W. Kantelhardt, T. Penzel, J.-H. Peter and K. Voigt Correlated and uncorrelated regions in heart-rate fluctuations during sleep, Phys. Rev. Lett. 85, 3736, 2000.

[10] M.V. Berry; Z.V. Lewis, On the Weierstrass–Mandelbrot Fractal Function, Proc. R. Soc., A 370, 459–484, 1980.

[11] Koscielny-Bunde, E., A. Bunde, S. Havlin, H.E. Roman, Y. Goldreich, and H.J. Schellnhuber, Indication of a universal persistence law governing atmospheric variability, Phys. Rev. Lett., 81(3), 729–732, 1998.

[12] Koscielny-Bunde, E., J.W. Kantelhardt, P. Braun, A. Bunde, and S. Havlin, Long-term persistence and multifractality of river runoff records: Detrended fluctuation studies, J. Hydrology, 322, 120–137, 2006.

Author Index

Coastal Engineering VII

Modelling, Measurements, Engineering and Management of Seas and Coastal Regions

Edited by: ***C.A. BREBBIA****, Wessex Institute of Technology, UK and* ***M. DA CONCEIÇAO CUNHA****, ISEC, Portugal*

Recent disasters have confirmed the importance of adequate protection of coastal areas. The cumulative effect of development in these regions needs to be better understood in order to ensure that they are managed in a sustainable fashion. Given the dynamics of the processes taking place on shorelines and the multitude of aspects that need to be considered, the problems involved are extremely challenging.

Bringing together contributions from researchers and professionals engaged in the development of modern computational and experimental tools, this book addresses many subjects relevant to the successful management of such areas and offers new insights. The papers featured were originally presented at the Seventh International Conference on Modelling, Measurements, Engineering and Management of Seas and Coastal Regions.

Topics and techniques covered include: Shallow Water Studies; Estuarine Problems; Pollutant Transport and Dispersion; Water Quality Issues; Oil Spills; Coastal Ecosystems and their Sustainability; Coastal Erosion and Sedimentation; Coastal Geomorphology; Sea Defence and Protection Systems; and Case Studies.

WIT Transactions on The Built Environment, Vol 78
ISBN: 1-84564-009-8
Published 2005 / 368pp / £155.00

WIT*Press*
Ashurst Lodge, Ashurst, Southampton,
SO40 7AA, UK.
Tel: 44 (0) 238 029 3223
Fax: 44 (0) 238 029 2853
E-Mail: witpress@witpress.com